Imperial Unknowns

In this major new study, the history of the French and British trading empires in the early modern Mediterranean is used as a setting to test a new approach to the history of ignorance: how can we understand the very act of ignoring – in political, economic, religious, cultural and scientific communication – as a fundamental trigger that sets knowledge in motion? Zwierlein explores whether the Scientific Revolution between 1650 and 1750 can be understood as just one of what were in fact many simultaneous epistemic movements and considers the role of the European empires in this phenomenon. Deconstructing central categories like the mercantilist "national," the exchange of "confessions" between Western and Eastern Christians and the bridging of cultural gaps between European and Ottoman subjects, Zwierlein argues that understanding what was not known by historical agents can be just as important as the history of knowledge itself.

CORNEL ZWIERLEIN is Professor at the Department of History at Bochum University, Germany. He completed his PhD in 2003 at the University of Munich. From 2013 to 2015 he was a Fellow at the Harvard History Department (Henkel/EU) and in 2014 a Fellow at Centre for Research in the Arts, Social Sciences and Humanities (CRASSH) in Cambridge, UK.

Imperial Unknowns

The French and British in the Mediterranean, 1650–1750

CORNEL ZWIERLEIN
Bochum University, Germany

CAMBRIDGE
UNIVERSITY PRESS

CAMBRIDGE
UNIVERSITY PRESS

University Printing House, Cambridge CB2 8BS, United Kingdom

One Liberty Plaza, 20th Floor, New York, NY 10006, USA

477 Williamstown Road, Port Melbourne, VIC 3207, Australia

314-321, 3rd Floor, Plot 3, Splendor Forum, Jasola District Centre, New Delhi - 110025, India

79 Anson Road, #06-04/06, Singapore 079906

Cambridge University Press is part of the University of Cambridge.

It furthers the University's mission by disseminating knowledge in the pursuit of education, learning and research at the highest international levels of excellence.

www.cambridge.org
Information on this title: www.cambridge.org/9781316617502

© Cornel Zwierlein 2016

First published 2016
First paperback edition 2018

A catalogue record for this publication is available from the British Library

ISBN 978-1-107-16644-8 Hardback
ISBN 978-1-316-61750-2 Paperback

Contents

Figures

Acknowledgments

I am extremely grateful to all institutions and people who have supported me and who have invested their trust in me and in those ignorant paths. The European Commission (Marie-Curie-Actions) together with the Gerda-Henkel-Stiftung (m4human) financed twenty-four months free from teaching and committees. At Harvard, I was warmly welcomed within the History Department, first of all by the Early Modernists: Ann Blair, David Armitage, Daniel Smail, Joyce Chaplin, Tamar Herzog, James Hankins, the late Mark Kishlansky. Several old and new colleagues and friends have read entire chapters and have given me extremely valuable suggestions and critical feedback, Mark Kishlansky, actually some months before his death, so I hold his suggestions specially precious, Malcolm Snuts, Jacob Soll, Albrecht Fuess, earlier parts Christian Windler; Alastair Hamilton, Martin Mulsow, Bernard Heyberger; the chapter on history had been precirculated at the Western Ottoman Workshop at UC Davis, January 2015, organized by Baki Tezcan, combined with the Mediterranean Seminar organized by Brian Catlos and Sharon Kinoshita, and at the Mediterranean Crossings Conference at Yale, organized by Francesca Trivellato and Alan Mikhail, where I received a wealth of helpful questions and comments by Baki Teczan, John-Paul Ghobrial, Alexander Bevilacqua. The chapter on science has been read and commented upon by Joyce Chaplin and was discussed with much profit in the Harvard Early Sciences Workshop organized by Devin Kennedy and Allyssa Metzger in February 2015. William O'Reilly has done the work of a friend to read and comment on the whole twice. I profited immensely from presenting parts of the book on other occasions at Harvard, Fribourg, Amherst, Brown University, University of West Virginia, receiving helpful suggestions by Mario Turchetti, Yves-Charles Zarka, Sara Miglietti, David Hershenzon, Will Smiley, Adam J. Kosto, Rachel J. Weil, Sam Boss, John Moreau, Matthew Vester, Wolfgang Kaiser, Guillaume Calafat,

Cemal Kafadar, Bernard Heyberger, Clifford Beckmann. Some answers to specific requests were generously given by Antoin E. Murphy, John Shovlin, William Deringer, Diego Quaglioni, Daniel Stolzenberg, Abigail Krasner Balbale, David Carrasco, Aristotle Papanikolaou, Maksim Astashinskiy, already years ago by Eric Vallet and Françoise Micheau. For more general questions of the history of ignorance I had extremely helpful conversations with Daniel Smail, Anthony Grafton, David Bell, John Hamilton, Ann Blair, Sheila Josanoff and the whole group of the double conferences at Harvard and at the GHI Paris on the History of Ignorance of which selected contributions have appeared in parallel to this book (*The Dark Side of Knowledge. Histories of Ignorance, 1400–1800*). I also profited from the several circles and working groups at Harvard as Mary Lewis' French History circle, the Early Modern History Workshop, the Early Sciences working group, the Intellectual History workshop, and perhaps most of all on several occasions from a free exchange of ideas with those who were also studying, reading and writing at the same time, as the Harvard graduate students Devin Kennedy, Louis Gerdelan, Joshua Ehrlich, Stuart McManus, Rowan Dorin, Devin Fitzgerald, Taylor Cowdery, Paolo Savoia, Allyssa Metzger, Joe La Hausse de la Louvière, Michael Tworek, Meredith Quinn, Honora Spicer, Taylor Cowdery.

A special thanks goes to Richard J. Evans and his Leverhulme Trust project on "Conspiracy & Democracy" at Cambridge University and the CRASSH where I experienced, as very-short-time-fellow, an extremely warm and intellectually conspiring welcome and exchange, "paying" just with one lecture. They enabled me to do some research in the Manuscript section of the Cambridge University Library, but I also profited from linking older topics of research (Conspiracy) with the new ones (Ignorance). I received very helpful remarks by Richard Evans, David Runciman, Hugo Drochon, Rachel Hofman, Andrew McKenzie-McHarg, Alfred Moore.

Older roots of this book are still the intellectual environment of the alma mater LMU Munich. Topics discussed there, sometimes some ten years ago, with friends and colleagues as Andreas Höfele, Tobias Döring, in the entourage of Winfried Schulze, Annette Meyer, the late Karl Eibl, Katja Mellmann, Jan-Dirk Müller, Thomas Duve, Oliver Primavesi, Claudia Märtl, Friedrich-Wilhelm Graf have left their enduring marks. Similar is true for related discussions at Bochum on "transcoding" with Reinhold Glei, Christian Frevel, Stefan Reichmuth, Rudolf Behrens, Marion Eggert, Nikolas Jaspert, and on the Bochum conferences on piracy and human security with the late Daniel Panzac, Salvatore Bono, Georg Christ, Leos Müller, Joachim Östlund. The ongoing conversations with the former PhD student and now colleague, Magnus Ressel, on Barbary and Mediterranean affairs since 2007 was a decisive stimulus. One has read nearly every word twice or three times: Stephen Walsh, who, while graduating himself at Harvard, has done a great job in editing the English. At a later point Miranda Critchley (Cambridge/UK) and Monica Hoesch (Salt Lake City) have helped editing. Diarmaid McCulloch, with whom I share an interest in the East Frisian and London first Puritan Johannes a Lasco which seems to have not much to do with this book, cleared the way to the Bodleian Manuscripts. I thank all staff of the French, British and German archives and libraries for their support. I could not have written this book in such a concentrated and quick way without the access to all books and resources at Harvard, of the special collections (Ernst Mayr, Countway, Houghton, the Map Collection and the University Archives), but even more important of the "normal" collections of the Widener, the Fine Arts, the Law School, the Business School, the JFK School, and even the never closing Lamont. Checking out those 600 books of your bibliography for years and having them at your disposal while writing was a privilege. Thanks to all those who work in those libraries and centers and who have helped.

A final word of thanks goes to Cambridge University Press: Michael Watson, Lucy Rhymer, Teresa Royle and the two anonymous reviewers of the manuscript for having taken it seriously, Robert Judkins, and the whole editing team at Integra for the final production, orchestred by Karthik Orukaimani, and copy editor Brian Black.

Note on Conventions

Seventeenth- and eighteenth-century French has been cautiously modernized following the rules given by Bernard Barbiche (theleme .enc.sorbonne.fr/cours/edition_epoque_moderne/edition_des_textes). For English, German, Italian and Latin, no modernization has been adopted, but abbreviations have been tacitly dissolved and Latin u/v have been standardized as vowels or consonants, respectively. Accentuation and Aspiration of Greek has been standardized if missing in the sources. The rare Arabic and Ottoman names and book titles have been transliterated following the rules of the *International Journal of Middle East Studies*, but with some exceptions concerning names returning very often and which were already familiar and westernized in early modern times, such as *Abulfeda* instead of *Abū'l-fidā*. Different spelling and transliterations in titles of literature cited have been kept.

Abbreviations

ADM	Kew, National Archives, PRO Admiralty
AE and place	Archives des affaires étrangères, name of site (La Courneuve or Nantes)
AN AE	Archives nationales, fonds Affaires étrangères
AN MAR	Archives nationales, fonds Marine
BL	British Library London
BNF	Bibliothèque nationale de France Paris
Bodleian	Bodleian Library Oxford
Chalmers	G. Chalmers, *A Collection of Treaties between Great Britain and Other Powers*, 2 vol., London 1790.
CO	Kew, National Archives, PRO Council of Trade and Plantations
CTB	Calendar of Treasury Books, 32 vol. (1660–1718), London 1904–1957.
CTBP	Calendar of Treasury books and papers . . . preserved in Her Majesty's Public Record Office, 5 vol., London 1897–1903.
CTP	Calendar of Treasury Papers, 1556–[1728] preserved in Public Record Office, 6 vol., ed. Redington, London 1868–1889.
	CTB, CTP and CTBP are fully searchable online with BHO (www.british-history.ac.uk); page numbers have been verified with the printed originals.
Dumont	J. Dumont et al. (eds.), *Corps Universel Diplomatique du droit des gens*, 8+5 tomes, Amsterdam 1726–1739.
MD	Mémoires et Documents
Ms. fr.	Manuscrits français
NAF	Nouvelles acquisitions françaises

ONDB	Oxford National Dictionary of Biography (online version, the author of the article is named)
PC	Kew, National Archives, PRO Privy Council
SP	Kew, National Archives, State Papers
UB	Universitätsbibliothek
UL	University Library

Introduction

Empires are built on ignorance. This is the assumption that this book takes as its starting point to compare the history of the French and the British trade empires in the Mediterranean between 1650 and 1750. It is a premise whose slightly paradoxical appearance can be tempered by elaborating that empires are built on ignorance as much as they are on knowledge. The historiography of empires has often insisted on the importance of the gathering, management and use of information and knowledge for administrative purposes. A whole branch of historiography on "empire and science" has developed that focuses on the infrastructure, aims and impacts of knowledge accumulation.[1] Asking differently for the unknowns within imperial communication does not intend to assert the unimportance of knowledge and of the results of that research. The point made here is to shift the focus to the other side of knowledge – its absence – and therefore to the enduring relationship between ignorance and knowledge. Such a change of perspective also implies that the very category of "empire" is put in second place. I will instead concentrate on the relevant *actors* in the Mediterranean who worked for or under the protection of France and Britain: the ambassadors, consuls, chaplains, merchants, translators and travelers on the one hand; and the involved actors in the English and French centers on the other, from the ministers and secretaries of state to the councilors, experts and advisors in commercial, cultural or religious affairs. How did these imperial actors cope with

---

[1] For the early Spanish-Portuguese empires cf. Barrera-Osorio, *Experiencing nature*; Cañizares-Esguerra, *How to write*; Bleichmar et al. (eds.), *Science*; for the passage from the eighteenth to the nineteenth century cf. the special issue MacLeod (ed.), *Nature and empire*; Raj, *Relocating modern science*; Gascoigne, *Science*; Drayton, *Nature's government*; Stern, "Rescuing"; McClellan III and Regourd, *Colonial machine*; Kumar (ed.), *Science and empire*; Harrison, "Science"; Cook, *Matters of exchange*; Delbourgo and Dew (eds.), *Science and empire*. For the later empires cf. Stuchtey (ed.), *Science*; most of the volume Beinart and Hughes (eds.), *Environment and empire* follows similar questions.

the multiple forms of ignorance within the four epistemic fields of politics and economics, religion, general knowledge and history of the Levant regions, and science? If one looks at the average daily activities of a consul, these four fields cover roughly all of his "official" time. A history of ignorance concerning imperial communication in the period from 1650 to 1750 will not just highlight, in a general way, that something was unknown. Instead, such a history assumes that the very act of ignoring, different forms of specifying and coping with ignorance, and the communication of and reflection on ignorance in relevant fields and matters have a constructive power that can be historicized for given periods, cultural settings, institutions, and in its evolution and epistemic movements. In this case, the aim is to understand in a new way the shape of what one might call "early Enlightenment empires." To clarify the problem with an example, one can point to the well-known fact within the specialized field of the history of cartography that around 1750, with some precursors, within the circles of the royal cartographers and the *Académie des sciences*, a new rigorousness arose that led to the deletion of uncertain, unclearly identified or imprecisely measured locations and items from maps, and their replacing and explicitly exposing as "empty spaces." This is an easily recognizable phenomenon because of its visual form, that belongs to the narrower field of the history of cartography.[2] But the assumption of this book is that there was, between 1650 and 1750, *in general* a new and evolving way to become aware of ignorance, to expose it and to define the unknown across the four fields of politics and economics, religion, general knowledge and history, and science. The assumption is also that, despite the fundamentally different character and type of these epistemes, there were similarities and parallels between the epistemic fields concerning the sequence of exposition and awareness of ignorance, the forms of specifying it, the process of knowledge-gathering in response to it, and the endpoint of the movements, which are called here non-knowledge cycles. The question, therefore, is which epistemic forces caused those changes? In politics and economics, the way in which the category of the national was newly defined as *the* determining matrix of mercantilist political economics will be understood as

---

[2] Broc, *La géographie des philosophes*; Konvitz, *Cartography*, 1–31; Laboulais-Lesage (ed.), *Combler les blancs*; Safier, *Measuring*; Haguet, "J.-B. d'Anville" for d'Anville as the promoter of "empty space" cartography.

a specification of unknowns. It was only through those specifications that a mighty process of nationalization was set in motion. In the field of religion, a quite sudden awareness of ignorance about the "present state" of even major Oriental religions and churches, like the Greek Church, creates the central complex of unknowns with which the actors tried to cope and which led to the massive communication of knowledge gathering and to unexpected entanglements and disentanglements between East and West. In the field of general knowledge and history, consuls remained in a state of unconscious ignorance for a long time about major parts of the history of the countries where they worked and lived in – unconscious, because one even does not find explicit mention of a lack of something. This will lead to the question of how one organizes one's own world, the structures of history and descriptions within such a state of nescience, and how that might have even served the functions of the empire. Finally, the consuls, voyagers, translators and chaplains were also active in exploring unknowns within the field of science, where their networks of the trading colonies on the Mediterranean shores intersected with the institutions and networks of the so-called "scientific revolution" centered in London and Paris.

Both empires were thus formed as a function of the continually ongoing work and communication of ignorance in these epistemic fields – and at the same time the empires shaped the form of that communication.

History of Empires, History of Ignorance

Is this, then, a history of empires? Is it a parallel history of epistemic developments? Is it a Mediterranean history? It is all of these three. To start with the last, as epistemes follow no border, this is more a history "in" than "of" the Mediterranean in a narrower sense.[3] The link between a history of ignorance and the communication within trade empire networks is not reserved to the Mediterranean and could

---

[3] Horden and Purcell, *Corrupting sea*. An exception where a history of ideas and of intellectuals can also be a history "of" the Mediterranean, are genealogies of early modern concepts of "the Mediterranean" as an geo-historical actor itself, blending, for instance, Peiresc with Braudel: Miller, *Peiresc's Mediterranean world*. For a (necessarily non-exhaustive) overview of Early Modern Mediterranean Studies cf. Zwierlein, "Early Modern History."

be transferred to other regions by taking into account that settler colonies would lend a different setting than that of the European merchant colonies within the Ottoman Empire. The nationalization of political communication concerned, in different forms, all European shipping on all oceans and seas. On the other hand, the Mediterranean and the Orient was usually the preferred location of the search for European civilizational roots and so there is a "material" Mediterraneanism to this history in parts of the chapters on religion, history and science. The realization of ignorance and unknowns concerning just those regions supposedly at the roots of European history was striking more so than ignorance of Asian and American lands. A history of the reflexive awareness of ignorance within the Mediterranean thus both echoes and transcends a history of the epistemic processes involved in the "discovery" of the New World. The former is something like a belated and paradoxical re-establishment of cognitive relationships and an ongoing clarification of the limits of knowns, and perhaps even of knowables, in the context of the region's political and cultural forms in its past and present states.

Including the Mediterranean within British imperial history is not very common, and in French historiography, a branch of "imperial history" is not established at all, since the common term for the history of former transatlantic possessions is "colonial history," in which the Mediterranean is not included, at least not for the period before 1830. Yet if one admits the existence or at least the historiographical notion of an English/British empire before 1750, there can be no question that the Mediterranean was a crucial part of the empire of trade – it accounted for roughly a third of the foreign trade of "England."[4] The acquisition of Tangier at the same time as Bombay is not merely a coincidence. There is no reason to regard the Mediterranean as less important than the Empire's informal trade empire in India, at least before the Bengal acquisition of 1757 and the 1783 laws. The British presence in India in the eighteenth century was largely confined to Bombay, Calcutta and Madras, constituting a total of about fourteen small locations along the coast of the subcontinent. This corresponds very closely to the situation in the Mediterranean with Aleppo, Constantinople, Smyrna and many

---

[4] If one takes a short look at the volumes of the *Oxford History of the British Empire*, the Mediterranean is nearly completely absent. The same holds true for the mercantile and economic histories of the Empire, cf. Chapter 1 on politics and economics and the broader discussions on mercantilism (notes 25, 26, 229).

minor places, foremost on the Barbary Coast. One should not therefore be misled by the late eighteenth-century importance of the American colonies, the post-1776 shift, and the developments in the nineteenth century by projecting this later situation back on the late seventeenth-/early eighteenth-century realities. *Ex ante*, it was not clear whether the economic and imperial future lay in the Mediterranean and/or a far larger "world."

As the aim of this book is a comparative French-British history, a common designation must be chosen. Using the framework of "imperial history" would tend to place the work within an Anglophone tradition; despite important differences in how the crowns and ministerial bureaucracies were organized, there are many analogous forms and tendencies that suggest the comparability of the French and British cases: the concentration of the organization of Mediterranean trade in a group of specialized merchants with their own "board" (*Chambre de commerce de* Marseille, Levant Company), the almost simultaneous formation of councils or boards of trade, of the Royal Society and the *Académie des sciences*, the synchronous regional differentiation of competencies within the ministries, and the establishment of a consular network. As many of these developments took place far away from the central state, and as, furthermore, no settlements and colonies in the formal sense could be built under Ottoman rule, "colonial history" is unfitting; but the "history of trade empires" seems an appropriate alternative. More important than the label itself is to reflect what historiographical burdens risk being unintentionally imported by putting the study at least partially under the cover of the Anglophone tradition of imperial history. Despite its richness and diversification, one may argue that within that tradition, studies typically organize their object according to a two-level approach: on the one hand, historians concentrate on narratives of ideas, concepts, ideologies and discourses of "empire"; on the other, on structures, institutions, commercial and economic exchange and data, and the political and diplomatic history and events. Often, this is an either/or classification, as studies seldom address both levels with the same priority. Granted, this simplification of a broad common juxtaposition of structure versus discourse/ideas does not do justice to the richness and diversity of the different approaches, both old and new. But accepting this simplification for a moment, these approaches tend to put the empire itself into the position of a logical a priori. This holds true not only for studies of the British empire's trade in the colonies, where "imperial" and "empire," as well as the competing nations in the Ocean are usually taken as given categories, but also for studies concentrating on political language analysis and the "ideological origins of empire." Once the first appropriation of imperial language is identified – as early as Henry VIII, for instance, and with the writings of John Dee in the reign of Elizabeth I –, the empire itself is presumed to exist as a framework.

The slight turn taken here by studying the communication and handling of ignorance within the epistemic fields as the base upon which empires are built, and vice versa, by analyzing the empire's function for shaping those epistemes, leads to a different setting: if one combines the study of mercantilist commercial and political communication with their central category of the "national" as just one epistemic type with the other epistemic types – religion, history and science –, one folds the bifurcation of ideas versus structures into each other. This consciously deposits the empire into co-existing epistemic fields.

The already mentioned younger branch of "empire and science" research has developed as a merging of the post-Foucaldian expansion of the history of discourses/knowledge or "science" in its older narrower sense with imperial history. Here, the epistemic approach has been the norm, and often the histories either concentrate more on the institutional dimensions of the organization of scientific research and knowledge management within the empires or on the contents of the science/knowledge in question. This is, at first glance, quite close to the approach chosen here. Those studies often place emphasis on networks and circulation, and authors try to abstract from the notion of empire or state to demonstrate how communication within the republic of letters spread through imperial networks on the one hand, and how state-biased and institutional communication organized science for the purposes of the state on the other, thereby transcending a narrower political and institutional history of empires. But in that research, the political, economic and other structures and institutions of "the empires" are mostly assumed to be pre-existing and the epistemic "liquidization" only concerns the part of its scientific functions and entanglements associated with the republic of letters. The goal here has been to totalize that point of view and find a way to treat all as a parataxis of coevolving epistemes. Politics/economics, religion, history *and* science are observed on equal terms. There is no "science *and* empire": imperial communication is a fourfold communication within several epistemes. The "empire" is *in* all this and produced by it, rather than above or before it. This is in no way a radical revelation, but instead an attempt to synthesize both traditions. And it is those actors between London/Paris and the factories or *échelles* who constantly kept these processes in motion, by asking, by not-knowing, and by partially answering their own unknowns.

In the context of the four cycles and epistemic developments observed, one might think in a first quick assumption that the totalizing of the epistemic approach and transfer from the narrower field of the history of science to the parataxis of all four epistemes important for imperial communication leads to something akin to the export or generalization of the "structure of scientific revolution(s)," in a refined Kuhnian sense as the structure of all epistemic developments and movements.[5] The four developments could be conceived as: (a) the birth, climax and change or decline of mercantilist trade, (b) the entanglement and disentanglement of the Western and Eastern churches during the former's transition from confessional conflict to

---

[5] It is not my purpose to integrate here a detailed analysis of the concepts of "(scientific) knowledge," "structure," "paradigm-shift" and "context" inherent to Kuhn's work nor of the decades of discussion about the work, nor of all attempts already made to transfer the Kuhnian theses to other realms of "sciences" and of knowledge. Mostly, only the concept of paradigm and the question of how one paradigm is accepted within competing worldviews is transferred to fields such as economics, politics, religion or "humanities" in general. While for my purposes, the notion of "paradigm" is far less important, within the more recent discussion and revisitations of the work around the fiftieth anniversary of its publication, one should underline that the concept of the "structure" or the shape of larger epistemic movements as such has attracted little attention for decades. On a still more general level, the question, if we must first distinguish different types of knowledge before transferring concepts about knowledge change from the history of science, seems not to be very sharply articulated within those debates. From the myriads of contributions one might refer to here, rather as an introduction into that bibliography than being itself one, see the issue devoted to *The structure of scientific revolution* of *Historical studies in the natural sciences* 42, 5 (2012), and Mayoral, "Five decades"; for an example of narrowing "religious conversion" and acceptance of scientific paradigms Drønen, "Scientific revolution"; for a reasoning about if and how the epistemic shifts around 1800 within history, philosophy and other fields combined as "humanities" can be modeled in analogy to Kuhn's model for "science," opposing by that in a dichotomous way those two fields, Solleveld, "Conceptual change."

the notion of natural religion and other trans-confessional forms, (c) the transformation of early modern *historia* into empirical history, geography and natural history due to standardization, spatialization and deepening of historical thought, and (d) the scientific revolution. The challenge, for a more general history of ignorance, is then to observe and assemble the similarities and differences of the shapes of those movements and to characterize "non-knowledge cycles" as a general pattern that is not restricted to the very specialized epistemological framework of the emerging scientific empiricism. Whether the model(s) for scientific developments of phases of knowledge accumulation followed by revolutionary shifts of paradigms and similar reflections about the general form of epistemic processes can just be transferred from the realm of science, or whether it must be altered, will be tested throughout this book. In fact, the question is not to transfer a model of Thomas Kuhn conceived in 1962 for the field of science on other realms. The question is rather about the internal form and behavior of the epistemic movements themselves, how they were realized in history, how we can describe them for each field and for all fields compared. It is not a question of models, but of the historical reality itself. The results of those observations will be collected in the final conclusion.

Actors, Institutions, Places, Period

The French consular network in the Mediterranean developed in the sixteenth century, with a different starting date for each individual location. Nevertheless, the monarchy's occupation with internal problems from the last Huguenot war (1629–1631) to the Fronde in 1648/49 to its involvement with the Thirty Years' War after 1635 led to the neglect of Mediterranean commerce and navy forces, a disregard which was strongly felt in its major ports, particularly Marseille. As far as can be judged, complaints and calls for reform in the 1650s/60s were not only rhetoric, but reflected the reality that French politico-economic engagement in the Mediterranean had reached a very low point after the Richelieu era. Colbert's reforms,[6] which I will address in the

---

[6] Soll, *The information master*; Dessert, *Argent* and Dessert, *Colbert*; on his
    successor cf. Dingli, *Seignelay*; on the internal French development of Colbert's
    reforms cf. Minard, *La fortune*; the transition to the later Maurepas
    administration is characterized in Pritchard, *In search of empire*, 230–263.

first chapter from the point of view of norms regulating commerce, were in fact a real starting point, as economic history since Masson has shown.[7] Only from these years onwards does the archival documentation emerge as a steady flow of dispatches and letters between the Secretary of State, the newly founded Chambre de commerce in Marseille (1650), and the consuls.[8] The reforms predate the concentration of power in Colbert, who accumulated between 1665 and 1669 the effective control of the navy, direction of commerce and the office of a sécrétaire d'état de la marine du Ponant – all in addition to the office of controleur général des finances. This has been identified as the "origin of the Ministry of the Navy."[9] At the beginning of the sixteenth century, there was just one French consul in the Levant (in Alexandria); in 1600 there were five (Alexandria, Tripoli/Syria, Chios, Algiers and Tunis), and in 1715 there were fourteen major consulates, all in the Mediterranean: Algiers, Tunis, Tripoli/Barbary, Cairo, Seyde (Saïda, Sidon), Aleppo, Chania, Lanarca, Smyrna, Saloniki, Nafplio (Peloponnes or "Morea"), Prevesa (Larta), Zakynthos (Zante) and Durazzo. Several new consulates were founded during the eighteenth century, and some vice-consulates were established.[10] Marking a new era, in 1669 authority over the consuls was switched from the Département des affaires étrangères (Hubert de Lionne) to Colbert's *Marine*. At about the same time (1664), the first *Conseil de commerce* was established, which prefigured the 1700/1 Conseil or Bureau.[11] While the medieval roots of the consular office were situated in trade and commerce, during the seventeenth century, royal authority came to place the office and institution more and more tightly under its control.[12] The interdiction, from 1691 onwards, against conducting commerce confirmed that development.[13] The type of writing in use between the state center and the consulates became very similar to

---

[7] Masson, *Commerce XVII*; Bergasse and Rambert, *Histoire*, 204–214; Paris, *Histoire*, 3–43; Carrière, *Négociants*.

[8] The Chambre de commerce was a municipal institution but, despite the usual struggles over competencies and interests, was quickly integrated into the central state's system.

[9] Dagnaud, "L'administration," 332f.

[10] Poumarède, "Naissance," 66; for the later developments Mézin, *Les consuls*; Windler, *Diplomatie*; Ulbert and Le Bouëdec (eds.), *La fonction consulaire*.

[11] Schaeper, *French Council*; Kammerling Smith, "Le discours économique."

[12] Poumarède, "Naissance," 128.

[13] The offices of the Secrétairerie d'Etat de la marine were first in Versailles under Louis XIV; in 1715 they moved to Paris, and they returned to Versailles in 1723. Cf. Dagnaud, "L'administration," 714 n. 1.

the diplomatic forms of instructions, dispatches and reports (*mémoires*) that had first been introduced during the Renaissance diplomatic revolution.[14] The consuls, helped by their chancellors, by dragomen, and in collaboration with the elected representatives of each échelle's nation, were now "an instrument serving the politics of penetrating the Ottoman Empire," and it is therefore fair to see them as imperial agents.[15] One of the most important sources for several chapters of this study are *mémoires*, or advisory reports, which can be found in many different archival locations. Many were written for the Chambre de commerce's internal process of decision-making and even more for negotiations between the Chambre and the Secretary of state in Paris/ Marseille; others were written by consuls in the Levant and sent to Paris, or by councilors and advisors in Paris/Versailles, within the Bureau of the Marine Ministry or within the Councils of Commerce. It is not always possible to determine their precise institutional origin, or to ascribe authorship, because they were frequently written anonymously.[16] There are also *mémoires* concerning the scientific activities of the consular network, which can be found in the archives of the Marine ministry and the *Académie des sciences*.

The English/British[17] organization was somewhat different because the consuls depended on the Levant Company whereas the French never succeeded in putting the Levant trade under the control of a chartered or privileged company. It is a somewhat surprising fact that all four major institutional studies on the Levant Company for our period were written in the 1930s, three of which remain unpublished.[18] Later works have

---

14 For those origins cf. Mattingly, *Renaissance diplomacy* and Frigo (ed.), *Politics and diplomacy*; Senatore, *"Uno mundo de carta"*; Zwierlein, *Discorso und Lex Dei*, 193–294.

15 Poumarède, "Naissance," 128.

16 Major archival locations of *mémoires* that circulated within the Ministry or between the Chambre de commerce and the Ministry are the series *Mémoires et documents* in the *Archives des affaires étrangères* (La Courneuve), the series F12 in the *Archives nationales* (site Pierrefite) and the series AE B III 234ff and several cartons of the AE MAR B VII series in the *Archives nationales* (site Paris) and the series C, E and H in the Archives de la Chambre de Commerce, Marseille. But one finds them also in many other places, mostly in the consular and the ambassadorial correspondence.

17 On the use of "British" and "English" cf. already a contemporary voice: John Toland in Hoppit, *Land of liberty?*, 242.

18 Ambrose, *Levant Company*; Russell, *Levant Company*; Matterson, *English trade*; Wood, *Levant Company*.

concentrated more on the London/England-based class of merchants than on Mediterranean affairs or on either individual merchants or mercantile families, but a renewal of interest for the whole Company and group has just started.[19] The institutional combination of the secretary of state,[20] the board and body of the Levant Company, and the admiralty and customs officers in the state center[21] together with the ambassador to Constantinople, the consular network in the Levant, and the English merchant nations in the periphery,[22] made up the structure that corresponded with the French setting just described. But in the English case, complications ensue due to the administrative division between the eastern and the western Mediterranean. The former was the charter zone of the Company, but the latter was not covered by its privileges and was hence open, despite many discussions, to non-Company traders, particularly the "Italian merchants," operating mostly to and from Livorno.[23] The Levant Company had the three major factories, in Aleppo, Smyrna and Constantinople, but consuls were also installed in Algiers, Tunis and Tripoli.

The main archival records thus include state papers containing consular correspondence and the records of the Levant Company. The Privy Council and the Board of Trade, which would be a parallel to the French *Conseil de commerce*, were also often occupied with Mediterranean affairs, even if the Board of Trade, at least in its 1680s/90s form, was instituted to concentrate on the plantations in the American colonies.[24] However, as the Levant Company papers show more clearly than the records of the Board of Trade, the Board repeatedly became involved with Levant matters such as conflicts between the East India and Levant Company.[25] The universities of Oxford and Cambridge as well as the

---

[19] Davis, *Aleppo*; Anderson, *An English consul*; Pennell, *Piracy*; Brenner, *Merchants*; Gauci, *Politics of trade*. A very lively picture of the Levant merchants' world is drawn by Mather, *Pashas*.

[20] For the institutional settings cf. Braddick, "English government."

[21] Cf. still Hoon, *Customs*.

[22] For the chaplains cf. more below, chapter "Religion."

[23] Pagano de Divitiis, *Mercanti inglesi*. On the Mediterranean and even global Jewish merchant networks that were central also for the English merchants cf. Trivellato, *Familiarity*.

[24] For the sequence of (re)establishing and dissolving councils, committees and the Board of Trade cf. remarks and literature in Chapter 1 on Politics and Economics before note 230 and note 310.

[25] Cf. the discrepancy between what seems to be a concentration on the "plantations" of the Board of Trade in CO 391/9ff and the many traces of the

Royal Society and the English Church's network were extensions of that apparatus as far as there was an exchange and overlap between those persons and circles.[26] For intellectual affairs, the Anglican chaplains to the Levant Company were the most important actors in the Levant. Concerning the sources, a clear difference is visible between the French and the English. In the French case, the forms of longer narrative descriptions and advisory reports were integral aspects of routine consular and diplomatic work. These descriptive activities were expected, controlled and directed by the center, and the archives in Paris and Marseille are consequently teeming with this type of source. The English forms of diplomatic and politico-economic correspondence were different, apparently more dedicated to the humanist virtue of *brevitas* and less shaped by diplomatic forms. The effect is that, with some noteworthy exceptions – for example, Paul Rycaut, Samuel Martin and Samuel Baker – most consuls did not write longer narrative reports in either manuscript or published forms. Conversely, the chaplains – and former chaplains – of the Company regularly produced extensive reports, as if there were a tacit division of labor. On commercial affairs it was "learned merchants" themselves who wrote more, and more often, than the consuls.

In general, the first chapter takes most of its examples from the consular correspondence regarding the Barbary Coast. The chapter on religion focuses on Constantinople and parts of the Holy Land because the exchanges with the Greek Church always went through Constantinople and because the challenges and problems of religious protection were much more frequent in those *échelles* than on the Barbary Coast, where Christian communities were few and the level of missionary activity was comparably low. For the history and science chapters, the spotlight returns to the Barbary Coast. As the epistemes are not strictly confined to one region however, many aspects of those chapters also apply to a wider spatial dimension.

The period chosen, from roughly 1650 to 1750, might seem simply to correspond with the conventional post-1648 second half of early

Board/councils being addressed by the Levant Company, e.g. SP 105/144, f. 33v–36v, ibid., 1657, f. 87v-88r, ibid., 1664, f. 143v; ibid., 1702, SP 105/145, pp. 304–306. Letwin, *Origins*, 24–31.

26 Cf. Hamilton et al. (eds.), *Republic of letters* which is not reduced to English affairs. For more literature concerning the Royal Society as far its connection to the Levant is concerned, cf. below chapter "Science."

modern European history. In fact the periodization is here less a choice of pragmatic top-down framing according to an uncritical post–World War II historiographic tradition; instead, it is part of the question and the results of the study itself. From the perspective of a history of ignorance that encompasses non-knowledge operative within everyday administrative and economic practices on the one hand, as well as within other epistemic forms less directly operative for daily "uses" on the other, an "early Enlightenment" shift beyond the Hazardian recognition within the field of the history of philosophy still has to be proven. The developments within the epistemic realms will suggest that there is a material substrate to that epochal shift. It is less important to discuss whether the Colbert administration would already fit under the term of early Enlightenment, as the Maurepas administration firmly does; instead, the shift of period within the administration and the overall forms of communication that are studied is a major focus on its own. In this regard the period from 1650 to 1750 reveals itself to be coherent, not via deductive labelling, but by inductive research.

Terminology and the Questions a History of Ignorance Asks

This book is not intended as *prolegomena* to a general theory of historicizing ignorance; instead, it focuses on the forms of ignorance, its exposition and reflection, and tries to identify empirically the precise historical forms active around 1700 within the chosen Mediterranean context. To achieve this, a terminology and grammar are necessary, along with some clarification of the problems in historicizing ignorance and the forms of its communication. The sociology of ignorance immediately suggests itself as a branch that might lend such a grammar. It was pioneered by Georg Simmel around 1900[27] and later Robert

---

[27] Usually one refers here to the chapter by Simmel, "Secrecy," English translation in *American Journal of Sociology* 11, 4 (1906), 441–498, but while the terminus technicus "Nichtwissen" is coined in this chapter, its concentration on "secrecy" within interpersonal relationships limits its dimensions. On the translator's use of "not knowing" or "nescience" as translation of "Nichtwissen" cf. Gross, *Ignorance and surprise*, 55; post–WWII sociological research on ignorance developed a far broader concept. For the germs of a notion of "Nichtwissen" already present with Simmel that points in this direction cf. the idea of "entanglement between knowledge and non-knowledge [*Mischungen und Verwebungen des Wissens mit dem Nichtwissen*]" (Simmel, "Lebensanschauung," 303) in a passage where he reflects on the devaluation of

Merton around 1950, but found its real impetus within the tiny niche of environmental and organizational sociology from around 1975 onwards. It largely focuses on the question of how political and social decision-making takes place in circumstances of ignorance and is largely ahistorical. But it is perhaps worthwhile to remember that for much of the necessary language of description, one could use the thirteenth-century distinctions developed within moral theology by Thomas Aquinas on the basis of Aristotelian concepts of action. In contrast to sociology today, every act of ignoring is considered to be a voluntary act related to the problem of sinful action. While, in this framework, nescience (*nescientia*) is simply the complete absence before and beyond the moral discourse about its sinfulness (*simplex negatio scientiae*), ignorance is the privation from knowledge (*privatio scientiae*) in a defective form of will. This can be an unwilling form of ignorance (*involuntaris*), or it can occur in several forms of wilful ignorance according to the degree of active voluntary concentration on the not-knowing: (a) *ignorantia voluntaris directa*, (b) *ignorantia voluntaris indirecta* – the latter meaning ignorance through negligence – *ignorantia per negligentiam contingens voluntaria*, and (c) purely accidental ignorance, which can nevertheless entail sin: *ignorantia voluntaria per accidens*.[28] Current sociology certainly does not reason in terms of "sin," but questions of conscious, unconscious, willed, unwilled and of "negative (non-)knowledge" – something ignored that someone knows to be knowable but decides is unimportant – as discussed today are all quite easily translatable into Thomist terms, and opinion will be divided between "traditionalists" and "progressive" readers about which choice of terminology is more elegant. Although I do not identify with progressivism, I chose the terms more commonly in use today; the others would create too antiquarian an impression.[29] The comparison between Thomism and the sociology of ignorance reminds us that, as Tarsky has taught us, a language of description is needed to conceive and identify sharply what can be found in sources;

scientific knowledge from medieval to contemporary times. Cf. Wehling, *Im Schatten*, 43–46.

[28] Cf. Thomas, *De malo*, quaestio III, art. 7 et 8. Cf. for a slightly more comprehensive attempt Zwierlein, "Towards a history of ignorance."

[29] Ravetz, "Usable knowledge"; Collingridge, *Social control*, 23–43; Smithson, *Ignorance*; Ravetz, *Merger*; Luhmann, "Ökologie des Nichtwissens"; Gross, *Ignorance and surprise*.

but that this can only be a first step to historicize ignorance, because the medieval Aristotelian terms are, as such, as ahistorical as Merton's "specified and unspecified ignorance." Beyond medieval moral theology and legal thought, the major emergence of a reflection not on a terminology of how to judge absent knowledge in eventually culpable or sinful voluntary actions, but on the dimensions, shape and status of "the abyss of ignorance" as the "dark side" of knowledge within the framework of an empiricist epistemology, does not occur prior to Bacon, Descartes, Locke and Pascal.[30] This is thus in synchrony with the historical starting point chosen for this study. That should not necessarily narrow the focus to the revolutions in seventeenth-century science and cognitive philosophy. The history of ignorance may have, in its own approach, a genealogical root in those reflections, but its objects are far wider and more general than the history of natural sciences. If this were not the case, the distinct shape of knowledge gaps and ignorances within premodern administrations, politics, religion and many other areas would be neglected.

The most basic definition of ignorance is the absence of knowledge, so the one term seems to necessitate a good definition of the other. Within the already broad field of a history of knowledge that has tried to get beyond a narrow concept of a "history of science," several terminological debates have been conducted, for instance, about the difference between "information" and "knowledge" as basic categories.[31] I will not enter here into the scholastic, definitional work of distinguishing both terms: briefly, I use (non-)knowledge instead of information to cover small pieces of information, such as the nationality of a given ship or product in one moment, as well as whole discourses like the Greek Church's teaching about transubstantiation. In addition to distinguishing the epistemic types, it is necessary to see what they have in common as ideational objects.

"Nescience" then is used in the following as a synonym for unconscious ignorance, when a person or a collective does even not know that they do not know something. In contrast, "non-knowledge" is used for the result of a specification of ignorance, that is, for the transformation of nescience into a known unknown. Now it is known only that

---

[30] Locke, *Essay*, IV, 3, § 22, 24, pp. 553, 555; Bacon, *New Organon*, Aphorism III, p. 33; Pascal, *Pensées*, nr. 77, 185; Descartes, *Meditationes*, III, 1.

[31] Burke, *A social history*.

something is unknown, and perhaps how and to what extent it is unknown; the item itself remains unknown, but the subject of non-knowledge is aware of his ignorance. Yet another form of ignorance, which is certainly very present in the early modern period and in our own era, is negative knowledge and willed ignorance. This means that one does not know something and one decides that it is not important, not necessary or even bad and forbidden to know. Case studies within historical agnotology have mostly concentrated on this kind of ignorance.[32] Most of the material treated in this book does not deal with such forms of implicit or explicit censorship in a metaphorical way. One could ask if the long-enduring, conscious state of ignorance on the part of European merchants on a mere cognitive level concerning (many parts of) the culture of the "non-Europeans" for centuries can be considered as something like willed ignorance. One might further inquire to what extent the perpetuation of such willed ignorance was functional for the very maintenance of the merchant colonies as European proto-imperial bridgeheads in their distinctiveness from the local environment.[33]

Two other distinctions are sometimes used: The first is that between operative and epistemic non-knowledge,[34] which allows us to consider the subjects of the first chapter and those of the third and fourth together, on the one hand by treating them as comparable, on the other hand by distinguishing them according to their basic potential of usability. Scientific knowledge about laws of nature and the universe has a mainly cognitive function for most users; historical knowledge about the roots of one's nation or about the founder of one's religion typically has largely emotional and cohesive functions. Both types may be categorized as predominantly "epistemic knowledge," while one may label as "operative" the knowledge that can be used in concrete decision-making. All three (or more) functions – for description and cognition, for stimulating cohesion and identification, for action and decision-making – are always co-present, as the functional types of knowledge are likewise never exclusive. Operative knowledge can become purely epistemic, for example if an institution is shut down and its whole body of knowledge is transformed into historical remains

---

[32] Proctor and Schiebinger (eds.), *Agnotology*, especially Proctor, "Agnotology."
[33] Zwierlein, "Coexistence and ignorance."
[34] Zwierlein, "Diachrone Diskontinuitäten," 426.

along with all the usable content about how to operate its administration. And an item of epistemic knowledge is nearly always linked to, or transformable into, an operative one. One can reuse and re-enact knowledge by, for instance, attributing a normative validity to it. What is true for knowledge also applies to the other side, to ignorance, even if it may sound strange to speak of "operative non-knowledge." What I mean by this term is that the specification of ignorance which leads to non-knowledge is done with regard to a type of knowledge that would be, if possessed, applicable and operative.

A second distinction concerns the epistemic objects and to what kind of *res* one is referring. Sometimes the knowledge possessed or not possessed concerns something presumably or really "out there," which can be found or understood in the Orient. These are natural unknowns and their laws on the one hand, and unknowns about human productions and artifacts like texts, discourses produced in the past or contemporary Oriental context on the other. One could call these "heteronomous unknowns." Even if the term is not very elegant, it serves to make clear that the object of knowledge or ignorance does not belong to the sphere of the observer and does not obey the rules the observer creates. In contrast, if the French and British inquired as to which nation a ship at sea or a given individual on the Mediterranean shores belonged, the unknown would not concern something that was simply "out there" in "the Orient." Instead, human normative definitions such as nationality always concern something that comes only into being while one is requesting it. The national emerges at the same time as the question is uttered and then answered: a "nation" only starts to exist through the use of schemes of inclusion and exclusion that compare normative ideals with empirically unknown realities – but that unknown is, strictly speaking, unknowable *tout court*, as "the nationality" of someone is not a *res* just to be found. Therefore, there are paradoxical processes of searching for auto-referential or auto-reflexive unknowns in the Mediterranean which do not have much to do with "the Orient" and the observation of "the other." The question is how the process of coping with ignorance was different in both cases.

Beyond that basic terminology, several questions will arise: first, how can one conceive of the difference and the relationship between the level of momentary everyday realization, statements and coping with ignorance in the fields chosen, and that of more developed discourses and topical communication? This will be reflected upon with regard to the relationship between the ongoing processing of national unknowns and the level of different – Grotian, proto-liberalist, ... – imperial discourses: both are linked and influence each other, but one follows normative rules, is ephemeral and momentary; the other evolves more slowly and is far more detached from everyday work and decision-making, but still serves as a general frame of thought. Another question is that of the dominant direction of the currents engendered by the newly emerged awareness of collective ignorance. Entanglements as studied in the chapter on religion are apparently based on such currents and selections of knowledge according to a given setting and obeying the form of the specification of ignorance. The way that religious and theological knowledge is requested and how it wanders from the West to the East and from the East to the West, differs significantly according to the starting trigger and impulse and contributes during the seventeenth-century to the establishment and later, around 1720 to 1740, to the erosion of an entanglement between Anglican, Jansenist and Orthodox theologies and groups in one of the examples chosen here. This leads, finally, to the general question of the dynamics of non-knowledge communication and the shapes of non-knowledge cycles: these shapes will be described and analyzed in each chapter and then summarized and compared in the conclusion, which asks whether some common characteristics and differences can be identified, for instance according to the epistemic types in question, but also merely due to specific developments in individual fields.

Disclaimers: The Author's Ignorances

This is not a book by a French citizen or a British subject, nor is it by someone who has biographical links to the Mediterranean. It is not a book by a trained Ottomanist or Arabist or specialist of Greek Orthodoxy or any other Oriental Christian theology. Instead, this is a book written by an early modern historian trained in central and parts of (south-)western European and environmental history whose personal experience is the experience of ignorance – concerning, for example, many languages spoken in the Mediterranean – who has tried to cultivate the awareness of those ignorances as a hermeneutic prudence regarding the object studied. This study only works toward an investigation of the English and French forms of acting and perceiving in the

Mediterranean. In so doing, and in stressing several times the existence of an epistemic hiatus between "European" and non-Western inhabitants, I do not want to reintroduce the conception of a Mediterranean divided into a dichotomous civilizational conflict between "the West" and "Islam," which was the pre-Saidian state of discussion between the Braudelian and Hessian concepts of a strongly interconnected world or one sharply divided into two parts by an invisible border. This is not the concern of the more general question of a history of ignorance. It is rather the attempt to introduce a third level of reasoning into the scene of the Mediterranean as well as to imperial histories. Beyond the long and reiterated disputes between different macroscopic approaches – from Braudel/Hess to Horden/Purcell – and beyond the ongoing implicit or explicit forms of reciprocal discomfort between partisans of microhistory and those of imperial and macro-history, I think an epistemic history of Mediterranean relations and proto-imperial communication is a worthwhile undertaking. The dense and different forms of interaction between merchants, dragomen, men of politics, European and Barbary corsairs, multiple converts in many directions, and women and men within those places and regions should not obfuscate the – yes, invisible borders of – ignorance that were present in those societies, as they are, in all societies. Moreover, the macro-historical narratives of neo-Braudelian, economic Israelian or Horden/Purcellian stamp might gain some hermeneutic sensibility by adopting an epistemic perspective, even if it is just to encourage awareness of what their actors *are* – their nations, civilizations and the forms of interaction – and of the fluid ground of asking, ignoring and coping with ignorance on which the entities were built and formed. Being and doing through ignoring, one might say. Finally, it is clear that there will necessarily be blind spots in this book and in my own approach, as it is impossible not to be at least partially peering through, or out of, the shadows of nescience. This is fair enough; otherwise, no new books would need to be written.

1 | Politics and Economy: Nationalizing Economics

The Constructive Power of Non-Knowledge

The British and French Empires in the Mediterranean were trade empires. They were mercantilist empires. On the one hand, this is obvious. On the other hand, however, to describe a form of prevailing economics as "mercantilist" does more to start a series of questions than to clarify matters. Without entering into the ongoing and renewed debate about the term "mercantilist" itself and its usefulness, I will define it here as the "nationalization of economics."[1] And I will treat the question of nationalization as an epistemic one: It is the crystallization and hardening of the distinction between "internal" and "external" that defines this form of economics. But while most research on mercantilism concentrates either on trading practices or, within the history of ideas, on theoretical treatises that discuss matters such as bullionism or the balance of trade, here I take a step backwards. I first start with the central perceptional structure organizing all mercantilist communication, the distinction between the trade of "our" nation and that of others. Without that, Thomas Mun could never have calculated a balance of trade, nor could any import/export regulation have functioned. The hardening of that distinction, and its exposition in everyday trading communication, is a distinctive phenomenon of the period, as comparison with the Middle Ages will demonstrate. Only in the second stage, will I address ideas and discourses, investigating the general frames of thought of Empire that governed and directed the

[1] For classical works on "mercantilism," cf. Heckscher, *Mercantilism*; Cole, *Colbert*; Cole, *French mercantilist doctrines*. For the current renewal of the discussion cf. the special issue of *The William and Mary Quarterly* 69, 1 (January 2012); Isenmann (ed.), *Merkantilismus;* Stern and Wennerlind (eds.), *Mercantilism reimagined*. Here, Sirota, "The church," 197, has already pointed to the concept of "nationalization." More strongly concentrated on economic language and ideas are Magnusson, *Mercantilism*; Finkelstein, *Harmony*, but the question of the "national" does not play a role in that literature and the link to practice is missing.

differing French and British conceptions of rule in and of the Mediterranean. In so doing, I follow the heuristic assumption that those empires themselves emerged via a bottom-up process that involved the continual specification of "nation non-knowledge," through asking and answering questions about the national. This happened in an osmotic relationship with framing and circumferential imperial discourses, but this imperial thought was changing more slowly and it remained detached from everyday practice.

The national form of distinction that began to dominate Mediterranean trade was a question of operative (non-)knowledge, while imperial discourse was moving toward epistemic knowledge. On a very basic level, one had to know the nation to which a given ship, sailor, passenger, cargo or captive being ransomed by pirates belonged. The nationalization of economics meant, first of all, the transformation of something that had hitherto been in a state of nescience into a specified unknown. From the highest level of imperial bureaucracy – the royal courts, the admiralties – to the London port officers and the *Chambre de commerce* in Marseille, the question "what nation is he or it from?," was a constant traveling companion for each man on a ship and each consul in the Mediterranean port cities, and it dictated everyday decision-making and politico-economic planning. The British and French did not ask about the national in the same way, however, and that national distinction was embedded in different general frames of thought. In the following, I compare both trade empire mercantilisms from the perspective of "non-knowledge about the national." This is an approach different and complementary to macro- *and* microhistorical research on imperial economics in general and on Mediterranean commerce in particular. Macro-historical approaches tend to presuppose the category of the nation in their narratives: "The Dutch," "the French" and "the English" conduct trade; but how those categories were themselves new, and to some extent arbitrary, and how they created paradoxes and were an object of continual discussion and interrogation, is not taken into account.[2] Microhistorical works, on the other hand,

---

[2] To give just one prominent example: the category of the nation is a blind spot in the narratives of Jonathan Israel, where the nation is a preformed category and not an object of historical investigation itself (cf. Israel, *Dutch primacy*; Israel, *Conflicts of empires*). For research more considerate of the dimensions of intellectual history cf. Hont, *Jealousy of trade*; Cheney, *Revolutionary commerce*; Reinert, *Translating empire*.

are familiar with how to look into the concrete realities of investigation into the national, but they are usually less interested in administrative standards and practices in addition to the overarching imperial concepts that were still the rules of the game. Even for a single actor, "the national" was intrinsically important in a myriad of interactions. The emphasis is put here on the osmotic relationship between practice *and* theory. I combine the macro and the micro, and focus on the mercantilism of empires. Because of that, other figures, groups and institutions play a minor role here, even if, in purely economic terms, they were very important. For example, the Greek, Jewish and Armenian trading diasporas (among others) were many things, but not imperial actors. There were no mercantilist norms, ports, or institutions that inquired in a comparable form into, say, Greek nation-non-knowledge in the Mediterranean.[3] What one may hope to learn by this third approach beyond the micro/macro opposition is, at first, somewhat tautological. It is how these empires, by defining and searching for the unknown national, were searching and finding themselves by defining what they are. I am interested in the constructive power of non-knowledge, something that might seem to be a paradox. The void of unknowns seems to be the least firm ground to build an empire upon. Yet it was precisely through the continual consideration of the question about the national and the nation abroad that the limits of the empires in question became visible at all. In addition, we must also consider the extent to which the category of "state" was connected to those of "nation" and of "economy."

This has to be seen within what one can define as a two-level system of Mediterranean trade. On the first level, European merchants were, and saw themselves as, competing against each other. The second is the parasitic corsair economy. As the corsairs gained most of their whole societies" wealth from piracy or its functional equivalent, maintaining the threat of piracy but allowing its replacement by regular payments according to international peace treaties, states started to protect "their" merchants in different ways against their European and corsair competitors. The protection of merchants – in the Mediterranean cities as well as at sea – was thus an important economic factor on the first level, a transaction cost, shared between the merchants themselves and

---

[3] Cf. Trivellato, *Familiarity*; Eldem, *French trade*; Aslanian, *From the Indian Ocean*; Greene, *A shared world* and Greene, *Catholic pirates*.

the states. It was also possible to borrow or "buy" a nation's flag, or even use it without formal permission, and therefore take advantage of a given nation's protection. This was actually in the interest of the states themselves because they obtained valuable duty payments from each shipowner or captain flying their nation's flag. This could also be detrimental for a state, however, if there was abuse or the unauthorized use of a nation's protection. From this, we see that the two-level system was transforming into a three-level system: competing European merchants, competing states/nations, parasite corsairs. How these various circles interacted with each other will be seen in the following.

Norms as Specifiers of National Non-Knowledge

In theory and in practice, the English normally distinguished between the particular interests of merchants and a general interest of "the nation," while the French usually used "the state" in that second position.[4] That seemingly small, but fundamental difference in wording ("nation" vs. "state") has to be kept in mind when studying the meaning of the national in the Mediterranean Empires' trade organization and competition. The French increasingly conceived of their trading houses in the Mediterranean, protected by their consuls, as a part of

[4] Cf. as examples for the English case: Petty, *Britannia languens* (1689), 10f: "Trade is either National or Private ... Private Trade hath regard to the particular Wealth of the Trader, and doth so far differ in the scope and design of it from the National, that a private Trade may be very beneficial to the private Trader, but of hurtful, nay of very ruinous Consequence to the whole Nation"; Cary, *Essay* (1695), 1: "It being possible for a Nation to grow Poor in the Main whilst private Persons encrease their Fortunes"; Praed, *Essay* (1695), 51. Cf. in contrast to the French case: Éon, *Commerce honorable* (1646), 3: "le commerce est une des principalles & des plus essentielles parties de l'État. Car comme l'État consiste dans l'assemblage de diverses personnes, le Commerce & le Gouvernement sont les deux parties qui le composent."; Pottier de La Hestroye, *Restablissement* (1715), 117: "il faut scavoir demesler l'interest général qui s'accorde toujours avec l'interest de l'État et l'interest particulier qui est presque toujours opposé à celuy de l'État." Because the French production of more general treatises starts only later in the eighteenth century (aside from Montchrestien etc.), the *state* centered perspective is obvious in all prior publications such as the *Advis*, the *Testaments politiques*, the state finance projects like that of Vauban or Gueuvin de Rademont, and in the texts of John Law (*Œuvres*). Pierre de Boisguilbert, for example, always used "la France," "l'État," "le roi" as point of reference in his late seventeenth- and early eighteenth-century texts, but nearly never "nation." I checked virtually all works before the physiocracy 1750 watershed as listed in *Économie et population*.

the state's extensions overseas. The internal/external distinction took the form of an invisible appendage of state borders abroad. The English mercantilist conception of trade did not subordinate merchants' activities to the *state* as much, but it did integrate forms of state power into their commercial network. English merchants acted more as agents of their nation than their state. While this is a difference encountered throughout all sources in the following, in a striking parallel the fundamental guiding standards emerged for both England and France around 1650/60.

Defining the Unknown

A very important process of reform and legislation around 1660 provided the pivotal moments for England and France, when respective shifts occurred, turning economic activities in a state of nescience about "nation" into one where the nationality (of merchants, sailors and ships) became *the* central specified unknown.

England

The 1660 Second Navigation Act[5] and the 1662 Act of Frauds,[6] together with the system of peace treaties and sea-passes, marked a decisive point of the nationalization of English seafaring in the Mediterranean. The first 1651 Navigation Act had been an "experimental law" to some extent, and, even though it had been strongly influenced by the lobbying of the Levant Company, only with the 1660/62 combination of laws did legislation achieve enduring decisiveness and incorporate important clauses concerning the southern trade.[7]

This occurred through the transformation of a state of nescience embedded in former practices into specifications of non-knowledge about the nationality of sailors. Those regulations required that English merchants who wanted to import from or export to the Mediterranean "beyond Malaga," had to provide an English ship with a minimum of two decks, armed with sixteen guns with at least thirty-two

[5] "An Act for the Encourageing and increasing of Shipping and Navigation," *Statutes of the realm* 5 (1819), 246–250.
[6] 'An Act for preventing Frauds and regulating Abuses in His Majesties Customes,' *Statutes of the realm* 5 (1819), 393–500 = 14 Car II c. 11.
[7] Harper, *Navigation laws* remains unsurpassed for the history of the legislation itself (citation on p. 53).

men, with the master and at least ¾ of the crew needing to be English.[8] The Navigation Act was rigid insofar as it allowed the seizure of foreign ships and their goods; the Act of Frauds dealt with the discipline and fine-tuning of the English ships themselves. The Mediterranean clause of the Act of Frauds only concerned the merchants who mostly conducted trade between Livorno, Spain, Portugal and England; the Levant Company – founded first as Turkey Company in 1580/81, and provided with a renewed charter in 1662 – was not affected. The Englishness of the company's trade had already been secured by virtue of the company being closed to both foreigners and naturalized merchants until 1753. Even beyond the question of nationality, the company could only be joined by "meer merchants," a restriction which remained firm despite frequently recurring complaints.[9] This stabilized the "Englishness" of the factories' personal in the Levant (Constantinople, Aleppo, Smyrna) probably more effectively than the French did.[10]

As for incoming ships until the 1740s – mostly between 1675 and the early eighteenth century – there were numerous cases when merchants applied to the Treasury to be freed from the one percent duty as they had lost men during the voyage due to several problems. These applications demonstrate how rigorously the surveyors of the Navigation Act and the Customs Commissioners controlled the ships in the English port cities. Mostly the problem was that the overall number of men was too small.[11]

The one percent duty of the Act of Frauds concerned the character of the ship to be armed and suitable for defense, an armed condition that had to be maintained by English men for their English ships. To better understand the meaning of those norms, one has to take a short look at its pre-1660 history. Following 1617, when the Barbary corsairs attacked English ships and port cities on the Western English coast for the first time, the English government raised £40,000 over the next three years, in order to finance warships and men against this new

---

[8] 14 Car II c. 11, § 33. [9] Schulte Behrbühl, *Deutsche Kaufleute*, 226–233.
[10] Cf. Wood, *Levant Company*, 136–140; Matterson, *English trade*, 222–242.
[11] Cf. cases from the 1670s to 1742: CTB V, 99, 1133; CTB VI, 616f., 644; CTB VII, 305, 349, 364, 375f., 533, 645; CTB VIII, 2135, 2147; CTB IX, 1, 2166f.; CTB IX, 485, 1247f., 1259f.; CTB XI, 252, 268, 332, 366; CTB XII, 155, 249, 271; CTB XIII, 145, 312, 343f., 360, 384; CTP II, 73, 108; CTB XXXI/2, 108; *Journals of the Board of Trade and Plantations* IV, 379–385; CTBP II, 223; CTBP III, 161, Nr. 20; CTBP V, 137; CTBP V, 148.

threat.[12] The government seized the money through fixed sums demanded from the port cities in proportion to the amount of their trade which the London administrators had calculated from past customs records.[13] Some of the cities, apparently first of all London, but also Weymouth, decided to raise that money through a one percent duty on import and export customs. This was probably a repurposed technical practice that the port's financial administration had utilized before.[14] Others simply collected the required money from their merchants. Nearly all complained that the London center's pretended knowledge of local trade was false and outdated, not least because of the current losses caused by the corsairs. The local character of the duty also created several unintended problems within the inner-English competition of the outport cities.[15] Perhaps because of that experience and due to intense discussions about the similar and related ship money (1635–1640),[16] the solution of a duty on imported and exported goods – even if still remembered[17] – was not chosen during the 1620s

[12]   Hebb, *Piracy*, 21–42. The Merchants of the East India, the Turkey, Spanish, Barbary, French Eastland, Muscovy, West Country and Flanders Companies all wrote a petition to Sir Thomas Smith asking for help and defense against the corsairs (March 9, 1617, PC 2/28, f. 581). The idea was to hold "a continued Force and strenth [sic]" (John Digby, April 30, 1617, SP 14/91, f. 78), and for that purpose £ 40,000 should be collected by the City of London and other port cities by a "proportionable contribution" (the same, April 30, 1617, SP 14/91, f. 79).

[13]   City of Southampton to the Council, February 22, 1619, SP 14/105, f. 195; " ... wee suppose that the other Ports of the Kingdome doe contribute according to the proportion" (City of Bristol, February 28, 1619, SP 14/105, f. 222; Dartmouth, March 6, 1619, SP 14/107, f. 12).

[14]   Exeter, March 20, 1619, SP 14/107, f. 65v: the first step was a "ticket ... certifying that everyone ... have paid their due uppon this collection" before being allowed to "receave ... goode or marchendizes ... in or out"; Weymouth, March 10, 169, SP 14/107 f. 23r mentions explicitly the "Customes ... of ... one upon every hundred which is the charge as they have heard that upon the like occasion is taken in London and elsewhere."

[15]   For example, the Dorchester merchants withdrew their trade from Weymouth because of the local one percent duty. See Weymouth to the Council, June 8, 1619, SP 14/109 f. 153 and May 30, 1620, SP 14/115, f. 85. Other merchants complained about being charged twice for the same purpose if they conducted trade in two cities (Barnstaple, June 17, 1620, SP 14/115, f. 137).

[16]   State of research: Langelüddecke, "Ship money."

[17]   The members of the Algiers Commission, Paul Pindar, Kenelm Digby and John Wolstenholme recalled on March 15, 1631 that "About 12 years since ... order was taken to leuie one percent of the merchant goods to raise such a some" for the purpose of suppressing the pirates, but the commissioners judged that now there would be "no hope of raising money in that way" (SP 71/1, f. 111r).

and 1630s when the corsair problem was growing.[18] Only twenty-three years later, in 1642, just a year after the Long Parliament had prohibited Charles' ship money, was the so-called Algiers (*Argiers*) duty adopted for that solution. It was, however, moved to the national level: the one percent was now to be levied in every English port city. While paying ship money for a royal navy was unpopular, such a duty to deal with the problem of piracy was accepted.[19] The 1642 solution decentralized the necessary knowledge about the amount of trade by ordering that local customs officers assess the levy according to current circumstances instead of calculating in London a proportion from past data meant to be valid for the present and through its national character; the unintended problems involving increased inner-English port competition were resolved. The 1642 duty act was extended several times.[20] While the money not used for ransoming captives was finally allocated to financing the navy in general, the 1659 overview of England's revenues still listed the one percent duty.[21] Those solutions prior to 1660 were not linked to the rules of nationality concerning the ships and their men. It was first (in 1617–19) an answer based instead on the old feudal concept of the defense of the realm to which the cities had to contribute. The second step, the national tax of 1642, still had its roots of legitimacy in this concept of the defensive obligation of the king against the realm's enemies and of his subjects to contribute the financial means to this aim.

The 1660/2 standards represent the sublimation and projection of the earlier defensive character of state violence against foreign threats into a mercantilist internal/external distinction by inquiring into and controlling the national character of commerce. Paragraph 33 of the

[18] Gray, "Turkish piracy"; Barnby, "The sack"; Hebb, *Piracy* and Matar, *Britain and Barbary*, 38–75.

[19] Matar, *Britain and Barbary*, Appendix 1, 173–176 for a recent print of the Act. Hebb, *Piracy*, 27f. was the first (and nearly only) to see a parallel between financing the navy against the threat of piracy and the ship money, but his study stops before the Algiers tax and the continuity of the one percent duty from 1617 to the 1660/62 Acts is not seen.

[20] Prolongations: January 28, 1644/45 Firth and Rait (eds.), *Acts and ordinances*, vol. 1, 609–611; July 7, 1645, ibid., 731–732; May 11, 1647, *Journal of the House of Lords* 9 (1646), 182–185; March 26, 1650, Firth and Rait (eds.), *Acts and ordinances*, vol. 2, 367f.; June 26, 1657, Firth and Rait (eds.), *Acts and ordinances*, vol. 2, 1123–1130.

[21] 'The income of England,' April 7, 1659, in: *House of Commons journal* 7 (1659), 627–631.

Act of Frauds set minimums on the type of English ship capable of being a "swimming defence machine" on its own. Now the duty worked by forcing merchants to use such "swimming little exclaves of England" in the Mediterranean – if not, the duty served as a contribution to necessary convoy shipping sponsored by the crown. The distinction between foreigners and Englishmen, present in the ports and – at least theoretically – in the whole Mediterranean, pointed in an abstract manner back to those older roots of the defense of the realm. The economic and prohibitionist impact of the 1660/62 regulations was high. Transport between the Mediterranean and Britain was nearly completely monopolized by British ships.[22]

From an epistemic point of view, the watershed of 1660/2 transformed the state of nescience about nationality into a central specified unknown. Non-knowledge about the nationality of each person on each ship in the Mediterranean was now of importance. It was specified as a problem and formed *the* central directive rules of Mediterranean commerce.

Eighteenth-century merchant handbooks transmitted those norm specifications as they had developed and practiced during the seventeenth century and following the 1701 Union. According to this, "British-built ships" were:

Ships of the Built of Great Britain, Ireland, Guernsey, Jersey, or the British Plantations in Africa, Asia, or America, and whereof the Master and three fourths of the Mariners are British, that is, his Majesty's Subjects of Great Britain, Ireland, and his Plantations, and three fourths of the Mariners such during the whole Voyage, unless in Cases of Sickness, Death, etc.[23]

A "stranger" was someone:

born in a foreign Country, under the Obedience of a strange Prince or State, and out of the Allegiance of the King of Great Britain; or a British Man born, who has sworn to be subject to any foreign Prince; though if such British-born Person, returns to Great Britain, and there inhabits, he must be deemed as British, and have a Writ out of Chancery for the same: And likewise the children of all natural-born Subjects, though born out of the Allegiance of his Majesty, etc. and all Children born on board any Ship belonging to, or in any Place possessed by, the South-Sea Company, are to be deemed natural-born Subjects of this Kingdom.[24]

---

[22] Cf. below n. 69.   [23] Crouch, *Guide*, 131, 142.   [24] Ibid., 145.

Evidently, similar to the Navigation Act itself, handbooks like that of Crouch reflect an already strong Atlantic orientation, but at the time of the handbook's publication, the Southern-European trade still represented a good third of Britain's foreign commerce and non-European trade only another third. London remained the uncontested British center of Mediterranean commerce,[25] and even more so, "as much of a third of New England's adverse balance of payments with the mother country came from available returns from the Spanish, Portuguese, and Mediterranean markets."[26]

Probably nowhere else besides the kingdom's naturalization records do we find more precise definitions of "Englishness/Britishness" and "strangers" than in these foreign trade records and merchant handbooks.[27]

France
The French parallel to the English combination of the Navigation Act, Act of Frauds, as well as war and convoy shipping, were the almost exactly contemporaneous French reforms of the 1660s regarding Mediterranean shipping and the central port of Marseille. Most significant for our purposes was the prominent edict of March 1669.[28] This edict laid the ground for the status of Marseille as a free port and its monopoly over the Mediterranean for French imports and (less so) exports. If one reads its text, especially the first section, one sees how the edict uses the old concept of *commercium*, as exchange between peoples and "even the most opposite spirits who become conciliated

---

[25] Imports and exports to and from Southern Europe and the Mediterranean – the Barbary risk zones – each made up between 26 and 30 percent of all foreign trade in 1663/69, 1699/1701, 1722–1724, 1752–1754. Exports only make up 26.6 percent. While its absolute volume remained quite stable throughout the eighteenth century, its share in the overall growth foreign trade sank to 19.4 percent in 1752–1755 (imports) and to 14/17 percent (imports/exports) in 1772–1774, cf. Davis, "English foreign trade 1660–1700," 164–165; Davis, "English foreign trade, 1700–1774"; French, "London's overseas trade," 482; French, "London's domination," 29. More recent survey articles are usually neglecting the southern and Mediterranean commerce, cf. Engerman, "Mercantilism."

[26] Morgan, "Mercantilism," 183. Cf. Lydon, *Fish and flour*, 8 and passim.

[27] Cf. for comparison with the Atlantic perspective Zahedieh, "Economy," 55; Braddick, "Civility and authority," 128. Kidd, *British identities*, 250–286 focuses on Gothicism for its Atlantic dimensions.

[28] For its text cf. Julliany, *Essai*, vol. 1, 221–228, and *Lettres instructions et mémoires de Colbert*, vol. 2, 796–798.

through a good and mutual correspondence." It stressed that, by royal grace, the act rescinded all former duties levied upon foreigners, specifying each old half to one percent duty annulled, and announced a message of liberty for all foreigners to come to Marseille.[29] Nevertheless, under the umbrella of that gracious liberty, the second part of the edict introduced a heavy 20 percent duty on all goods not shipped in French vessels. Works on economic history from Masson to Carrière have consistently stressed the prohibitive character of the edict due to that 20 percent duty.[30] Masson and Rambert called it "a kind of [sc. French] Navigation Act,"[31] and Carrière devoted a chapter to evaluating the economic rationality of the prohibitive 20 percent.[32] One of the best informed contemporary historical accounts of French-Anglo-Dutch commercial competition from the sixteenth century until around 1700, Pottier de La Hestroye's *Mémoire sur le restablissement du commerce*, highlighted the anchorage duty and the 20 percent duty as the only French means of protection, endangered by new peace treaties of Nijmwegen and Utrecht which allegedly granted the Dutch too many liberties. Colbert's 1664/69 reforms and legislation was placed here in exact parallel to the English and the Dutch prohibitive measures.[33]

Without doubt, this is the perspective of the state-centered discourse of political economy. Today, scholarship less interested in inter-state competition than in the complex realities of merchant networks, sometimes hidden by the perceptional framework of *state simplifications*, has come to appreciate the edict's impact on attracting foreigners,

---

[29] Cf. the beginning: "Comme le commerce est le moyen le plus propre pour concilier les différentes nations et entretenir les esprits les plus opposés dans une bonne et mutuelle correspondance . . . déclarons le port et havre de nostre ville de Marseille franc et libre à tous marchands et négociants . . . à cet effet nous avons supprimé et supprimons [sc. the following duties . . .]" (ibid.). Cf. for that most general notion of commerce close to "communication" and "relationship" Steiner, "Commerce," 182.

[30] Masson, *Commerce XVII*, 160–177; Bergasse and Rambert, *Histoire*, 204–214; Paris, *Histoire*, 3–43; Carrière, *Négociants*, 309–330.

[31] Masson, *Commerce XVII*, 166; Bergasse and Rambert, *Histoire*, 208.

[32] Carrière, *Négociants*, 319–330.

[33] The treatise, first written in 1698, was revised and expanded until 1715. I follow here Rothkrug, *Opposition*, 435 against Harsin. Cf. also Faure, *La banqueroute*, 57f.; Murphy, *John Law*, 8–11. On the context of the establishment of the *Conseil de commerce* on June 29, 1700 cf. Kammerling Smith, "Le discours économique," 31–37. On Pottier de la Hestroye as advisor of the crown cf. McCollim, *Assault on Privilege*, 143–145.

foremost Armenian and Jewish colonies – but there is a risk of over-looking the second, prohibitive part of the legislation.[34] Many of the foreigners would probably not have come to Marseille if French trade had not been monopolized through the 20 percent duty. Their immigration was only partly voluntary, prompted by Colbert's promises of liberty and freedom. This Janus-faced character of the French regulations of prohibition *and* free port establishment was different from the English Acts, and they likewise produced different results on the epistemic level of ignoring/knowing nationality.

The 20 percent duty had been advocated by Marseille merchants as early as 1658.[35] Their main interest was to exclude both foreigners and other French merchants, in addition to monopolizing the Levant trade in Marseille hands. In October 1662, Henri de Maynier de Forbin, baron d'Oppède, the first president of the Parlement de Provence, general counsellor to Louis XIV, and Colbert's close collaborator in reforming the port of Marseille,[36] called together, by royal order, the aldermen and deputies of Marseille, as well as the deputies of the cities of Toulon, Antibes, St. Tropez, Fréjus, La Ciotat and Martigues in the *Chambre de Commerce*. After having collected several *mémoires* and *advis* on the "restablissement du commerce," the deputies assembled between October 9 and 14 in the presence of the Duke of Mercœur, they deliberated over propositions and d'Oppède produced minutes of their discussions to be sent to the king.[37] Royal protection of commerce, they argued, should be granted by the renewal of the capitulations with the sultan and the installation and strong empowerment of the office of ambassador to Constantinople. Levantine commerce, they maintained, could not flourish without the state's protection. The central measure of reform in this important 1662 collective *mémoire* was a proposal by the assembly and d'Oppède to introduce the 20 percent duty. Its aim would be to distinguish between the commerce of the Ponant and of the Levant

---

[34] As already Cole, *Colbert*, vol. 1, 392–396, Takeda, *Between crown and commerce*, 31–36 completely omits the 20 percent duty. But cf. Trivellato, *Familiarity*, 116f.

[35] Bergasse and Rambert, *Histoire*, 72.   [36] Masson, *Commerce XVII*, 141–146.

[37] This central document in AN AE B III 234 n. 13 bears the autograph signature of Oppède. Masson, *Commerce XVII*, 160–177 remains the best account of the discussions between the crown (Colbert), its representatives d'Oppède and Arnoul, and the *Chambre de Commerce*, but is mistaken to believe that only in 1667 "prit corps le projet d'affranchissement du port de Marseille."

and to prevent the English and French from conducting Levantine trade under "their own flag" instead of the "French flag."[38] The whole compromise – 20 percent duty on the one hand, affranchissement and cottimo on the other – was already worked out at that time in collaboration with all other Levant port cities. In its general framework, the dual character of the standard was present from the beginning of the year-long decision-making process. The matter still took some time and discussion.[39] Once established, the duty formed a firm basis for Marseille's Levantine monopoly to which the *mémoires* concerning Levant commerce addressed to Colbert and Seignelay in the 1670s and 1680s always referred as such.[40]

The norms were successful. Of the 371 ships that departed between 1680 and 1683 from Marseille to the Levant, only ten were not French. The monopoly was enduring. In 1753, of 439 ships, there were only seven ships classified as foreign. This is even more decisive if we look just at France's northern competitors (the English, Dutch and the French "Ponant"): Of the 16,210 ships that entered the port of Marseille from the Levant between 1709 and 1792, only 199 vessels (0.12 percent) were from Northern France or Northern Europe.[41] Vice versa, Marseille merchants were largely unable to participate in the Northern trade, owing to their ships not being competitive with those produced by the Northerners. From the point of view of Marseille, commerce had been partitioned between the passive Northern and Atlantic trade on the one side and the concentration on their active Levant trade on the other.[42] But one has certainly to distinguish shipping *in* the Mediterranean from shipping to and from the Mediterranean and the respective home country. Duties such as the 20 percent had a decisive impact on the latter, but less so on the former.

As opposed to the English standards, the French rules were at first rather imprecise about the "Frenchness" of a French ship. The 1669 edict expressed the norm in a quite complicated manner, articulating

[38] AN AB III 234 nr. 13.
[39] Masson, *Commerce XVII*, 160–164; cf. *Mémoire* 1669, inc. "Trois choses ont ruiné la ville de Marseille et son commerce," CCM H 7.
[40] CCM H 7 (1678, inc. "La chambre du Commerce de la ville de Marseille"); AN B III 234, nr. 30 (1682, non fol., last page); ibid., nr. 34 (1684, non fol., point 4); cf. however, the letter of Choiseul to the *Chambre*, Versailles, August 2, 1762, CCM E 148; and CCM C 143–162 for frauds and debates concerning the duty.
[41] Carrière, *Négociants*, 584–594.   [42] Carrière, *Négociants*, 500.

foremost the idea that all ships should sail directly from their outgoing Levant port to Marseille without stopping "in Livorno, Genoa or elsewhere," and by charging all goods transported on foreign ships, even if French-owned, with the duty. It also formulated rules of registration with the French consuls in the Mediterranean port cities. The edict always used the legal distinction between "our subjects" and "foreigners [*estrangers*]" or between "foreign" and "French ships," but it did not define what was to be considered a "foreign" as opposed to a "French" ship.[43] Only through evolution in practice and through further royal decrees – a 1681 ordinance and an elaborated 1727 declaration – did those definitions become more specific. Now, to be officially counted as French, a ship had to meet a quota of being at least ⅔ "really French." By this qualification, French administration allowed foreigners to hold as much as a ⅓ interest in a French ship. This was different from the English case, where the body of the ship had to be not only British-built but also completely British-owned. Nevertheless, the question remained, what was to be counted as the corresponding "thirds" of a ship? Its sailors? The value of the ship itself? The goods it carried? From the 1681 Ordinance until the second decade of the eighteenth century this remained rather unclear, and therefore probably also not seriously debated or inquired into, either by the merchants, the state, or the *Chambre de commerce*.[44] The mixed character of free-port politics and of protectionism led to a higher degree of fluidity in the French case.

The French were also undecided about the best way to control the "floating Frenchness" in the Mediterranean. Early ordinances opted for centralizing administration and procedures within Marseille. Passports and *congés* were only to be issued by the Admiral and his officers in that port city, no consul, and not the ambassador in Constantinople were permitted to issue such passes, something that had been a common

---

[43] Julliany, *Essai*, vol. 1, 224–227.

[44] There were some attempts to gain higher precision as by the *Intendant des galères* Pierre Arnoul (on him Dessert, *La Royale*, 46f.) who proposed in 1715 to form a register in which the *Commissaire de la Marine* should duly note the first name, last name, age, place and address of residence of each sailor who had lived there for at least five years to be then recognized as "French" in the *bureau des entrées et des sorties* (CCM H 7). This corresponded with the number of years for naturalization as stated in the 1669 edict. Cf. Sahlins, *Unnaturally French*, 96 for a similar but later 1718 rule.

practice.[45] Several subsequent orders and *mémoires* relied solely on the consuls' authority and their, by definition, decentralized administrative power. Consuls were called upon to examine ships and use their chanceries for the registration process.[46] This was in fact a question of the practical organization of epistemic processes: At what location can one best cope with ignorance? Where should knowledge crystallize? This is foremost a question of deciding between centralized and networked organizational structures for knowledge.

The differences between the rigorous demarcation of Britishness and the more fluid, evolving Frenchness of the merchant vessels might be explained by the different spirit and roots of the French legislation. Referring to the basic model of Mediterranean economics, defined by the two levels of inter-merchant and inter-European competition on the one hand, and the parasite economy represented by piracy/privateering on the other, the Colbertian standards were predominantly formulated from the perspective of the first level, while the British ones had largely been developed from the perspective of the second. While the French had known well of the Barbary problem for a long time, the 1669 Marseille edict evinced no genealogical roots in concepts of the kingdom's defense, as the English legislation did. In fact, in that same year, 1669, Colbert transferred navy warships from Toulon back to Marseille and the French trade ships were likewise armed with cannons like the English. From the point of view of how the conglomerate of Versailles/Paris/Marseille and the London centers of administration thought and ruled, there was a different logic at hand. While we usually conceive of British trade politics as the spearhead of modern economic development, paradoxically, English mercantilism seems to have been far more "war-born," while French mercantilism – despite being

[45] The ordinances from between May 22, 1671 and March 24, 1686 are imprecise concerning the consul's power, but the ordinance of December 22, 1686 explicitly prohibited consuls from issuing passports. Still, the consuls were allowed to issue a passport to a captain who had bought a ship in a foreign country. He then had three months to come to Marseille with this provisional passport to exchange it for one issued by the admiralty. Intelligent merchants sometimes used that "provisional" passport much longer and thus undermined the admiralty's authority (CCM E 146, and Pierre Arnoul, "Mémoire sur les abus que font les Nations Etrangères de la Bannière de France," December 14, 1715, CCM E 147).

[46] Cf. at latest the Art. 17, 25, 31 of the Maurepas Ordinance December 9, 1727, CCM E 147 specifying the consul's role in registering the *rôle d'équipage* and embarking of ships.

so decisively envisioned from the perspective of state – seems "trade-born."

Mercantilist Paradoxes: Known Flags, Unknown Nations

From the point of view of inquiring into the epistemics of nationality, the interdependency between the two-level system and the emerging systems of security production – Ottoman-European and Barbary States-European treaties, sea-passes and armed naval support –[47] created several further administrative procedures. These were in constant interplay with the duty and customs system and the definition of the nationality of a ship, its cargo and its passengers.

National competition did not just need to play the protectionist card. It also pursued an expansionist agenda in terms of its nationality, and both lines of reasoning were partially contradictory. This becomes evident when analyzing the carrying trade and its gradual merging with practices of (ab)using a foreign flag. The carrying trade, the transport of other ("foreign") merchants' goods was of importance as was also the practice and possibility for "foreign" merchants to sail under the protection of one of the large naval powers.[48] The English plied the carrying trade in the Western Mediterranean to some extent, mostly along the Italian coast or along routes from Livorno to non-European ports beyond the Mediterranean. The French did both on a much larger scale, as the attractiveness of their protection grew from the late seventeenth century until a period that was once termed the "French reign" of the Mediterranean after 1740.

The traditional sign of being under the protection of a given "nation" was the ship's flag. The stronger the risk of piracy was, and the more resources that had to be used for the protection of the ships flying one's flag, the more it was in the interests of a given state to ensure that a ship flying a French flag was also a real French ship – whatever a "real French ship" might be. And as the treaty system between the Barbary

---

47 Cf. Paris, *Histoire*, 188–193 for how the *Chambre de Commerce* paid large parts of the Royal navy expeditions; Villiers, *Marine royale*, vol. 1, 64f. for the growth of the French navy in general; Hebb, *Piracy*, 136–143 for an estimation of the costs of piracy attacks directly affecting England or English ships 1627 to 1640 (£1,000,000 to £1,300,000).

48 Panzac, *La caravane* is now the starting point, but cf. Heywood, "Ideology" 18: "A study of specifically Anglo-Saxon – better, British – participation in the Mediterranean caravane maritime remains to be written."

states and the European powers had evolved and the corsairs became more or less willing members of an international maritime law system, they became increasingly interested in being certain what *was* a French or English or Dutch ship and what was not, if they met one at sea. This was a fairly simple logic, but it resulted in continual communication about ignoring, want and the need for knowledge about those criteria.

From the given interdependency between inter-European competition and the parasitic piracy economy, one can derive, grosso modo, a rule that in times of significant pirate activity, when the transaction costs for security were high, it was more advisable to keep the number of those who profit from a given state's valuable protection small, so as to act in a protectionist manner against the practices of flag borrowing. In times when pirate activity was low, it became profitable instead for the flag of a given state if many merchants from other nations sailed under its protection through increased duty revenues, the concentration and attraction of flows of merchandise and the symbolic "branding" effect of apparent domination at sea. No early modern state and port accounting calculated those relationships in a mathematized form in our period, but the general rules of relationship and interdependency were clearly perceived and explicitly articulated.

The Normative Framework
The first English-Algerian peace treaty that contained detailed provisions concerning encounters between Algerian and English ships at sea and the control of the pass was concluded for the Crown by John Lawson in the same year as the Act of Frauds was issued, on April 23 / May 3, 1662. Under its terms, Algerians were to let every merchant ship whose captain could provide "a pass, under the hand and seal of the lord high admiral of England" sail in peace. If such a pass did not exist, the ship was, nevertheless, supposed to go free if the "major part of the ship's company be subjects to the King of Great Britain."[49] Following treaties repeat that latter clause.[50] The first order by which Charles II instructed the Lord High Admiral and the farmers of customs about the procedure is

---

[49] Art. II (executive instructions), Chalmers II, 364 (date old style) = Dumont VI/2, 419f. (date is Gregorian).

[50] Treaty October 30, 1664 (Art. II executive instructions), Chalmers II, 32; Treaty November 29, 1672 (Art. IV), Dumont VI/1, 205; Treaty April 10 (old style), 1682 (Art. IV), Chalmers II, 367.

from November 23, 1663.[51] Following those orders, when a ship wanted to leave the port of London for a destination in the Mediterranean beyond Malaga, several necessary steps had to be taken: "The Surveyor of the Port where the Ship lies must go on board, and examine and survey her, and muster the Seamen; then he must certify in Writing under his Hand to the Collector of the Port, the Burthen and Built of the Vessel, the Number of Men, distinguishing Natives and Foreigners, the Number of Guns, what sort of Vessel she is, &c." The Collector then prepared an affidavit in which the Master's oath to the truth of all noted particulars was testified.[52] This affidavit was next transmitted to the secretary of the Admiralty who checked if the master of the ship had returned all past passes previously granted to him. The secretary had to ensure that the ship was English built or foreign but "made free," that its master was the king's "Naturall Subject" or "Forreign Protestant made Denizon," and that "two thirds of the marriners" were the king's subjects. The secretary then issued against the payment of a bond – in 1682 that tariff was £50 for ships up to 100 tons, £100 for larger ships[53] – the sea-pass which the master had to give back after returning to the port.[54] The first register of sea-passes issued by the Admiralty dates from 1662, but the extant series is interrupted between 1668 and 1683, and again between 1689 and 1729.[55] The first register 1662–1668 and one may probably infer by that also the passes themselves, did not mention the destination of the ship.[56] In 1682, the practice changed; the officers of the Admiralty were now to use

---

[51] Keppel, *Keppel*, vol. 1, 158.
[52] Cf. "The Forme of the Oath to bee made by the Master of an English built Ship" according to the 1682 rules, in ADM 7/76, f. 3r (for foreign-built ships, ibid., f. 4r).
[53] Cf. "The forme of the Bond directed to bee taken" according to the 1682 rules, ibid., f. 2v.
[54] *Ductor mercatorius*, 36f.; "The severall Rules, now in Force for the Granting Passes, made since the Peace with Algire in Aprill 1682," ADM 7/76, f. 1–5.
[55] ADM 7/630 (1662–1669) and ADM 7/75–76 (1683–1689), ADM 7/77 starting with 1730.
[56] ADM 7/630: The first 600 entries are numbered (until f. 43), then this practice seems to have been lost again, while later, at least since 1729, each pass had an individual number in uninterrupted chronological order. This verified the identity of each pass. The 1660s entries contain (1) name of the ship, (2) home port, (3) name of master, (4) burden in tons, (5) number of men, specifying mostly "all English," (6) number of canons, (7) construction (Dutch, English . . .), (8) identificatory picture on the ship's body or figurehead. The "forme of the Pass" (ADM 7/75, f. 2r) does not indicate the destination.

a table that contained the same scheme of entries. Still, the destination was not mentioned. Instead, an entry indicated the current location of the ship ("Place shee lyes at").[57] After July 10, 1683, the alternative that it was sufficient if the "major part" of the men were English – which opened a door to case sensitive consular negotiations in favor of English ships without passports – was made invalid. Indeed, corsairs often rigorously seized ships in the early eighteenth century if they had no "proper Pass ... altho she evidently appears to be a British ship" and confiscated their cargos.[58] The pass system had stabilized at least by 1682/83; the passes remained in use until the middle of the nineteenth century.[59]

In 1717, a typical sea-pass issued by the commissioners of the Lord High Admiral read:

Suffer the Ship Royall George of London John Levett Master, Burthen about Two hundred & fifty Tuns, mounted with Eightin Guns and Nauigated with Twenty two men, seventeen his Majestys subjects, British Built, Bound to Affrica to pass with her Company Passengers, Goods & Merchandizes without any Lett, Hindrance, Seizure, or Molestation, The said Ship appearing unto Us by good Testimony, to belong to the Subjects of His Majestie, and to no Foreigner.[60]

This example also shows how the captain respected the rule of at least ¾ English men of the Navigation Act at the lowest possible limit – 16 of 22 would not have been enough – as foreigners were cheaper. The ⅔ rule of the sea-pass issuing instructions was overridden here. The difference

---

57 ADM 7/75: the current location refers to the dock in the London port or the riverside place along the Thames, to an outport city or even to a place abroad (Newfoundland, the Straits, Barbados, Jamaica). Thus passes could be issued *in absentia* during those early years. Also added were entries for the date of pass issue, the name of the signing officer, and a space for a memorandum note.

58 Duke of Newcastle, Whitehall, November 27, 1732 in response to the Algerians' requests, SP 71/7, f. 617–622, 619.

59 The sea-passes were engraved printings with blank spaces for the individual ship and master. They were also cut into two parts along a scalloped line. The upper, smaller part was sent to the consuls who gave them to the Bey/Dey of the Barbary cities. The lower part was given to the captain. Only the perfect match of both sides on the sea or in Algiers and Tunis, when a ship was brought in, granted free shipping and secured from captivity. For reproductions of English, Dutch, Danish, Swedish sea-passes around 1800 cf. Gøbel, "Danish," 171; Müller, *Consuls*, 145; Ressel, *Zwischen Sklavenkassen*, Abb. 3, 4, 6.

60 "Original English Mediterranean Sea-Pass," October 21, 1717, AN MAR B7/474 nr. 21. In 1717, a typical sea-pass issued by the commissioners of the Lord High Admiral read (Fig. 1.1.)

between those several quotas remained. It also shows that quite often the ships left England with fewer than two men for each of at least sixteen guns, which would be here thirty-six men. If they entered like that London on the way back, they would have to pay the one percent duty of the Act of Frauds. Often, they hired still more men within the Mediterranean.[61] The problem of how to handle "English ships" without passes remained, as is evident already from the later treaties.[62] As those early eighteenth-century passes contain the destination of the ship, so did also the registers restarted in 1730, obeying a different notation system, following an order of December 18, 1729: a first destination ("Whither bound" directly from the place the pass is received at) and a second destination ("Whither bound from thence") was noted in two columns. So, passes were now issued to ships for instance "of London," currently lying in the "Thames," heading for New England by passing the Straights of Gibraltar and then back to Lisbon. Or first to "Lisbon, New England," counted as one first destination, and then to the West Indies. All ships now enrolled in the London Admiralty registers had to be currently anchoring at a port of the British Isles.[63]

One may ask why the northern nations relied on the strongly formalized sea-pass system, while the French congé documents remained far less developed. The higher interpenetration of Atlantic and Mediterranean trade reveals a reason for this. As the Atlantic trade grew, it was first not necessary to demand a sea-pass from the Admiralty in London when coming from the plantations; a certificate from the respective governor or his representatives was considered sufficient[64]. This, however, was a simple document written in English like a French congé without the haptic element of two parts that had to fit together. A ship, coming from there "and trading to Portugal, the Canaries, Guinea and the Indies"

---

[61] For the requests of exemption from those rules cf. the references above n. 11.

[62] Treaty April 10 (old style), 1682 (Art. IV), Chalmers II, 367; "Treaty concluded by Admiral Arthur Herbert and ambassador William Soames," April 6, 1686 (Art. IV), Chalmers II, 381. "Additional articles agreed with Captain Munden and consul Cole," August 17, 1700, Chalmers II, 387.

[63] ADM 7/76. Only from that time, they can serve as supplementary source to the Port Books where the customs officers noted the foreign outgoing shipping and which are preserved for 1686 and again since 1709 (Davis, *The rise*, 380).

[64] "Additional article agreed with George Byng," October 28, 1703, Chalmers II, 389. Registers of those early American Mediterranean certificates are to be found in the records of the provincial governors, see for Pennsylvania the entries excerpted in Linn and Egle, *Pennsylvania Archives*, vol. 2, 628f. from 1761–1764. Cf. Lydon, *Fish and flour*, 54f.

Figure 1.1 and Figure 1.2 Two English Sea-Passes, left of 1717 without the upper part, on the right of 1719 with the upper part still attached (AN MAR B7/473). On the left are visible the two blue 6 pence duty stamps for issuing, which are not the bond to be left, the stamp of the Admiralty on the left and the number of the passport (Nr. 17675) as it was registered in the passport register. The only printed matter is the ship and title in the wreath of leaves; but the text below is nevertheless a form already written by the officers with blanks left to be filled in (names of ship and captain, dates etc.).

Figure 1.1 and Figure 1.2

without any such pass would be taken by the Algerians despite its clear appearance of British property.[65] While the Romance languages were strongly present among the Barbary corsairs – themselves often renegades

[65] James Wisham to Lord Bolingbroke, Algiers, August 8, 1719, SP 71/5, f. 203r.

from Mediterranean countries – and as their _lingua franca_ was itself a fluid mix of Castellan, Catalan, Italian and Arabic elements,[66] documents from Mediterranean countries were more likely to be understood at sea, even if illiteracy was a problem. English was not a frequently spoken or read language on the North African shore. Consul Wisham recommended, in 1714, the issuance of real Mediterranean sea-passes and not only the said certificate to ocean traders coming from the Americas, Asia and Africa, because the Barbary corsairs "being not only entirely ignorant of the English Tongue, but even of the Characters of any Christian Language, can make no Judgment whether such Certificates are true or falses, and that the only mean whereby they can know <a> ship belonging to Her Majestys subject is by comparing the Indent of Passes with the Passes themselves."[67]

The Admiralty and customs officers in British ports not only sought to determine the "Englishness" of the ships and their men, the administration also ensured itself by the typical early modern "last step" of assurance, the ship master's oath. A promise under oath that the ship was English-built, belonged to the English, and that the necessary amount of men were English, served as a convenient capstone for the construction of the account books' correctness. To a certain extent, it also replaced reality through its sworn statement. If it later proved to be wrong, the administration could always blame the master, but for the pass registers, the exact number of outgoing English and foreign men was saved. According to those rules, English nationalizing was successful. Nearly all importing and exporting into and from the Mediterranean was conducted by English ships as early as 1615. The English had lobbied with energy against any plan by Italian firms to be established in London.[68] The nationalizing of the ships themselves took longer. Between 1654 and 1675, the share of foreign-built ships was between a third and the half of all ships leaving English ports, a proportion that fell from the 1680s to less than 10 percent.[69] This

---

[66] Changing concerning the region: In Morocco the _lingua franca_ was more "spaniard," in Tunis more influenced by the Italian, cf. Cifoletti, _La lingua franca_.
[67] SP 71/5, f. 203r. For the triangular trades between Britain, the Mediterranean and the "rest of the World" cf. McCusker, "Worth a War?"; Richardson, _Mediterranean passes_, reel 1, 10.
[68] Davis, _The rise_, 232, 296; Pagano de Divitiis, _Mercanti inglesi_, 151f. (case of 1666).
[69] Davis, _English shipping_, 48f. For the steps of "naturalization" of a foreign ship bought by an English or to free a ship caught legitimately as prize cf. Crouch, _Guide_, 132f. The Treasury Books and Papers contain numerous records on that.

indicates that the naturalization of ships – or, "making them free" – was of high importance; a large number of ships taken from the Dutch was "freed" in 1668/70.

The elite in Algiers and Tunis, the Dey, Bey, the milice and the Divan, was well aware of the European legislation[70] and knew what the treaties stipulated. When the English government communicated the new rules[71] for granting sea-passes to its consul in Algiers, Samuel Martin, in 1675, he transmitted this information to the Dey, provoking discussion in the Divan. The new Algerian ruler Bobba Hassan complained about the "many forraine shipps sayling under English Collours" but admitted that this could very probably not have been by permission of the English themselves as it was not in their own interest as the "true [sc. English] subiects found the lesse Employment for their shipping."[72] This was a time when the French were attacking their Northern European rivals – Hamburg, the Dutch – which rendered the British flag especially attractive. Many ships used it even if many of their sailors and even their captains were not English.[73] Such extensive use of a flag formally protected by the treaty induced the corsairs to bring ships into the Algerian harbor, and to distinguish there by closer scrutiny between "English" and "non-English" passengers. Applying in 1677 the clause of the 1662 treaty requiring a "major part of the ship's company" to be English, they classified a ship like the English-built *Susannah* sailing from Livorno to Palermo with its English captain Walther as not English "because his number of passengers, & strangers exceeded the number of the English Men."[74] The rule that ¾ of the men on board had to be English was apparently quite successfully implemented on vessels arriving in and departing from England. But English ships within the Mediterranean, such as most of

---

[70] Cf. below n. 123.

[71] There is no evidence for those specific rules (cf. the gap between 1664 and 1683, n. 55). It seems that there was in fact an interruption of the use of sea-passes between 1664 and 1683 (cf. SP 71/2, f. 172v).

[72] Samuel Martin to Whitehall, Algiers, May 6, 1676, SP 71/2, f. 103v.

[73] "The Narrative of Samuel Martin Consull of Algiers" (1676/7), SP 71/2, f. 172r: "those sea's swarmed with English shipping & indeed of all Nations under our Bandera."

[74] Samuel Martin to Whitehall, Algiers, May 31, 1677, SP 71/2, f. 183r. Martin then fought for the ship, referring to the main Art. 2 of the 1662/64 treaties, willingly ignoring the executory instructions which had just formulated that "major part" rule.

the Italian merchants based in Livorno, could and did depart more easily from those rules.[75]

The French were affected by the phenomenon of the borrowed or abused flag far more than the British. Before a ship left Marseille, it had to be checked by the *commis* of the *lieutenant de la Marine*. At least after 1681, there should have been a *rôle d'équipage*, a list of all men on board. The French counterpart to the sea-pass was the *congé* for the ship that was likewise issued against a certain payment according to a fixed tariff. The *congé* could, in theory, only be issued if two thirds of the men on board were French "currently resident in France."[76] From 1696, however, because of the system's frequent abuse by foreigners, a different *congé* form was distributed with the inscription "Etranger."[77] The higher frequency of flag abuse was due to several causes: the partial free-port status of Marseille, the effective nationalization of the Levant import and export shipping by the English and their Levant Company's monopoly, and the more fuzzy French legislation developing in a back-and-forth manner.[78] A 1671 order prevented French merchants and ship owners from lending their name to strangers ("prêter le nom"), an old practice of foreigners using the "Frenchness" of a Marseille merchant for their own purposes. Immigrants to cities often tried to use the names and the signs of the privileged burghers and guild members. This forbidden practice was also called "prêter les noms ou marques."[79] Between 1671 (Ordinance of May 20) and 1727, this practice, usually exercised by way of counterfeit contracts, was sometimes defended absolutely, sometimes it was allowed again partially as between 1684 and 1717. Other regulations concerned restrictions of naturalization standards. Foreigners who had been granted a letter of naturalization but who had not really abandoned their domicile in their former homeland were deprived of the privilege of naturalization, a rule designed foremost

---

[75] This concerns the *small vessel* shipping and carrying trade conducted for Italian, Jewish and Armenian commissioners. On its rise cf. Pagano de Divitiis, *Mercanti inglesi*, 80–91; for the cross-cultural collaboration of the Sephardic Jews of Livorno cf. Trivellato, *Familiarity*.

[76] Art. VIII, "Ordonnance du Roy," Strasbourg, October 24, 1681, CCM J 59.

[77] "Ordre de Louis Alexandre de Bourbon, comte de Toulouse," Versailles, May 21, 1696, CCM E 146.

[78] These *déclarations, arrêts* and *ordonnances* are filed in CCM E 146 and E 147.

[79] Cf. Art. 415 of the "Ordonnance générale," registered to the Parlement of Paris in 1629, cited by Éon, *Commerce honorable*, 234.

against Genoese merchants who had apparently established a practice of dual nationality and residence, and enjoyed the privileges of flag use, protection and tax reductions of both places, and the ability to flexibly choose the best conditions for each given freight and voyage (royal declarations of August 21, 1718; February 1720).[80] The 1727 declaration linked rules concerning the property of ships and the use of flags with very specific regulations about how to keep a register of the ship lists ("rooles d'equipages") containing the names and nationalities of all men on board, from the captain to the passengers and every sailor.[81]

The ways to undermine the rules relating to the distribution of *congés* were apparently much more multiform and frequently applied in the French case. Deviation from the ideal started right in Marseille, while for the English, it seems to have been more a question of difference between the London homeport and shipping realities in the Mediterranean.

Nationality in the Practice of Shipping and Slave Ransoming

This steady processing of "the national" through the observation and the breaking of rules, and through communication about them can be shown through several examples from the Mediterranean.

Regarding the English, it was not uncommon to find foreign-built ships, belonging to strangers with only the captain and a few English men using the English "Bandero." For example, the treasurer of the Bey of Tunis, a Jew, wanted to use a non-English built ship with only three English men on board. Old sea-passes were used[82] which had not been returned to the Admiralty with "scratch't names & put others in their stead."[83] Sometimes a consul issued a pass to merchants to let them deliver their goods to a North African city, and when departing, because they had no passes, these mariners obtained protection from privateers, either European or Maltese, or even the Algerians.[84]

---

[80] On Genoese naturalizations in that period cf. Sahlins, *Unnaturally French*, 96, 170, 175f. and mostly 199f.; in general Dubost and Sahlins, *Les immigrés*, 182f.

[81] "Déclaration du Roy, concernant la navigation des vaisseaux François aux Côtes d'Italie, d'Espagne, de Barbarie, & aux Échelles du Levant" (October 21, 1727), CCM E 147.

[82] The term of validity for passports was negotiable. For the British ships sailing to India, it was important to have a validity longer than one year, while the Algerians preferred a six month term of validity (1730/31, SP 71/6, 97–99).

[83] Consul Goddard to James Vernon, Tunis, December 30, 1700, SP 71/27, f. 105r.

[84] Consul Lawrence to James Stanhope, Tunis, February 17, 1716, SP 71/27, f. 213r.

New subjects of the British king were not always included under protection granted by peace treaties without some difficulty. The inhabitants of Gibraltar, Port Mahoney and Minorca, for example did not appear to be very "English" to the corsairs.[85] They obviously feared that those Catalan-speaking British subjects would be hardly recognizable and "that the Mayorkeens Cattalans and other Spaniards may hereafter clandestinately make use and navigate under Brittish Colours" if the door was opened once to such an un-northern Englishness. The consul proposed himself as the authority to "distinguish who are his Majestys right subjects and who are not" by inspecting Mediterranean sea-passes in question.[86] If ships came in now from Gibraltar to Tunis, it could be a matter of life or death if the consul and other representatives of the English nation, requested by the Bey, accepted the papers presented to them.[87] Could an English Mediterranean pass from Gibraltar protect a Catalan tartan and could a bill of health produced there testify to the health of the men on board, or neither of these things?[88] Ships which were even less clearly British – a Genoese captain coming from Gibraltar leaving a ship without men and protection anchored offshore Oran – could fall into the hands of corsairs if the English men were not on board and the British embarkation pass from Gibraltar was with the crew and captain at a tavern ashore.[89] While here Catalan-speaking British subjects and ships, or enslaved German-speaking subjects of the British Crown from Hanover, such as one Albrecht Wilhelm Forstmeier[90] had to be protected, English speaking Irish Catholics naturalized in Spain could likewise complicate the usual

---

[85] For the British discussion about the inclusion of Gibraltar into the British Empire cf. Plank, "Making Gibraltar British," especially 351–358; Constantine, *Community and identity*, 11–92.

[86] As n. 84.

[87] Cf. the case where a British Gibraltar privateer was brought in to the harbor of Algiers "because he had St. Georges Colours, which they mistook for Genoese and because the Captain & ships company spoke little English, it was with the greatest difficulty they got her released; but the poor Captain and all his Men lost all their Clothes." (Thomas Bolton to the Duke of Bedford, memorial of ca. 1748, SP 71/8, f. 301–333, 308).

[88] Consul Lawrence to the duke of Newcastle, Tunis, June 19, 1733, SP 71/28, f. 354r; similar case 1730/31, SP 71/7, f. 97–99.

[89] Duplicata [of a Letter of Robert Cole to the Earl of Dartmouth?], Algiers, October 8, 1712, SP 71/4, f. 204r (relating to affairs of May 1710).

[90] Recommendation of Forstmeier to Consul Hudson, Algiers, June 11, 1724, SP 71/6, f. 273r (cf. f. 483ff.).

patterns of recognition.[91] Far away in Northern Europe, those Mediterranean affairs with the corsair economy could trigger merchant migration and therefore the fluctuation and repartition of nationalities, as traders moved from a neighboring territory some miles across a border to become inhabitants and later naturalized subjects of a British dominion, such as Hanover, in order to obtain access to British sea-passes.[92]

The British protection of a ship could also be obtained by paying a consulage duty to the Levant Company in the Mediterranean. In those cases, completely "un-British" ships sailed under that flag, and if some problem occurred, the "national" was affected as the consul of the place was typically involved and had to resolve not only a technical problem, but also defend the reputation of the flag. A Greek, George Aptall, a former member of the Oxford Greek College,[93] who apparently pursued the career of ship captain, sailed with his ship from Alexandria to the North African ports "with goods belonging to Turks and with 30 Turkish passengers," but Aptall made a stop in Crete, murdered six of the Turks and sailed away with their goods. As the consulage was paid to the vice-consul of Alexandria, the Turks tried to get recompense from the English Nation, obviously thinking of the British protection as something like insurance.[94] Cases like this put the consul and "the nation" in serious conflict against the Ottoman authorities. A Turk of Smyrna who had chartered a pollacca of a certain Mariano Julia who enjoyed British protection as being from Port Mahoney, while of quite un-northern appearance (with a "poor old Turk, & a negro girl"), as the consul felt obliged to remark, dissolved his contract because of the French threat. In October 1741, a ship under British protection sailed from Malta to Tripoli with "Ninety Moors, men, and Women, all Pelgrimes who were coming from Alexandria with a Sweedish ship to Malta." The French took the ship and enslaved the passengers, putting some on the galleys in Toulon. In response, the pasha of Tripoli now appealed to the English, as the ship was under their protection.[95] The freeing of

---

[91] Consul Lawrence to the duke of Newcastle, to Tunis, August 18, 1735, SP 71/28, f. 463.

[92] Harding, "North African Piracy."

[93] See the next chapter for the English-Greek connections.

[94] Cf. Matterson, *English trade*, 156–158, the affair took ten years to be resolved.

[95] Consul Lawrence to the Duke of Newcastle, "Account of the severall insults offer'd by the French Cruizers to His majesty's Colours in these Seas," Tunis, August 2, 1742, SP 71/29, f. 215–217. A similar case: eight Algerian Mekka

enslaved Algerians, passengers of a ship under British protection taken by a Spanish cruiser, could become a state affair that did not just involve the captains and the consul in Algiers, but the Dey, and in London the Lord Justice, the Secretary of State, as well as the British ambassador to the Spanish Court in 1722/23.[96]

It is generally argued that the Europeans had their different ransoming institutions while the Moroccans and the Maghreb had not.[97] The cargo trade and the lending of protection to foreign ships, as in these cases, could produce paradoxical situations where the Muslims gained access to that infrastructure of the consular and diplomatic system, pitting Europeans against Europeans.

Despite *and* because of the 20 percent duty, and because of the steadily growing importance of the French market, the French flag was highly attractive in the Mediterranean.[98] Aside from the more complex naturalization process, the already mentioned counterfeit contracts were the most frequently used way. That those contracts were simulated was obvious to the locals in Marseille, cognizant that the modest fortunes of a given master could not have permitted him to really buy an expensive ship.[99] In reality, the ship master acted as employee of a foreign merchant, but before the French port administration, he presented himself as owner and, being himself of Marseille and falling under the other rules of the royal ordinances, he could obtain the *congé* of the Admiralty and the right to use the French flag for a ship that was completely foreign in economic terms. There were other ways too; the simplest was just to take several flags on the ship. This meant there were English ships with French and Spanish flags on board, and a ship with only Irish men on board sailing under English flag, but with passports of the French Admiralty in Brest, meaning Irish Catholics switching between Britishness and French protection.[100]

---

pilgrims were taken by the French before Tunis, sailing on an English ship, cf. the request of the Divan of Algiers to Louis XIV, Algiers, July 26, 1696 and response of Seignelay: Plantet, *Correspondance*, 252, 260.

[96] Cf. only Carteret to De la Faye, Göhre, November 22/11, 1723, SP 71/6, f. 223r.

[97] Ressel and Zwierlein, "Ransoming"; Hershenzon, "Plaintes et menaces," 453.

[98] Panzac, *La caravane*.

[99] "Coppie de la lettre du Conseil de Marine a M. Arnoul Intendant des Galères," December 2, 1716, CCM E 147.

[100] Grammont, *Correspondance*, 18, 45, 59.

A case in some way comparable with the English intervention for captured Muslims occurred in 1717 when Algerians took a ship commanded by a French captain with 119 Spanish passengers, sailing from Barcelona to Valencia, which, because of its French passport, the French consul in Algiers had taken under his protection. The consul ended up hosting the Spanish for three years in his own house, because the Algerians were only willing to free them in exchange for 130 "Turks and Moores" taken in 1716 by the Sicilians of Syracuse – then ruled by Savoy after the treaty of Utrecht – who had likewise traveled on a ship with a French captain. This put the French in the position of an "arbiter" between enemies, but the whole affair was perceived as causing dishonor to the French flag. While on the one hand the French could flatter themselves as having something like precedence and hierarchical dominant rule among the Europeans and even over the corsairs, they were also challenged to prove the efficiency of their protection.[101]

The deviations from the French norms about who was allowed to sail with the French flag were perceived by contemporaries as a mass phenomenon instead of single cases. This is also true for how the French perceived the matter themselves. During the regulatory period of the regency from 1715 to 1717, the *intendant des galères* Pierre Arnoul, and the *Chambre de Commerce* unsystematically gathered data about the abuse of the French flag at sea through the interrogation of incoming sea captains. Five captains gave accounts of many ships from St. Remo, Naples, Malta, Sicily, Genoa and Messina that they had encountered, providing incomplete lists of some thirty or forty ships whose names and captains they remembered. All had flown the French flag, but often it was at best the captain alone who had been French or naturalized, while all the men on board had not been French.[102] Similar

---

[101] Cf. the correspondance of Jean Baume, 1717 to 1719, Grammont, *Correspondance*, 138–152. Another case where the honor of the white flag was endangered: Denis du Sault of Tunis, St. Mandrier, August 27, 1720 to the Secretary of State, AN AE B I 1130.

[102] "Memoires donnez à la Chambre de Commerce de Marseille par les Capitaines cy après denommez des Batimens Étrangers qui naviguent sous la Bannière de France," sent as attachment to Arnoul's "Mémoire sur les abus que font les Nations Étrangères de la Bannière de France" (December 14, 1715), CCM E 147. Pirates sailed with a set of several flags to fly as was convenient in a given situation ("There was on board the sloop four sorts of Colours English, French, Swedes and a Black Ensign which had in it a Death head on one side and an Arm with a sword in ye hand on the other side," consul Lawrence to the duke of Newcastle, Tunis, October 2, 1720, SP 71/27, f. 301).

notes, gathered less systematically, are to be found in reports of how the Deys and Beys of Algiers, Tunis and Tripoli complained about counterfeit Frenchness.[103] Sometimes massive fleets, as in 1699 some 200 ships, passed the North African coast with French flags and the Algerians let them pass without examining them. But impossibly, the consul suggested, all of those 200 ships were really French.[104] The use of the French flag by ships that had in fact never visited a French port evidently occurred much more often than with other European nations from at least the 1680s, however.

Ransoming communication was, from the perspective of the epistemic of mercantilist political economy, a secondary phenomenon. The national belonging within captivity and ransoming correspondence was distinctly connected more to the state level than the merchant economy. Literature on captivity and ransoming is expanding, and sometimes the issue of identifying captives of one's own nation has been addressed explicitly.[105] The impression gained from what the archival records reveal and what scholars have discovered until now is that the sources do tell a great deal about the need to identify the nation of captives and about how eager both consuls and corsairs were to do so. Nonetheless, there is little evidence about how such identifications were conducted in early modern times. Seldom do we have consular correspondence or the ransoming orders really sent to a captive's country of origin, requesting entries in the parish records. For those that do exist, it remains unclear how that entry could have helped to confirm anything other than that there was an entry for a person of that name in that parish. The application of that knowledge gained for the captive claiming to be that person hundreds of miles away in North Africa still remained, in the end, arbitrary. Some slave lists (infrequently) contain rudimentary forms of captives' physical descriptions,[106] but they usually only offer

---

103 Plantet, *Correspondance*, 120 (case of 1686); AN AE B I 1088, no fol. (case of 1686/87); Du Sault, "Réponses aux griefs du Dey," Tunis, July 1, 1693, AN AE B I 1088, no fol.

104 Consul Durand to the *Chambre de Commerce*, Algiers, January 12, 1699, Grammont, *Correspondance*, 65.

105 The literature on ransoming Christian captives in the Mediterranean has exploded and regionalized since the 1964 monograph by Salvatore Bono. Instead of surveying that field here again cf. Ressel and Zwierlein, "Ransoming."

106 Kaiser, "Vérifier"; AE Nantes, Corr. Consulat Algier, A II, fol. 93–104; later examples with physical descriptions in CCM G 41, but this is not the usual case.

lists of names with the place of origin.[107] In all likelihood on shore, the language spoken by the captive and who he claimed to be, if believed and supported by the consul, was usually the only communication that reified a nationality, as many passengers and sailors possessed no passports or letter of conduct.[108] Usually corsairs, as well as Europeans, simply referred in writings to, for instance, "5 French and 7 Dutch slaves" or something similar. Research on early modern identification processes *within* the proto-national states has shown how complex and unclear the standardization of descriptions and modern record keeping was in times before photography.[109] Matters were still more complicated in a zone of constant travel and assimilation among merchants, sailors and passengers such as the Mediterranean shores and at such great distances from the metropole. Boubaker has underlined the high "importance attributed to national belonging" by religious as well as state institutions and that "mostly, both sides kn[e]w the identities of the persons,"[110] and this is true. Yet one must add that we do not know much about how that past conviction "to know" was achieved and if we, according to current measures of what might be deemed knowledge about personal identities and nationality, would call that "knowing."[111]

Where Does the Nation Start and Stop? Irritations and Reflexivity
It is evident from what has been shown that the mercantilist communication created the paradox of an ongoing query for the nation and for national attributions, but that it often ended up in an awareness of ignorance. Most interesting in this context are those testimonies of experts and administrators of the regulations to be executed that betray a reflexive analysis of those paradoxes and of the functioning of the norms themselves. Some of those observations show that, to some

---

107 "Rolle des Esclaues François qui ont esté remis au Sieur Le Maire par le Divan de Tripoly le 27e Septembre 1686," AN AE B I 1088 [unfol]; CCM G 39–48, specially the early *rôles* in G39, cf. two entries from a 1615 *rôle*; printed examples: *Les sources inédites de l'Histoire du Maroc*, IIe sér., tom. 4 (1693–1698), 376–383, 630–632.
108 Conseil de la Marine to Arnoul, April 6, 1718, CCM G39.
109 Caplan and Torpey (eds.), *Documenting individual identity* and for France especially Denis, *Une Histoire de l'identité.*
110 Boubaker, "Réseaux," 27.
111 This seems to be still close to the medieval situation, Kosto, "Ignorance about the traveler."

extent, what has been widely considered a major insight of the twentieth century – the constructivity of nations, or their status as imaginations, as that of other community discourses – was already available to the constructors themselves.

In 1748, the chaplain to the factory in Algiers, Thomas Bolton, wrote a memorandum to the Secretary of State, the Duke of Bedford, in which he explained the attractiveness of the British flag. He conceived of all Barbary/European intercourse in the Mediterranean as essentially a story of imitation of the initial English war/treaty solution established in 1662 after Admiral Blake's 1647 victory, even that of Ludovician France. Bolton attributed the rise of the Swedish, Danish and Hansa cities during the first half of the eighteenth century to the "inactive reign" of the former Dey who died in 1745.[112] The result was that all of a sudden the smaller neutral powers and Italians became the preferred carriers of trade, even by English merchants themselves.[113] Evidently then, and not without surprise, there was a precise awareness about the causalities between flag use in the carrying trade, war, peace and the opportunities of neutrality. The reflections of Robert White from two years later are even more sophisticated: The Act of Frauds of 1662 included explicitly under the term "English men" all the king's subjects from England, Ireland and the colonies.[114] The 1660 Navigation Act and the administrative practice similarly understood as "British-built Ships" all "Ships built in Great Britain, Ireland, Guernsey, Jersey, or the British Plantations in Asia, Africa, or America."[115] But during the last Jacobite upheavals in Britain and their repercussions on the shores of the Mediterranean – where fear of Jacobite conspiracies among the English nation and employees of the Levant Company was frequently mentioned before and around 1750 –,[116] Imperial administrators reflected upon the restrictions of that definition of "English." White creatively interpreted the renewed treaty clauses with Algiers of August 1700 in retrospect, trying to prove that neither of the contracting parties, at that time before the Act of Union between England

---

[112] SP 71/8, f. 307 – The Dey was Ibrahim Kheznadji Pascha (reg. 1732–1745). Bolton's analysis corresponds with current research, cf. Ressel, *Zwischen Sklavenkassen*, 467–469, 481.

[113] SP 71/8, f. 305.   [114] Act of Frauds 1662, Art. V – cf. above n. 6.

[115] Crouch, *Guide*, 131; Navigation Act 1660, Art. VII (above n. 5).

[116] Cf. Bolton to Whitehall, Algiers, s.d. (rec. December 15, 1748), SP 71/8, f. 343; Robert White to Whitehall, September 1, 1748 had to clear his "Charac'ter from the infamous imputation of Jacobitism" (SP 71/24, f. 62–63).

and Scotland that formed Great Britain in 1707, could have intended to include "Scotland, Ireland, or any of His Majestys other Dominions" under the term "England."[117] Nor could the term "British Ships" reasonably have been thought to "cover or imply all Country Ships belonging to His Majestys Dominion." Finally, White deconstructed the very core of the processes of identification and national attribution, the moment of matching the upper and the lower parts of the sea-pass:

Besides, this whole article seems to render the Security of Passes very uncertain and precarious, Because the vague term, *Fit*, which is the Ruling Term in the Article, may admit of many Different Meanings. It may be understood for the Scollops fitting, for the Lines fitting, for the Length, Breadth and Thickness of the Pass fitting, It may be constructed to mean One or all of these, or any thing else.[118]

Beyond the particular situation – five ships had been brought into Algiers because the scalloped Passes did not correspond with the corsairs' counterpart – the reflections show an awareness of the somewhat unreal moment of testing nationality, how the fitting (or not) together of some pieces of parchment was the code for being British/not-British. It was the code for freedom/captivity, and even for life/death.

White's reasoning, obviously based in his training in Common Law legal language analysis, was dominated by the aim of delegitimizing the existing treaties with respect to a precise political situation. But it shows nevertheless an astonishing amount of awareness of the problem of "how to know the nation." It even shows the availability of relativist thought about the constructedness of nationhood as it was defined by the "fitting/non-fitting" of two pieces of parchment. If, like White, one starts to reflect on definitions, they become fluid. The fifty-year distance since the Union had added an additional amount of relativity to that; it made the past political situation unsuitable to dictate the present. This magic moment of the nationalizing "fitting" of sea-passes was deconstructed like rationalist reformed theologians were trained to deconstruct the mystical moment of transubstantiation. While one analyzes today sometimes the past discourses of belonging or not to a nation in a Schmittian way as secularized forms of belonging or not to the *corpus mysticum Christi*,[119] proto-constructivist thoughts about national

---

[117] SP 71/8, f. 489v.   [118] Ibid., f. 490r.
[119] Cf. Bell, *Cult of the Nation*, 38. On the *communio cum Christo* as central category of forging communities and political alliances in pre-national

communion building were apparently possible just by reflecting on the crude materiality of identification processes such as pass matching.

For the French, one can find similar forms of reflexivity in *mémoires* concerning the functioning of the whole Mediterranean seafaring and trading system between 1715 and the 1730s in response to the abuses of the flag, mostly by help of the counterfeit contracts. In those cases, once issued, captains used the admiral's *congé* for several years and the port administrators tolerated all this when the ships returned to a French port. As a result, according to a memorialist in 1715, foreign ships had effortlessly taken over shipping from the French, but still under French flag, as the "real" French ships now remained without work in their home ports. This also created something like a low-cost market for French sailors who were present in foreign port cities to be hired to fulfill, at least partially, the French equipment rules. Paradoxically, the French realized that after fifty years of Colbertian legislation, French protectionism – at least if run too laxly – could create economic exiguity for the French and profits for foreign (Italian) merchants.[120]

Only fifteen years later, the situation changed once again. With pride, an anonymous 1731 memorialist of the *Chambre de commerce* remembered that they had flattered themselves by the French flag's attractiveness and its "high reputation with foreigners" until that date, while Dutch and English ships in the Mediterranean could find almost no tonnage besides their own commerce for decades.[121] Now, the peace had changed the situation. Competition had risen and the French now risked losing their share of the carrying trade market. At that point, this memorialist warned about the ill effects that would result from severe protection that followed too rigorously the "maximes d'Etat" instead of an economic rationale. He therefore openly recommended what seems to have even been the usual practice for a long time, a somewhat laissez-faire enforcement of laws that would re-allow foreigners to take up to a third of the stake in the property of French shipping as had been the rule in the glorious times of Louis XIV. In so doing, he justifiably criticized the decision taken by the *Régence*

---

reformed thought Zwierlein, "Les saints de la communion"; Zwierlein, "Consociatio."

[120] "Mémoire sur les abus que font les Nations Etrangères de la Bannière de France," communicated to Pierre Arnoul, December 14, 1715, CCM E 147.

[121] *Mémoire* of the *Chambre* directed to Maurepas, December 24, 1731, CCM E 147.

administration in 1716/17 as being too rigid and perhaps too informed by state theory, in an attempt to convince Louis XV to return to the political economics of his great-grandfather. Indeed, Maurepas responded to Lebret at the end of January 1732 by accepting the necessity to tolerate the abuse of the flag "by not always observing strictly the 1727 declaration" in order to not lose foreigners as important subcontractors of French navigation. The "inclination to our nation," he declared, was of high importance for France. If the advantages of the (ab)use of the French flag by the foreigners had been "more than reciprocal," the nation would have profited infinitely from it.[122]

The protectionist perspective on the second level of Mediterranean economics, piracy, in 1715 was that the corsairs were well acquainted with the rules requiring a ship to be manned by at least ⅔ Frenchmen. If they captured one of those counterfeit French ships with old passes and nearly no Frenchmen on board, they normally enslaved the whole crew to the great dishonor of the French nation. However, if the waters of the Mediterranean were plied only by ships flying the French flag, the corsairs would have had, in the end, no possible target for their main business, piracy. This would force them to resume attacks against French ships. The complete dominance of the French flag was therefore dysfunctional within the given system.[123]

It seems that the 1715 "white flag overflow" argument was not the winning one. After the Maurepas administration had adopted the policy of tolerant enforcement in the 1730s and following a favorable capitulation with the Ottomans in 1740, the French had found their balance.[124]

What is interesting here is how, within fifteen years, the arguments could completely change direction. It is remarkable how the French realized and reflected upon the economic functions and dysfunctionalities of their own regulations, how they recognized a quite simple everyday practice of simulating Frenchness. While naturalization procedures and belonging to a nation were, on the one hand, so restrictive and bound to blood and soil, as the research of Sahlins and Dubost has shown, the undercover use of nationality as a mere sign and currency

---

[122] *Mémoire*, December 24, 1731 and response letter Maurepas to Lebret, January 31, 1732, CCM E 147.

[123] *Mémoire* to Arnoul 1715, CCM E 147.

[124] The records in CCM E 148 show that after the 1730s there were only case-to-case decisions which increased a little after 1768.

within the Mediterranean shipping context betrays the other side of the same coin: the national became an element of steady processing from the 1660s, and because of that, its status as an attribute and a sign, that is, its constructedness became likewise visible. Nationality became objectified as the content of specified (non-)knowledge, but objectifying it also meant exposing its partial arbitrariness.

Those moments of high reflexivity concerning the sign status of the national show that, at the same time as the central mercantilist distinction of what was internal or external was sharpening, it also became underdetermined. Terminologically speaking, this was no retreat to the prior state of nescience about the "national," no return to the more fluid and much less specified situation before the 1660s. It was rather, in a spiraling way, a new form of reflexive awareness about the usability of the national created by undermining the valid norms or by the skilled use of them – all this long before the modern nineteenth-century heydays of nationalism. Just as how the post-Reformation plurality of confessions and the micropractices of playing and use of these confessional boundaries in a pluralizing manner were not the same phenomena as inter-religious exchange in the Middle Ages, the perforation and deaggregation of the specified national was different from the earlier state of unconscious ignorance about it.

A Comparative Look at the Medieval Conditions

To test the historical specificity and the new character of the 1660s regulations, it will be helpful to have a comparative look at earlier situations in the Mediterranean. As the sixteenth century was, from roughly 1492 (the Christian conquest of Granada) to 1574 (the Ottoman reconquest of Tunis) a period of unsettled circumstances along the North African coast, it is more enlightening to consider the late medieval period, before the Ottomanization of the region. Many caveats of "incomparability" may be brought forward, but on the other hand, the continuities of the corsair activities of the North African cities/kingdoms are striking.[125] The Berber kingdoms of Fez, Tremecén (with Algiers, Oran), Tunis (Bugía), Granada (Almería) and their corsair

---

[125]   For authors who trace the lines of continuity between the North African Arabic and the Ottoman period cf. Abun-Nasr, *A history of the Maghrib*; Heers, *The Barbary Corsairs*.

attacks on European – foremost Aragonese – merchant shipping form the direct precursors of the early modern situation. The sixteenth-century Ottomanization meant mostly the implantation of a Janissary elite in the city states which took over much of the Berber seafaring and corsairing traditions.

The different ways of obtaining ransom from Berber captivity in late medieval times were (a) ransoming through individual friends, merchants, families,[126] (b) ransoming through the religious orders, above all the Trinitarians (founded in 1198) and the Mercedarians (founded in 1218), but also through the military orders, (c) ransoming by help of corporative and municipal institutions, (d) ransoming through a monarch's direct diplomatic intervention. In Spain, the *alfaqueques* had been professional ransoming mediators acting on behalf of the crown or of municipalities since the thirteenth century, but they mostly operated on the inner Spanish Christian-Muslim territorial border.[127] Trinitarians and Mercedarians were late order foundations linked to the challenges of the crusades, and their only reason for existence in terms of competing with the established military orders was their specialization in ransoming captives.[128] They were not very active in either the Holy Land or the Eastern Mediterranean, and if so, more as hospitallers than as ransomers.[129] Despite the early foundation of the Trinitarians in Marseille, which is probably attributable to specific *Provençal* interests,[130] their activity in the late medieval period is scarcely documented in either France or Genoa.[131] Consequently, following the spread of their monasteries in Spain, the Trinitarians formed, with the Mercedarians, whose origin and center had been in Aragón, a rather

---

[126] López Pérez, *La corona de Aragón*, 806–812 and Rodriguez, *Captives*, 107–118.
[127] Cf. Díaz Borrás, *El miedo*, 61–72.
[128] Cipollone, *Cristianità – Islam*; Cipollone (ed.), *La liberazione* (but no contribution on late medieval practice); Cipollone, *Marsiglia*, 105–135, 111, 115; Brodman, *Ransoming Captives*; Rodríguez-Picavea, "The Military Orders"; Forey, "The Military Orders."
[129] Friedman, *Encounter*, 203, 210.
[130] Le Blévec, "Le contexte parisien," 120f., 124f.
[131] In the massive study Deslandres, *L'Ordre des trinitaires* only in vol. 1, 324f. and in the documentary annex vol. 2, 61f. are some short allusions to fourteenth-century ransoming activity. Cipollone, "Contributi," 36–40 can only refer to testamentary legacies where persons attribute some money to ransoming purposes of the order, no proto-national specification is mentioned. For Genoa Heers, "Gênes," 238; Porres Alonso, *Libertad*, 187–198.

regionalized religious order concentrated on the problems of Aragón/
Castellan/Murcia connections with the inner Spanish Arabic and the
North African Berber corsair threat.[132] Even in this region of their
main activity, we do not possess much certain archival evidence of
their ransoming activities, even though largely retrospective, hagio-
graphic and advertising publications of the orders claim that they freed
several thousands of captives.[133] The non-hagiographic documents that
we possess from the fourteenth century show that the ransomed persons
were in fact all from the dominions of the Aragonese crown in the
Mediterranean, from Spain, Sicily and Sardinia.[134] But none of the
Trinitarian or Mercedarian sources spoke of their task other than as
freeing "Christian captives" (not "Aragonese captives").[135] Municipal
institutions are a different matter.[136] The most famous of these is the
Entidad Valenciana en pro de los cautivos, founded in 1323 and active
until 1539, which organized the charitable collections from a city's
parishes for ransom in the form of efficient city-state bureaucracy. This
became a hybrid between a municipal duty or tax and traditional church
collection. Here, the money was reserved only for Christians of Valencia.
Only if there were no more Valencian captives could the money be used
for Christians from the kingdom of Valencia, preferably from the city's
neighborhood. Here a tiny proto-national element was evident, but in
practice it was purely municipal.[137] Diplomatic negotiations between
the Aragonese kings and the Muslim kingdoms over peace treaties could
seem most similar to early modern realities. The *cedulae*, the royal
instructions to Aragonese ambassadors and the peace treaties themselves, show how the kings focused on the freeing of subjects from
their territory as an act of the lord's duty of protection.[138]

---

132 Brodman, *Ransoming*, 120.
133 For example the *Historia general* of Gabriel Téllez (1639), cf. Díaz Borrás, *El
     miedo*, 55f.
134 Brodman, *Ransoming*, 114.
135 Cf. Porres Alonso, *Libertad*, 277–284, 285–307, 425–429. There is only one
     privilege to the order granted by Juan II of Castille, April 6, 1448 that states "la
     dicha Orden [...] han fecho a mí muchos seruiçios e a los mis reynos mucho
     prouecho *en redimir mis vasallos e naturales y súbditos* de la dicha cautibidad
     [...]" (ibid., 289).
136 Brodman, "Municipal Ransoming Law."
137 "Ordinacions per a traure cautius christians de poder de infeels" (1324), § III
     and IV, cited in Díaz Borrás, "Notas," 346.
138 Jayme II to the king of Granada, Mahomad Aben Nacer, 1301: "fazemos vos
     saber que el fiel nuestro Bernart de Segalar debe fablar con Vos sobre feito de los

The normative form of the peace treaties is very different from the early modern treaties between the Barbary and the European states. The Aragonese treaties were formulated like conclusions of a peace following war. The liberation of captives was conceived in terms approximating the prompt exchange of captured warriors after a decisive battle. There were no precise regulations of future shipping or other exchanges, of ship control, and no pass system was inaugurated. The liberation of captives in those peace negotiations was not accomplished via ransoming either. Only the fact that the Aragonese envoys negotiated the number of freed captives in exchange for a proportional number of years of peace shows that both sides acknowledged a prolonged period of regular corsairing and the taking of captives instead of a situation of discrete war and enduring periods of peace.[139] The selectivity of freeing and protecting just *their* subjects was therefore not born out of the context of inner-European and inner-Christian economic competition and the captivity market.

With good reason medievalists stress that the practice of mercantilism has its roots not in seventeenth-century England or France, but in twelfth- or thirteenth-century protectionist trade policies, for which not just Aragón but also the French kingdom and the Italian imperial republics Venice and Genoa were spearheads of development.[140] Aragón's competition with late medieval Italian merchants is perhaps the most profiled example and anti-Italian Aragonese legislation,

Christianos *de nuestra tierra*, qui son cativos en poder vuestro" (Capmany y de Monpalau, *Memorias historicas*, vol. 4, 30); peace treaty between Jayme II and the king of Bugía "Alid Abu Zagri" (that must be the emir Abū al-Baqā'), 1309, Art. III: "Item: que tots los catius ó catives qui sien *de la terra ó Senyoria del Senyor Rey d'Arago*, è son en la terra o Senyoria del Rey de Bugía" (ibid., 40); embassy of Jayme II to the ruler of Tremecén (Tlemcen/Timlisān) Abd al-Rahmān I b. Mūsa I, Abū Tāshufīn, 1319: "vulats deliurar è soure tots los dits catius *de la nostra terra è Senyoria* qui son en vostre poder" (ibid., 67f.); the same language is used in 1309/19, Capmany y de Montpalau, *Antiguos tratados*, 73, 96, 98. An exception is the instruction of Jayme II to the Mamluk sultan of Egypt Al-Malik al-Ashraf Khalil, August 9, 1292, where the aim was to free not only Aragonese, but also Castillan and Portuguese subjects (ibid., 31). Cf. Rodriguez, *Captives*, 123–130; for the Hafsid rulers Bosworth, *The New Islamic Dynasties*, 43f.; Rouighi, *The Making*, 42.

139 Capmany y Montpalau, *Antiguos Tratados*, 101 and Capmany y Montpalau, *Memorias historicas*, 68f. (1319).

140 Heckscher, *Mercantilism*, vol. 1, 325–337. A clear "strategy of commercial politics on an international level" is not visible for fourteenth/fifteenth-century Florence, cf. González Arévalo, "Rapporti commerciali," 182 (Venice, Genoa).

culminating in King Martin's edict of January 15, 1401 can be seen in perfect continuity with the policies and practice of trade of the then leading seventeenth-century states.[141] Yet the lack of competitive proto-national semantics within ransoming correspondence suggests that the late medieval economic system of the Mediterranean as a whole cannot be understood as an already fully developed two-level system of inter-national trade competition on the one hand and parasitic corsair activity on the other. The smaller naval capacities of both sides meant that exchange between them was more regionalized, and that inter-Christian competition was far less state-defined in the Mediterranean as a whole and in interaction with the Levant and North Africa in particular.[142] If Aragonese politics in North Africa has been character-ized largely as an imperial economic enterprise,[143] it had not been so in competition with European powers at this point. If, too, Aragonese ships were competing with Castellans and Italians and, from the fifteenth century, with the Portuguese, this competition was not linked in a triangular way to the confrontation with the Berber kingdoms as was the two-level system of seventeenth and eighteenth centuries.

Even in quite similar structural situations and with actors such as Aragón, perhaps the "most modern" of all medieval kingdoms, one still cannot detect real genealogical precursors for the epistemic situation of the seventeenth century. National belonging remained, in medieval times, more in a state of nescience. If the processing of "the national" in all politico-economic communication did not first begin with the standards of the 1660s, this still was a moment of unprecedented enforcement wherefore it can be taken as an epochal turning point.

The Political Arithmetic of the Unknown: The French Nation

Attempts to shape the reified result of the investigation processes into national attributes were a final matter in which ignorance and specifi-cation of the unknown national played a crucial role in the seventeenth- and eighteenth-century Mediterranean imperial administration. At first, this concerned the nation in its narrower sense, which derived

---

[141] Del Treppo, *I mercanti catalani*, 163–173; Ferrer i Mallol, "Genoese Merchants"; Houssaye Michienzi, *Datini*, 378–384.

[142] Panzac, *La marine*, 88–92.

[143] Dufourcq, *L'expansió catalana*; Dufourcq, "Un imperialisme médieval"; López Pérez, *La corona*, 267 supports Dufourcq's interpretation.

from the medieval concept of naming a group from a given region or country within a plural, "multinational" context a *natio*: the *nationes* of a university, of a military order or, as here, of merchant colonies. By and large this only applied to the French side in our study as there were virtually no traces of the state dirigist organization of the "English nation."

There are several thorough studies of different French "nations" in the narrower sense for Tripoli in Syria,[144] Tunis,[145] Aleppo[146] and Constantinople,[147] but there is no study of how the Paris/Versailles center tried to organize the nations in the *échelles* as a whole.

The decisive period of the French state's empirical investigation into the *échelles* and of its major regulatory efforts was between 1685 and 1730. Since the late seventeenth century, the ambassador in Constantinople and the consuls regularly reported to the *Chambre de commerce* in Marseille and also to Versailles about the number and the character of the merchants in each place.[148] Sometimes a general numeric overview of all the French in the Levant was elaborated, but there was no established administrative practice of annually counting all French subjects.[149] But sometimes, as in 1732, the French Ministry tried to obtain a current overview "of all the Frenchmen living in the échelles." This was linked to the decision to send certificates of residence to all those who had not yet acquired one if they matched the criteria of bon-conduit.[150] A constant perception of deception, of not knowing the exact realities is apparent and a continual desire for central control was notable.[151] Many unforeseen travelers,

---

[144] Roux, *Les échelles*, 32–49. [145] Debbasch, *La nation*.

[146] Fukasawa, *Toilerie*, 71–109.

[147] Eldem, *French trade*, especially 203–283; Frangakis-Syrett, *Smyrna*.

[148] Cf. Eldem, *French Trade*, 205 for the census of 343 individuals of the Constantinople nation and Fukasawa, *Toilerie*, 78 and 97 with n. 22 for the references of the "états des Français résidant à Alep" (AN AE B III 290, CCM J 901–921, J 932–967, J82); some other examples from the consular correspondence in AN AE B I 630 for Cyprus 1691 and AN AE BI 320, f. 324r for the French nation in Cairo in 1730 (11 maisons, 17 marchands).

[149] The *état* of 1769 is in CCM J 59.

[150] Maurepas to the *Chambre de Commerce*, Compiègne, May 30, 1732, CCM J 59. /

[151] "[Send me an] autre État plus complet ... faire connoître que comme l'on veut sçavoir le véritable État des choses." (Lebret, Aix-en-Provence, May 20, 1724 to the *échevins* and *députés* of Marseille; Maurepas to the *Chambre*, Compiègne, May 30, 1732, both in CCM J 59).

"vagabonds," "français oisifs" – as they were called in the sources – were detected.[152]

The administration distinguished between several different units within their populationist regulation attempts. These included the number of whole *maisons*, i.e., firms in a city; the number of merchants; the number of foreign merchants under French protection; the number of craftsmen and servants who performed auxiliary services for the merchants; the dragomen, translators and enfants des langues; relatives and family members of the aforementioned groups; Frenchmen who were censured as not productive. The king and his ministers tried to regulate the number and even the quality of all those groups.

Regarding the most important group (the merchants themselves), after 1685 the Crown developed something like an ideal scheme of the population that it wanted to realize which one can summarize with the following parameters:[153]

- The overall number of merchants in each city should be "proportionate" to its economic trade balance.
- The merchants should not be younger than 25.
- They should be unmarried, preferably.
- They should be good Catholics and not Protestants.
- They should stay no longer than ten years abroad.

That implied that the sojourn should be made by a cohort of men aged around thirty, who would then return to France at a time when they were still young enough to marry, yet experienced enough to fill the full position of head of a Marseille firm. Excepted from this scheme were the consuls who were mostly older but who were also no longer real merchants after the arrêt of 1691.[154] Connected to this was the repeated prohibition against marrying in the Levant, aimed mostly against marriages between French merchants and Christian Ottoman subjects. The French state did not want the uncontrolled establishment and proliferation of a Levantine French population. For the most part,

---

[152] For letters and orders concerning the "françois oisifs" cf. Pontchartrain to the *Chambre*, November 21, 1714, Maurepas to the *Chambre*, September 9, 1725; July 4, 1731; to president Lebret, July 3, 1731 (CCM J 59).

[153] The ordinance of February 12, 1720 prescribes the necessity of the special certificate for naturalized strangers; the ordinance of March 21, 1731 fixed the time of residence in the *échelles* at ten years, CCM J 59.

[154] Debbasch, *La nation*, 182–194.

however, the administration feared the uncontrolled dispersion of French property in the case of death and inheritance. This problem found its way to the agenda most prominently during the *Regency* in 1716. There was first an ordinance of the *Conseil de la Marine* which decreed that every merchant who married an Ottoman subject would lose all his rights as member of the nation and to attend its assembly (August 11). An ordinance of July 20, 1726 followed that entirely forbade merchants and their family members from marrying any "daughter or widow, stranger or from that country, subject or not subject of the sultan or of the powers of the Barbary regencies, even those of French origin or who are born in the Levant and those regencies" during their Levant stay. Only the king could grant exceptions.[155]

The merchants tried to oppose sections of the rules. Under this regulatory situation, a French merchant had perhaps just five years of productive time in the Levant. Their European rivals, they argued, would gain advantage through the extra time they could devote to gaining additional experience, power, connections and roots in the Ottoman world. As a compromise, the ordinance of 1716 fixed that a French merchant might go to the Levant at the age of 18, but that he would be admitted to the assembly only at the age of 25.[156]

As a result of all those differentiations, one can distinguish at least five categories of the "national":

(1) Frenchmen who came to the Levant at the age of 25 and left after ten years. These had full access to the merchant nation's assembly if they were active in commerce.[157]
(2) Frenchmen who arrived in the Levant at the age of 18. No access to the assembly.
(3) Frenchmen and women who were born to other Frenchmen in the Levant and had stayed there.

---

[155] Debbasch, *La nation*, 128–136.

[156] Ordinance of October 21, 1685; *Mémoire* by the merchants to Pontchartrain; negative response by Pontchartrain (letter of June 15, 1701); Ordinance of March 17, 1716, CCM J 59.

[157] The intendant des *échelles* Isnard reserved the right of admission to the assembly to those merchants whose *per annum* economic productivity equaled at least 2000 piastres, on Cyprus, 1500 were enough: decision of the *Chambre*, June 26/August 1, 1731, CCM J 59.

(4) Foreign merchants naturalized in France (a) with and (b) without special permission to conduct trade under French protection in the Mediterranean.
(5) Foreign merchants not naturalized but enjoying French protection.

The most intensive discussion concerned the actual number of merchants in the Levant. The government's fear was always about the potential to lose productive subjects into the periphery.

The guiding principle of the French administration was the "proportionality" between the number of different kinds of individual Frenchmen (merchants, craftsmen [*artisans et gens de métier*]) and of the *maisons* on a corporate level, which meant the merchant factories and firms on the one hand and the volume of the trade done in a particular port on the other. The idea of proportionality developed from an embryonic stage to a quite elaborate version after the middle of the eighteenth century.[158]

The process of fixing the number took a great deal of time and apparently remained undecided for decades. Nevertheless, the Ministry had quite concrete conceptions of misbehavior and rule transgression, such as the founding of new firms "directement [ou] indirectement" by splitting up an older one. Maurepas aimed for a "general rule [*une règle générale*]" regarding that perennial question about population and several times ordered the chamber and the Levantine outposts to refrain from actions that would create unwanted facts before that rule had been established.[159] The chamber acted with some dilatory tactics on that "projet de la reduction,"[160] and Maurepas ordered that the *Chambre* should at least not distribute any new certificates to merchants other

---

[158] One had to fix the number "in proportion to the trade conducted during a normal year in each *échelle* [*proportionnement au commerce qui se fait année commune dans chaque Echelle*]" (Maurepas to the *Chambre*, Versailles, January 14, 1733, CCM J 59). The notion of "année commune" meant "normal year" in the sense of an average peacetime year, as war conditions always changed situations beyond the calculable. Earlier, Maurepas had stated that one should be attentive that the "nombre des négociants [would not grow] d'une manière disproportionnée au commerce qu'ils peuvent faire" (cf. Roux, *Les échelles*, 27).

[159] Maurepas to the *Chambre*, Versailles, November 29, 1736, CCM J 59.

[160] Maurepas to the *Chambre*, Versailles, April 30, 1737: "il peut y en avoir un trop grand nombre à proportion du Commerce qui s'y fait, cette matière ayant esté examinée, il a esté resolut de ne prendre quant à présent aucun arrangement général à cet Égard" (CCM J 59).

than to those who would join the ones already active in the Levant.[161] Finally, the inspector of the échelles, Pierre-Jean Pignon,[162] was sent to conduct census-like visitations of all the French nations of the Levant in 1740–1742,[163] and to negotiate the question within the Chamber of Marseille.[164] The "proportionality" of the number of French residents conducting trade or associated with Levantine commerce had its parallel in the French court's concept of the proportionality of the trade conducted as a whole.[165]

In 1740, several *mémoires* addressed this issue. The merchants typically argued against an artificial reduction of houses. They proudly reminded Maurepas and the King in Versailles of the historical process of their installation in the Mediterranean: they had achieved "a revolution" in Mediterranean commerce. While during the seventeenth century, the Dutch, the English and the Venetians had possessed a share of the overall commerce four times larger than the French, now it was precisely vice versa; the French had four times more. The merchants evaluated the overall size of Mediterranean commerce at 18 million livres (the proportion of foreign merchants sailing under French flag not included), and the net gain at 15 to 16 million. This was the result of their labors, and for that, they needed a stable number of merchants.[166] One memorialist

---

[161] Maurepas to the *Chambre*, Versailles, May 12, 1734: the practice would be "contraire aux veues que l'on a de reduire dans chacune les maisons des negoçiants à un nombre proportionné au commerce qui s'y fait" (CCM J 59).

[162] Pierre-Jean Pignon had negotiated in 1729 the French peace with Tripoli, had been consul in the important city of Cairo in 1729–1734, had been *premier commis du bureau du Commerce et des Consulats de Levant et de Barbarie* in 1738–1741, and was *inspecteur du commerce du Levant* at Marseille as successor of Icard in 1741–55 and 1757–59 (cf. Masson, *Commerce XVIII*, 7 n. 1 and passim; Mézin, *Les consuls*, 493f.; AN MAR C7/248). His economic thought, as revealed in his 1750 *mémoire* about the Levant Commerce from the Gournay papers, is analyzed by Meyssonnier, "Vincent de Gournay," 100–106).

[163] Masson, *Commerce XVIII*, 20–25; Paris, *Histoire*, 331.

[164] There is no study of the office of the inspector of the Levant commerce. One may transfer some elements from the several dozens inspectors of manufactures to its functions, cf. for that Minard, *La fortune*.

[165] Maurepas to the *Chambre*, Versailles, April 22, 1740, CCM J 59: "Si les résidents au Levant, Messieurs, proportionnoient les ventes qu'ils font à crédit à la consommation annuelle de leurs échelles, ou si les negociants de Marseille qui règlent leurs operations mettoient quelque proportion entre les envoys et la consommation [they might be correct …]."

[166] *Mémoire* April 22, 1740, CCM J 59. The number seems to be quite accurate: Carrière, *Négociants*, vol. 2, 1040, col. II for 1740 = 17,9 Mio. livres.

estimated the number of persons employed in the Mediterranean trade at 100,000. If one sought to reduce that commerce to less than half, it would not only affect the Mediterranean. As the current French commerce with Germany, Italy, the Netherlands, Spain and the Americas would function only "in relationship and in proportion" to that of the Levant, the implications of reducing the Levant trade would necessarily lead to the stagnation of all French commerce. By "fixing the number of trading houses in the Levant," he claimed, "one would also fix the [sc. all French] commerce."[167] The *Chambre* even evinced a form of self-perception that was similar to present forms of mercantile acuity: for a duration of thirty years it calculated an average of one percent economic growth which shows the use of a long-term memory and a correspondent regulatory idea concerning the amount of commerce as a whole.[168] Rather than cutting the number of merchants, one of the *mémoires* authors gave detailed advice about how the organization and the quality of several products of the Levant trade could be improved.[169] The danger of the planned reduction having deleterious effects was seen first regarding the international competition in the Mediterranean. Markets would be left to France's competitors. Resulting from the 1740/41 discussion and Pignon's visit, the number of the *maisons* was fixed in 1743,[170] but still, the rule of "proportionality" behind that limit was disputed and not rationalized in a mathematical way.

Only in a *Mémoire sur le commerce de Tunis* from 1765 does the mathematical calculation of proportionality become visible. This author implemented the idea of the "normal non-war year" expressed by Maurepas in 1733. First he calculated the overall average import volume for the ten years from 1755 to 1764, at 862,499 piastres per annum. A quarter of this was attributable to four French factories, and the rest to the community of 2,000 Jewish merchants from Spain and Italy. Export revenue was calculated to about 234,187 piastres. For the four years of peace 1754, 1756, 1763, 1764, disregarding the impact of the Seven Years' War, average annual French import revenues were higher, 359,529 and not 213,348.[171] The author judged that an

---

[167] "Mémoire sur la réduction des Maisons du Levant" (s.d., 1740), CCM J 59.
[168] "Mémoire des Négocians de Marseille ... 1742," CCM J 59.   [169] Ibid.
[170] Masson, *Commerce XVIII*, 26–28; Paris, *Histoire*, 331, 333.
[171] AE La Courneuve MD Afrique 9 n. 22 (1765). Public accounting in Italy was experienced by the sixteenth century in distinguishing between "normal" and war years, Zwierlein, *Discorso und Lex Dei*, 443.

increase of the number of merchant houses was advisable but remained undecided between six or eight as the optimal number and he displayed no idea of a formula or equation to decide the matter. In any case, the *mémoire* shows that the idea of proportionality developed from the early eighteenth century to the last third of the century from a simple wording or phrase into a form that necessitated the precise analyses of the economic past and future prognostics for a given *échelle*.[172]

The general political aim pursued is not surprising. During the first half of the eighteenth century, the usual populationist concepts for metropolitan France centered on increasing the population as far as the agricultural production could allow. After the expulsion of the Huguenots, in France one usually never finds the kind of overpopulation arguments by pro-colonialists that are known to have circulated in England around 1600, when overpopulation and depopulation arguments were in balance. Avoiding the loss of people to the *échelles* was precisely in line with that general idea.[173] One can practice populationist politics without political arithmetic, even without knowledge about the number of inhabitants.[174] It is not that general aim, but the *how* of administrative practice seen here in the *échelles* that can be called astonishing at this time (1720/40). What can be witnessed here is an embryonic form of applied quantitative political "laws" of arithmetic combined with an enduring political desire to also shape a given population qualitatively.

Thierry Martin once put forward the question of whether there ever really was French political arithmetic. The usual answer was that it could be found after 1750.[175] The findings here presented from the Maurepas administration seem to contradict this claim. Was there an earlier comparable form of applied political arithmetic in Europe? Political arithmetic itself was "invented" and proudly advertised by William Petty and Charles Davenant in England as early as 1690.[176] There are some passages in Petty and Davenant that show ideas close to those applied by the Maurepas administration. These concern the

---

172 AE La Courneuve MD Afrique 9 n. 22 [1765].
173 Usually, the increase of population was recommended for the main state, while emigration to the colonies should be regulated strictly for preventing France from losing population: cf. still helpful Spengler, *Économie et population*, 32–105, for the English case Campbell, "Of People."
174 Nipperdey, *Bevölkerungspolitik*, 119–121.
175 Martin, "Une arithmétique politique française?"
176 McCormick, *William Petty*.

delineation of a given population into productive and unproductive sections and the question of how to increase the number of the former and to decrease that of the latter,[177] a plan that corresponds with Maurepas' continual repetition of orders to get rid of the "français oisifs" in the *échelles*. Generally not much is known about the *early* reception of the English political arithmetic in France before 1750.[178] Petty was fascinated by the idea of proportion in the sense that he thought square roots to be inscribed into the nature of things. This was a reasoning not related to practical application.[179] In the flood of French political economy literature that started to pervade the public sphere after the Montesquieu choc and the year of rupture and takeoff in 1750,[180] one only occasionally finds echoes.[181] Vauban, often mentioned as an appropriate candidate for a real French political arithmetic in parallel to the classical English authors, offered in fact many

[177] Petty, "A treatise of taxes and contributions," 28; Davenant, "An essay," 202. Cf. Finkelstein, *Harmony*, 121–124, 221f. The early texts analyzed by Appleby, *Economic thought*, 129–157 did not contain further general reflections about the proportionality between commerce and the number of people (I checked William Goffe, Adam Moore, Leonard Lee, Samuel Hartlib, John Cook, Peter Chamberlen, Humphrey Barrow, John Moore).

[178] For the post-1750 reception cf. Reinert, *Translating empire*.

[179] For Petty's curious *Discourse concerning the duplicate use of proportion* (1674) cf. McCormic, *William Petty*, 190. Cf. for the English practice Deringer, "Finding the Money."

[180] Precisely around 1750, the overall print production in Europe, the number of published treatises on political economy, the overall number of translations, and the number of translations of economic theory, all exploded (Théré, "Economic publishing," 11; Lüsebrinck et al., "Kulturtransfer im epochenumbruch," 33; Reinert, *Translating empire*, 52–60). In 1750, the Gournay circle was formed: All those are clear moments of a new framework concerning the epistemic conjunctures, perhaps more important in this context than the outbreak of the Seven Years War or the 1750 export crisis in the Levant.

[181] Veron de Forbonnais, *Elemens du commerce* (1754), vol. 1, 56: "La population est l'âme de cette circulation intérieure, dont la perfection consiste dans l'abondance des denrées du crû du pays en proportion de leur nécessité"; Cantillon, *Essai* (1755), partie II, chap I, 151f.: "On a essaié de prouver, dans la Partie précédente, que la valeur réelle de toutes les choses à l'usage des Hommes, est leur proportion à la quantité de terre emploiée pour leur production & pour l'entretien de ceux qui leur ont donné la forme." But nothing similar in Melon, *Essai politique* (1734); Law, *Œuvres complètes*; Pierre de Boisguilbert; Huet, *Le grand trésor* (1713); Du Tot, *Réflexions politiques* (1738); Du Tot, *Histoire* (1716–1720); abbé de Saint-Pierre, "Utilité des dénombremens" (1733); abbé de Saint-Pierre, "Projet pour perfectioner le Comerse de France" (1733).

calculations about the regularities of population growth, and his taxation plans also contained ideas about just proportions. Once again, these calculations did not lead to considerations of practical implementation and there was no combined inductive-deductive generation of populationist laws in relationship to states of a given trade.[182] The general *enquêtes* performed to survey the population, in addition to the natural resources, monuments and curiosities of France in 1716–1718, 1730 and 1745, all with the participation of the country's intellectual avant-garde and its institutions, mainly the *Académie des sciences*, cannot be classified as being conducted according to principles of political arithmetic. They were counting the population, but not calculating in the sense that political plans relating to it were guided by certain rules derived mathematically from what had been tallied.[103]

Brian distinguished between three steps of development of political arithmetic: (1) Cartesian accounting as present with Vauban around 1700; (2) experimentation present in surveys like the one conducted under the controleur général Orry in 1745; (3) progressive abstraction from the stated facts, an analysis of the causes, the comparison, and finally the identification of regularities and abnormalities which he did not find in France before 1770 with the abbé Terray.[184] The characteristic feature of political arithmetic in the sense of step three is to consider a "population" as an object of nature that obeys rules similar to those physicists were searching for from the Padovan school to Galileo and Newton.[185] This was the case in early forms of probabilistic calculation of life expectancies in given cohorts of populations from Graunt to Huygens and de Moivre. Even there, however, it still took a long time to form mathematical functions that would approximate life expectancies and annuities instead of using and improving life tables. And it still took longer until mathematical laws like that were begun to be implemented in institutional administrative practice.[186] But what is visible in the Maurepas administration

---

[182] Cf. Virol, *Vauban*, 199–254 and the annexes 5 and 6 on pp. 415–420; Meusnier, "Vauban," 91–132, 98; McCollim, *Assault on Privilege*, 107, 141.

[183] Garner, "L'enquête Orry," 371. But cf. the circular letter of Orry, Paris December 17, 1744, Lecuyer, "Une quasi-expérimentation," 174.

[184] Brian, *La mesure de l'état*, 176f.

[185] Zwierlein, "Politik"; Daston and Stolleis (eds.), *Natural law*.

[186] Hacking, *The emergence*; Hald, *A history*, 513, 547; Krüger, Daston and Heidelberger (eds.), *The Probabilistic revolution*, vol. 1; Daston, *Classical probability*, 137, 172–174; Zwierlein, *Prometheus*, 202–208.

concerning the Levant trade is just fitting into that step of development: Rules active in the *nature* of trade and economy were to be found in the periphery. Next, normative rules which had the desired effect had to be formed accordingly.[187] Finally, the generated rules were supposed to dictate the realities, in movements backwards from the center to the periphery. Those were now half-normative and half-empirical rules, rules fitting the "macroeconomic" framework as one might call it today.

All that said, one would be tempted to depict the Maurepas administration's efforts to shape the French Nation in the *échelles* as a small but important "political arithmetic revolution" of French government. This was not just a speculative philosopher's abstract idea, but a practice. It was not just a rigid normative ruling;[188] it not only gathered data through queries, enquêtes, tables and columns of numbers, but it did so with the aim of generating socioeconomic rules like experimentators in the laboratory through empirical investigation and abstraction. It predated Brian's third step by roughly forty years. This means one may answer Martin's question in the affirmative. Yes, there was a French political arithmetic, but earlier in practice than in theory. And that happened in France's periphery or with respect more to the empire than to the core of the country. The tiny size of the population abroad, their high mobility and the absolute dependency of the merchants abroad on the crown's protection within the Ottoman lands presented a better object for "scientific" populationist experimentation than the homeland itself.

By conceiving of the *nations* in the *échelles* as a whole to be organized and shaped by such general rules, they were dismantled of their

---

187   It would be necessary to "establir une règle par rapport aux artisans . . . qui sont en trop grand nombre dans plusieurs Éschelles . . . " (Maurepas to the *Chambre*, May 30, 1732); "mais je prévois qui sera difficile de régler le nombre . . . il seroit convenable et aisé d'en régler le nombre proportionnement" (Maurepas to the *Chambre*, Versailles, January 14, 1733); the merchants themselves when they "règlent leurs operations mettoient quelque proportion" (Maurepas to the *Chambre*, April 22, 1740); the merchants responded with the same language: "toutte autre règle ou arrangement" would be against the necessary "Liberté" ("Mémoire des negoçians" 1742); or they claimed that "il convenoit mieux de laisser libres les negoçians, sans les assujetir à des règles générales" (*Mémoire* of the *Chambre* for the comte de Castellane, nominated royal ambassador, January 7, 1741 – all in CCM J 59).

188   As the extraordinary 1727 letters patent for colonial French America, cf. Pritchard, *In search of empire*, 241.

medieval corporative identity of merchant colonies from a given region (*nation* in the narrower sense) and transformed into parts of the allegedly well-regulated and controlled body of *the* nation, of France abroad. This becomes apparent in a late *mémoire* written by Pignon, who was perhaps the closest collaborator of Maurepas in the Levant. In this text, written during the 1750 grain crisis, Pignon shows that he conceived the work that had to be done from 1720 not only as a technical modeling of numbers. Instead, he now used all that language that is characteristic of the later eighteenth-century construction of modern nations that one usually associates with the times around the Seven Years War.[189] Political economy had to correspond to the "génie de la Nation" and one had to be careful to utilize commerce to augment the good and suppress the bad parts of "le caractère même de leur Nation." He reminded his audience about the necessary "union" among the merchants of the French nation, endangered by its own interests.[190] The seemingly materialist and technical political arithmetic reduction of merchant families to calculable numbers went now hand in hand with an early enlightened thought that aggregated the merchant nations into *the* Nation as the newly constructed subject of History.[191]

One can thus see a remarkable development. Around 1650/60, the hitherto nescient concept of nationality was transformed into a conscious specified unknown, in different ways in the English and the French case. It became a question and a matrix against which reality was continually measured and tested. In asking about the unknown nation, the national became more and more reified. It could become, around 1720/40 an object of abstract reasoning. An awareness of its constructedness grew. Finally, the reification of the national could lead, in the French case, to a new quasi-material object of political arithmetical calculation. Paradoxically, the aggregation of autoreferential unknowns could reach such a state of shapable, malleable concreteness. *Ab ovo*, there is no "nation" of anyone born, but the mercantilist practice of continually asking about and ascribing it started a process of administrative communication, building, in a manner of speaking, on

[189] Cf. Colley, *Britons*; Blitz, *Aus Liebe zum Vaterland*; Bell, *The cult.*
[190] Pignon, "Sur le Commerce" (1754), 75–80. For the identification of Pignon as author thanks to the (larger) manuscript copy in Gournay's library cf. Meyssonnier, "Gournay," 100–106.
[191] Stanzel (ed.), *Europäischer Völkerspiegel*; Bell, *The cult*, 140–168.

the void. At its end a second level was reached, a second-degree process emerged in which the reified nation, as result of the aggregated results of classification and attribution, could be planned, shaped and calculated regarding its arithmetic, physical and finally even moral character.

Baldus versus Grotius: Conceiving the Empires and Their Unknowns

Analysis has thus far moved in a bottom-up direction. This is because the central emphasis of the chapter is that the mercantilist trade empires, considered as institutional settings and as networks of humans that communicated in a specific way, were built on unknowns, on the question about the national once that category had emerged. The fluidity of that everyday microcommunication of the same or similar contacts and questions was even the socle of aggressive and violent mercantilist estrangements and competition. Its "fundaments" were, in the end, just reiterated forms of communication, of attributions, of words and signs. This brings us now to move from purely operative forms of (non-)knowledge to higher condensed epistemic forms. As was the case for the intersection between proto-political-arithmetic method and administrative practice, imperial discourses also had their impact on action and decision-making. Some of those moments of connection, which is not simply one between the sphere of "books" and the sphere of oral and handwritten "action," will be shown below – when a Grotian idea in an advisory text is put into practice for Colbert, when differences between free trade and regulated commerce were negotiated before the Council of Trade. There was also a complexity and systematicness to those discourses, as, for instance, the different approaches to Roman law by Grotius and Selden will show. This was detached from action, but it gave them their enduring impact and power to convince on the level of discursive tradition and as a firm point of retreat. So, the imperial concepts were present and more stable. They were developing more slowly, guiding, in some way, the general directions of decision-making processes in the long run, in other words, of the shape of the English and the French empires. If one puts the chicken-and-egg question for the "origins" of empires on the table, I would opt for the just mentioned bottom-up perspective and emphasize the relation between norms as specifiers of unknowns and their continual enactment in practice. But this does not mean that discourses or "ideology" were unimportant. Those two levels were in a relationship of interdependency, of mutual stimulus, and of "fitting" together. They were – to use a geological or Braudelian metaphor – like different discursive strata of fluid magma on the one hand and of more solid, more slowly moving tectonic elements on the other.

A growing body of literature has been investigating concepts of empire, perhaps more for British than for French history.[192] In this comparison of imperial concepts, one can exclude, for both France and Britain, the use and exploitation of crusader ideology around 1700. That does not mean its total erosion; France utilized those motifs, for example, in the context of its status as protector of the Christian faith. Nonetheless, it is of next to no importance for the question of nationalizing economics.[193]

From French Grotianism to the Property of the Mediterranean

There are few French theoretical politico-economic texts, however one might define them, *before* 1750. Some decades ago, Perrot demon-strated how the one important candidate for such an early theory, Montchrestien (published 1615),[194] was almost unknown and rarely cited by eighteenth-century bibliographers of the then newly estab-lished discipline of political economy. They all started with Melon, Boisguilbert or similar early eighteenth-century authors, while the extension of Bodinian thought into economics by Montchrestien was forgotten.[195] The one and nearly only author cited in several *mémoires* right until the times of Gournay is Richelieu.

---

[192] Colley, *Britons*; Wilson, *The sense*; Armitage, *The ideological origins*; Armitage, *Foundations*; for rather a later period Pitts, *A turn to empire*; Haran, *Le lys et le globe*. Usually, the Mediterranean does not play a great role in that literature.

[193] Charles V as the late medieval Iberian kingdoms still borrowed from crusader ideology during his Tunis and Algiers enterprises 1535/41 (cf. Poumarède, "Le voyage de Tunis"). Duke Charles II Gonzague-Clèves-Nevers who tried to establish a whole European military order still played on that until the third decade of the seventeenth century (Papadopoulos, Ἡ κίνηση, 148–196; Cremer, *Der Adel*, 144–168, but the ideological basis here was eclectic, mixing motifs from antiquity with references to Charles Martel, Byzantine rule and Godefroy de Bouillon). Crusader motives in late seventeenth-century attacks by British or French ships against the Barbary corsairs had become rather ornamental.

[194] Monchrestien, *Traicté*.   [195] Perrot, *Une histoire*, 67f.

Richelieu's *Testament politique* was published for the first time in Amsterdam in 1688. It had been known and certainly circulated in manuscript form among the French political elite before.[196] The parts of the *Testament* which go beyond all the political authors writing and publishing in Richelieu's own circle in the 1620s and 1630s, and even beyond most who wrote later until the times of Louis XIV, are the chapters on Economy, Navigation and Trade (second Part, Chapter 9, 5–7): all other prominent French authors and intellectuals of that time – Guez de Balzac, Boisrobert, Hay du Chastelet, Chapelain, Cardin Le Bret, Philippe de Béthune, Hersent, the anonymous author of the *Catholique d'État*, Louis Machon and Gabriel Naudé – think in quite purely political terms of reason of state and post-Machiavellian and post-Bodinian discussions of sovereignty; there is no "imperial political economy" to be found with them.[197] That is why the 1688 publication of Richelieu's *Testament* could make a "fresh" impression at a time when Colbert was already dead and many elements of the *Testament* had already been realized, and that is why this text could still be a reference in 1750, not only in veneration of the mighty cardinal, but also because it was still functional concerning the political economic foundations of French imperial expansion. There was virtually no other good text at hand.[198]

Richelieu started the fifth section of his work on "Sea Power" with the question of who possesses the empire of the sea.[199] Richelieu's own way of reasoning operated through examples rather than through systematization. The central example came from the times of Henri IV. When Henri IV sent the Duke of Sully to England in 1603 to honor James VI/I's accession to the throne, a patrolling English rowing barge stopped Sully's ship in the middle of the Channel and commanded that the French white flag be removed from the top mast in order to honour the English king as "the sovereign of the Sea." When Sully, who thought as ambassador he was exempt from that sign of humility,

---

[196] Richelieu, *Testament politique*, ed. Hildesheimer, 17f.
[197] Thuau, *Raison d'État*; Church, *Richelieu*; Cavaillé, *Dis/simulations*. For further references, cf. Zwierlein, "Machiavellismus / Antimachiavellismus."
[198] That is also the reason why the English authors always pointed to Richelieu and liked to cite the *Testament politique*: It was, ultimately, an authoritative source for "the" French politics with which one had to compete or which one should emulate. Charles Davenant was particularly strong on that, cf. Finkelstein, *Harmony*, 228, 337 n. 50, but cf. also Cary, *Essay* (1695), 141f.
[199] Thomson, "France's Grotian Moment?," 394.

refused, the English captain enforced his order with a barrage of cannonballs which "cut through the heart of all good Frenchmen." Starting with that example, Richelieu then reflected about possible solutions for such situations: French ships close to the English coast having to lower their flag, and vice versa, English ships close to the French coast theirs, or, a rather empiricist-statistical solution, that the ships of a smaller fleet would have to lower their flag upon encountering ships of a larger fleet anywhere at sea. Richelieu finished his discussion of this point by concluding that the king just has to "be strong at sea" regardless. Whatever might be a solution by what was later termed the international maritime law, the political answer had to lay in the expansion of naval armaments.[200] He proposed maintaining forty ships in the ocean and thirty galleys in the Mediterranean. As it is well known, the actual realization of those navy plans happened under Colbert and Seignelay.[201]

Bearing in mind what has been said in an earlier section on flags and signs of nationality, it is remarkable that the ship's flag was the principal iconic sign chosen by Richelieu to introduce the section of his treatise on international trade and naval competition in the two seas that he suggested considering separately, the Ocean (meaning the Atlantic) and the Mediterranean. In this example, which is followed by an analysis of competition between the English, Spanish, Dutch and French, Richelieu condensed the central distinctions of we/them, of internal/external and of foreign/home. The sixth section addresses "Commerce as depending of the Sea Power." Through that order, Richelieu made clear that commerce and commercial competition was a *filia potestatis*, and was subordinated to naval military strength, conceiving commerce as part of state power as was typical in the French tradition.[202] Following the major division between "Ocean" and "Mediterranean" which would later also be the principal division within the French Ministry of the Marine (Ponant/Levant), Richelieu first dealt with the (possible) colonial Atlantic and North European trade, then the Mediterranean Levant trade:

I have to add that I was mistaken for a long time concerning the commerce of the Provençaux in the Levant: I thought, with many others, that it is detrimental for the state, founded in the common opinion that it would tear

---

[200] Richelieu, *Testament politique*, ed. Hildesheimer, 323f.
[201] Dessert, *La royale*; Villiers, *Marine royale*, 1–120.   [202] Cf. above n. 4.

money [*argent*] out of the kingdom by importing nothing else than unnecessary merchandise which are only good for the luxury of our nation.[203]

Richelieu's views of the Levant has attracted less attention in recent years, but this passage is a clear indicator that the *Testament* is an Anti-Montchrestien and an Anti-Razilly on what concerns conceptions of imperial outreach into the Mediterranean as Henri Hauser demonstrated in 1944.[204]

Montchrestien had pointed out that the French Levant trade with the Ottoman Empire, Egypt and the Barbary Coast would stimulate the economies of those hostile countries. One could observe, he noted, that when French trade ceased, all the societies of that region always became destabilized, with the outbreak of "brouilleries, seditions, et guerres civiles." Relying on the basic mercantilist idea that it is detrimental for a country to let money flow out and foreign goods enter, he underlined that, in this particular case, the matter would be even worse as this trade would be "totalement ruineux" for the French and of great advantage for the enemies of Christians.[205] He then proposed complete European protectionism as a Europe/Orient blockade. French, Italian and Spanish manufacturers should answer the demand for silk, and the wool production of Southern France should be augmented so that France could withdraw completely from the Levant trade. Europe could stay apart from Muslim countries in complete economic autarchy. This done, "those excessive sums of gold and silver which are leaving France will remain there."[206] Montchrestien judged the Levant

---

[203] Richelieu, *Testament politique*, ed. Hildesheimer, 338f. Cf. that passage with that of Montchrestien cited below (n. 205), it's a direct response.

[204] Hauser, *La pensée*, 20, 74–107. After Hauser, it seems that Mousnier, "Le Testament politique," 137 and Louis André achieved to orient the discussion for decades on the themes of the "reason of state," the "grand dessein" and on the question of the *Testament*'s authenticity. Esmonin, "Observations"; Thuau, *Raison d'État*; Church, *Richelieu*, 480–495; André did not include Hauser's results in the critical apparatus of his edition, so there is no note on Montchrestien (cf. Richelieu, *Testament politique*, ed. André, 423). Thomson, "France's Grotian Moment?" correctly reminds that it is wrong to think of Richelieu as an ideologue who tried to form Europe like a demiurge according to the *grand dessein* (chapter in Hauser, *La pensée*, 108–120), but Hauser's results concerning Richelieu's sources – the *Mémoires*, today deposited in the *Archives des affaires étrangères* – should not be forgotten.

[205] Montchrestien, *Traicté*, ed. Billacois, 361. On Montchrestien cf. Clark, *Compass of society*, 10–14.

[206] Montchrestien, *Traicté*, ed. Billacois, 361s.

trade therefore to be completely damaging and superfluous for the French economy and, even worse, to Christianity as a whole, because it strengthened the hostile Muslim world.

The *Testament* developed its concept of the French trade empire in the Mediterranean in a completely opposite way, relying on several manuscript *mémoires* written for Richelieu on Levant commerce, one of them excerpted at some length. Against the idea of a European anti-Ottoman blockade, Richelieu argued that Levant goods were not luxuries for France but necessities. It followed then, that the import/export balance was positive for France and that the money (or rather literary, the silver) invested into the Levant did not come directly from France, but from Spain. From here French merchants obtained the silver merely through the sale of Levant goods. Marseille had grown with that commerce, Levantine merchandise was re-exported with a 100 percent profit margin, and therefore many craftsmen and sailors were fed by the Levant trade. Thus, Levantine commerce was necessary. While Montchrestien had argued for the complete reorientation of France's maritime trade to the Atlantic only about two decades after the Dutch and English entry into the Mediterranean, Richelieu's *Testament* had reinforced the Mediterranean perspective, a position which would prove to be realist in the long run. Also suggesting the promotion of France's ship building industry, the final paragraph of that chapter is devoted to the second level of the Mediterranean economy. To "clean" the Mediterranean of the Barbary corsairs, Richelieu argued, it would be sufficient to dispatch a squadron of ten galleys which would start patrolling each April from Gibraltar to Corsica, Sardinia and along the Barbary coast.[207] The bibliography of French printed works on political economy does not list almost any important work touching on maritime trade between the time of the *Testament*'s composition and Colbert, with the exception of Jean Éon's (Mathias de Saint-Jean's Ord. Carm.) *Commerce honorable* (1646).[208] Colbert knew the Cardinal's *Testament*[209] and his memorialists referred to Richelieu's Levant politics as exemplary even before the publication of the *Testament*.[210] But as the *Testament* itself relied

[207] Richelieu, *Testament politique*, ed. Hildesheimer, 342.   [208] Cf. above n. 79.
[209] Soll, *The Information Master*, 53–63, 116.
[210] The "Mémoire à Monseigneur Colbert sur l'establissement solide du Commerce en Barbarie" (October 1670), AE La Courneuve MD Alger 12, f. 172r–173r advocates the conquest of Tabarca from the Genovese Lomellini, a plan which was debated by the English and French for decades.

heavily on manuscript *mémoires* prepared for the Richelieu government, the Colbertian administration can surely not be understood as an "application" of the *Testament*'s "theory." It was instead a new osmotic cycle between more "theoretical" works, longer analytical *mémoires* and *avis* and everyday correspondence and orders. The many *mémoires* concerning the reform of the Levant trade written for Colbert's use from within and outside the first embryonic Conseil de Commerce in 1664 always thought it necessary to stress that the correct organization of commerce would prevent France from losing money and/or silver pouring out of the country.[211] The first point of d'Oppède's central 1662 *mémoire* for Colbert cited above[212] concerned the duties and tolls of Villefranche (Villafranca) requested by the Duke of Savoy and the similar coastal duties for the Grimaldi of Monaco. They hindered the free access of Marseille ships to their home port by forcing them to stay 100 miles off the coast. They were perceived as a sort of "tribute" of the king's subject to foreign princes and as an insult to the "authority of His Majesty and to the French name." D'Oppède's minutes argue:

[those duties of Savoy and Monaco are] of an insupportable inconvenience and danger for navigation. The sovereign princes are certainly empowered to impose whatever duties they think in their ports as everyone is free to go there or not. But they have not the right to impose duties within the sea which does not belong to them and which belongs to public and international law where there is no entitlement to property that ever could be valid.[213]

The sea was free. As a result, the 1662 assembly argued, said princes should not be allowed to levy tolls one hundred miles off the coast. Those formulations, betraying a degree of legal expertise (*droit public / droit des gens*) were very probably composed by d'Oppède

> The anonymous recalled "feu M. le Cardinal de Richelieu sur la fin de ces jours, entra en quelque traicté pour la tirer des mains de la famille des Omellins [sic] de Gennes."
>
> [211] "Le Transport de l'or et de l'argent a esté de tout temps déffendu en ce Royaume ... Les advantages que produira cest establissement sont de très grandes conséquences, le premier est qu'il ne se transportera pas aucun argent de France" (S. Correur, "Mémoire pour le negoce du Levant," AN Paris B III 234, n. 5).
>
> [212] Cf. above n. 37.
>
> [213] "D'une incomoditté et danger insuportable à la navigation; les princes souverains pouvans bien imposer dans leurs portz les droictz que bon leur semble pour ce qu'il est libre d'y aller ou de n'y aller pas. Mais non imposer sur la mer quy ne leur appartient pas et *quy est du droit public et des gens* contre lequel *il n'y a tiltre ny possession* que puisse valloir" (AN AB III 234 nr. 13).

himself.[214] It was, in fact, still the Grotian language of "freedom of the sea" used here as it had been of importance for Richelieu. As Thomson has shown, it was used in the 1620s as a weapon for forcing the strong re-entry of French maritime power and trade into the Mediterranean. Bearing in mind that the 1662 *mémoire* already incorporated all the measures of the 1669 edict, one may argue with good reason that the main "frame of thought" in which d'Oppède/Colbert acted was still very similar to where the Richelieu administration had left the issue around 1635, when the Thirty Years' War and the Fronde had interrupted a good deal of commercial politics in general. The combination of the language of the freedom of trade and of protectionism in the 1669 edict proves to be the extension of the Grotian legacy which, itself, embodied that bifurcate and somewhat dialectical framework.

Richelieu's *Testament* remained the fundamental reference for a long time. In 1700, Pottier de La Hestroye referred to Richelieu in praise of the Cardinal's support for the founding of merchant companies in 1642 and his homage to Marseille as where the Phoenicians had established their trading colony and where Caesar had entered France.[215] The donneur d'avis Adrien Cazier used Richelieu's *Testament* as his latest reference in his 1710 proposals for reform addressed to the King.[216] The Abbé de St. Pierre wrote a whole commentary on the *Testament*, albeit not on the Levant trade chapter.[217] Melon referred to the *Testament* in 1734, commending the Cardinal's efforts to build up the Navy.[218] And the Cardinal's legacy was well-remembered by French Levant specialists. When in 1731, the King sent Duguay-Trouin on a military and political-cultural expedition to the Levant, a member of the Lemaire consular dynasty wrote a *mémoire* (*Observations sur le Voyage des Eschelles de Barbarie*). Lemaire referred to Richelieu's *Testament Politique*, citing the central (hidden anti-Montchrestien) passage:

---

[214] Cf. Thomson, "France's Grotian Moment?" – for a more detailed analysis of Grotius cf. below the section on English imperial frames of thought.

[215] Law, *Œuvres*, 70–72, 133.   [216] McCollim, *Assault on privilege*, 177.

[217] Abbé de Saint-Pierre, "Observasions" (1741). But his judgment on Mediterranean trade conformed precisely with Richelieu: "De-là il suit que le reste étant égal, nostre Ministère doit porter la Nation le plus qu'il est possible au Comerse Maritime ... notre Nation peut faire beaucoup plus facilement, que plusieurs Nations d'Europe, la plus grande partie du Comerse de la Mediterranée." (Abbé de Saint-Pierre, "Projet pour perfectioner le Comerse de France" (1733), 204f.).

[218] Melon, *Essai politique* (1734), 37, 94.

No one doubts the utility and necessity of the Levant commerce which had been renowned as very advantageous even during times when we only could export money to buy their goods and despite the inconvenience of currency departing the kingdom, as becomes evident from the *Testament politique* of M. Cardinal Richelieu. He admits that he had been convinced by reason and experience, after thorough examination, and despite the contrary warnings of the people to whom the utility of commerce was little known in general, and in particular that Levant commerce would provide the state with very important profits.[219]

The chancellor of the Cyprus consulate and the future consul of Algiers, Lemaire, was again referring to competition with the English. The French were eager to achieve superiority over the English in the important *drap* trade with the Ottomans and North Africans, and Provence had recently surpassed England in terms of production. Following Richelieu, Lemaire stressed the state's interest in the trade, showing that economic competition was also competition for political predominance. The proximity of Provence to the Turkish lands gave an advantage to the French concerning their knowledge of transport costs and Ottoman tastes. If they proceeded in this manner, Lemaire argued, the French could totally exclude the "draps d'Angleterre en Turquie" and make themselves "masters of the Levant commerce." Already by 1731, the English were allegedly maintaining their position in the Mediterranean "rather by political reasons to compete over the terrain with us [sc. the French] than for their own utility." Besides the purely economic advantage gained, the French might also acquire "superiority

---

[219] "Personne ne doute de l'utilité et nécessité du Commerce du Levant, qui dans les tems même où nous n'y pourions porter que de l'Argent pour leurs marchandises, malgré l'inconvénient de la sortie des espèces hors du Royaume, étoit reconnu pour très avantageux, comme il paroist par le testament politique de M. le Cardinal de Richelieu. Il avoue qu'après un mur examen, et malgré la prévention de la pluspart des gens, auxquels l'utilité du commerce en général étoit alors peu connue, il s'étoit convaincu par raison, et par expérience, que le Commerce particulier du Levant procuroist à l'Estat un profit très considérable" (André-Alexandre Lemaire, "Observations sur le Voyage des Eschelles de Barbarie et du fond du Levant en l'année 1731," AE La Courneuve MD Alger 13, f. 127r-144v, 130r: this version seems to be enlarged later: "fait à Alger le 1er Janvier 1751" by Lemaire (autograph signature, f. 144v). The earlier version (without naming the author) in AN MAR B7/311 (unfol.) has as general title "Suite du journal de la campagne de 1731. Observations sur les Échelles de Barbarie" and for the subchapter with the Richelieu paragraph "Observations sur le commerce du Levant en particulier des Échelles que nous avons visitées."

and almost full ownership of the Mediterranean Sea." As the British parliament would only accept increased military spending when economic interests had to be defended, according to Lemaire, they would withdraw from the Mediterranean as soon as the French took over the *drap* trade. "To remain the masters of the Mediterranean Sea seems to be a considerable object for the State." The French as masters of the sea should remain the only option for trade with the Turks, and the only power to influence the decisions of the Ottoman Porte. As there were ten times more French subjects in the Ottoman Empire than of any other nation, the establishment of French hegemony in the Mediterranean would, Lemaire maintained, be only logical. The English in turn would, perhaps, even give up Gibraltar and Port Mahoney after some time if the costs of maintaining those possessions notably exceeded their commercial profits.[220]

Lemaire used the Grotian Richelieu to express an absolutely un-Grotian vision of the maritime trade, France's sovereignty and its imperial growth; there was no use anymore in camouflaging expansion with the language of free trade. It was now primarily a space to be conquered. The idea of "ownership, property" was applied to the whole sea. This explicitly contradicted the Grotian understanding of how the Romans had conceived of the sea: As will be discussed below, Grotius' central point was the re-enforcement of the original idea of the sea as a *res communis*, a thing common to all and indivisible, of which no one could claim and acquire as property. Lemaire's argument even went beyond the British and Iberian seventeenth-century adversaries to Grotius who had never claimed the property of the whole sea for their prince, but only the *dominium* over some coastal areas. Instead, Lemaire opted for the conquest of the "property" of the whole Mediterranean through commercial, followed by political, domination. This was a clear plea for a French Mediterranean maritime empire.

So, while French anti-Habsburg men of politics used the idea of a free sea in the first half of the seventeenth century as they supported the Dutch against the Spanish in the Mediterranean, by the turn of the eighteenth century, they tended to use a diametrically opposite argument stronger than Selden had formulated it. While it is usually acknowledged that the modern "imperialist" vision of foreign trade would, in the long run, follow a successful path from mercantilism to

---

[220] AE La Courneuve MD Alger 13, f. 131r–v.

free trade,[221] one sees here a different intermediate step during the eighteenth century. This was, perhaps, even a more imperialist policy based on a harsh opposition to free trade, and aiming at the monopolist domination of economic space.[222] That this vision was re-used during the 1751 crisis is telling, and, unsurprisingly, Gournay himself still referred to Richelieu in his *mémoire* concerning the rivalry between France, England and the Netherlands, just after citing the Ciceronian locus classicus on maritime empire ("qui mare tene[a]t rerum potiri necesse est," Cicero: *Letters to Atticus* X, 8 [Cumae, May 2, BCE 49], translated into French by the abbé Mongault in 1714 as "Celui qui est maître de la mer, le sera tôt ou tard de l'empire").[223] To those observations, one could add French projects to conquer parts of North Africa and to build colonies during the eighteenth century. There was a long-running discussion about which of the Mediterranean empires would buy the islet of Tabarca from the Lomellini. The small French possessions as the *Bastion de France* were then usually called "colonie."[224] This markedly imperial self-understanding of France in the Levant even reached, at some particular points, plans for terrestrial conquest of Egypt or the whole Ottoman Empire, as is known from Leibniz' famous *Consilium aegyptiacum*. Such grandiose schemes, however, remained unrepresentative for the predominant eighteenth-century imperial thought until 1830.[225]

---

[221] Mokyr, *The enlightened economy*, 156.

[222] That is not at all the *doux commerce* that one often associates with enlightened French economic thought, at least for the Gournay period (cf., after the classic account of Albert Hirschman, Larrère, *L'invention*, 144–172). Foreign trade planning was different from reasoning about internal commerce and agronomy.

[223] "On sçavoit pourtant dès le tems d'Auguste que ceux *qui mare tenent* [in Cicero: singular] *rerum potiri necesse est*. Le cardinal de Richelieu l'a répété depuis, et les Anglois nous le prouveront bientôt si on ne les empêche" (Gournay, "Moyens" (1755), 360). That Ciceronian *locus* was widely used in British-French discourses about maritime Empire. The passus from a French Navy *mémoire* in 1738 which Cheney, *Revolutionary commerce*, 35 quotes in his English translation ("whoever is master of the sea is master of everything") is obviously such a citation, apparently without indicating Cicero's name. Cf. Selden, *Of the dominion* (1652), 74; Evelyn, *Navigation and commerce* (1674), title page.

[224] Cf. only "Mémoire concernant le commerce des Colonies de Barbarie au pouvoir de la Compagnie des Indes" (1730), AE La Courneuve MD Alger 13, Nr. 22 f. 82r–85v.

[225] Cf. Drapeyron, "Un projet"; Bérenger, "La politique ottomane"; Leibniz, "Consilium Aegyptiacum"; Hirsch, *Leibniz*, 18–21; Dingli, *Seignelay*,

This synopsis shows that the everyday cognitive processing of the "national" largely corresponded with the larger mercantilist imperial frameworks. As long as the post-Grotian freedom-of-trade doctrine was endorsed by the French, the external/internal definition of the "national" remained more fluid than in the British case, as the 1669 Marseille edict was also more undecided and bifurcated – protectionism *plus* freedom. The more the practice developed in the direction of the hegemonic domination of the Mediterranean, the more the general frame of thought changed into opting for establishing the Mediterranean as "imperial property." As a consequence, the practice of defining and controlling the national increasingly insisted on a populationist modeling of *the* French nation abroad as has been shown, as if "France" already extended from Versailles to Cairo and Algiers. The national was still an everyday unknown, but its forms of specification changed on the microlevel at the same time as the broader visions of the French nation's possible empire in the Mediterranean changed on the macrolevel.

From Venetian dominium maris *to the Britons' Empire*

Since the 1990s, several studies have examined the development of an 'Imperial or Empire consciousness' by the British in the early modern

157–168; Leibniz' plans were not so far from other circulating projects as the literature on Leibniz as a philosopher might suggest sometimes; cf. the "Mémoire," February 6, 1664, AE La Courneuve MD Alger 12, f. 146r–150r: "Comme les forces de tout ledit royaume consistent en la seulle ville D'alger la Conqueste de cette place donneroit au Roy deux cent lieues destendue de pays que ladite ville faict"; for later proposals to conquer (Northern) Africa or Morocco cf. M. Taral de Montpellier: "[proposition] d'une expédition en Afrique," AE La Courneuve MD Afrique 5, nr. 36, f. 132–136v ("Il s'agit Sire de la conquette de leur pays, de cette partie que le Roy de Marroc, les Algériens, Tunis, et Tripoly occupent," 1729); a proposal to conquer vast regions in Africa, first in Morocco, but also reflecting the relationship to Tunis, Algiers, Tripoli, ibid., nr. 40, f. 141r–154v ("Projet pour jetter les fondemens d'un nouvel Empire ou Roiaume Chrétien, sur les Côtes de la Mer atlantique en Afrique, par l'établissement d'un ordre à l'instar de celui de Malthe, quant aux Chevaliers"). In 1785, the abbé d'Expilly proposed a new common Western diplomatic blockade of the Regencies, not to conquer them, but to force the replacement of the current governments of the regencies through "toute autre espèce de Gouvernement" to liberate the "divers peuple qui gémissent sous le joug, sous la tyrannie de ces Régences ... dans ce siècle éclairé" (Expilly, Genoa, August 16, 1784, AE La Courneuve, MD Alger 13, f. 297r–301v, 300r).

era. Wilson stated in 1995 that "ironically, anti-imperialist attitudes have been ably documented," while "the meanings and significance of empire in public political consciousness have only begun to be investigated."[226] This gap has been closed to some extent by numerous scholars who, implicitly or explicitly, have mostly focused on imperial consciousness relating to the Atlantic and Pacific dimensions of the growing British Empire.[227] Only two political affairs (Admiral Vernon's victory at Porto Bello over the Spanish in 1739 and Byng's defeat and the fall of Minorca in 1756) have been depicted as crystallizing moments when the British proto-national and proto-imperial public concentrated on the Mediterranean.[228] From a strict 1700-perspective, there is no reason to neglect the question of a "British Mediterranean imperial consciousness." For many in England, the politico-economic importance of the Mediterranean region was traditionally more evident. Samuel Pepys, the Secretary of the Navy, almost never mentioned the American colonies in his notebooks, but as member of the Tangier Committee he wrote a great deal, even if rather sarcastically, about Tangier and the Mediterranean.[229]

Moreover, concerning the concepts of maritime empire, for many centuries only one sea had been the reference point for the production of discourses and theoretical reflections by Europeans: the Mediterranean. If one wanted to formulate a theory of maritime imperialism around 1600, the only texts to base it on would have been legal or politico-juridical theories written by authors who thought about the Mediterranean, as opposed to other bodies of water, and problems to solve within it, beginning with antiquity. One ought to be wary of modernizing the mindsets and hierarchy of authors' priorities. The Mediterranean is significant here on *two* levels: as a point of reference for the reservoir of past discourses used by seventeenth-century authors to formulate their agendas of "modern" imperialism *and* as a scope of those agendas' application for the struggle for power within the present Mediterranean.

In order to understand the French imperial thought between 1615 and 1750, it is much more productive to analyze the huge amount of handwritten *mémoires* on commerce written for the monarchy, because

[226]   Wilson, *Sense*, 138.   [227]   Cf. literature above, n. 192.

[228]   Wilson, *Sense* with the many older studies, and recently, e.g., Kinkel, "The King's pirates?," 14.

[229]   Beach, "Satirizing English Tangier." Cf. Stein, "Tangier."

few texts on political economy and commerce were published. For the British case, however, quite the opposite holds true. Printed texts of various lengths about mercantilist, early populationist and commercial theory abounded, even if mostly in pamphlet size. This reflected the different structure of the relationship between the early modern public sphere and government. Certainly, as in France, many handwritten notes and treatises on political economy were also composed in England. However, it is telling that the transmission of papers from the early precursors of the 1696 (re)founded Board of Trade – the only 17 months active first Council of Trade 1651/52 and Charles II's Councils of Trade of 1660–1665, 1668–1670, the Council of Plantations, then Council of Lords of Trade from 1670 – is so nugatory. Before 1696, these documents are overwhelmingly dispersed in collections of private papers; they did not belong to the "heart" of the central state administration. The combination of King-plus-Houses as the center of decision-making produced other media for the transmission of ideas than in France. It was a process triangulated by "the public" where technical discussions about trade forms and company structures like that between EIC and Levant Company took place. In France, political economic debate was, at least before 1750, far less an object of the emerging public sphere – as were religious and other political matters.[230] This mirrored the structure of the French monarchy's power relationships, with its mono-centered orbit around the king. Due to this, thoughts circulated through the medium of handwritten *mémoires*. It made no sense to print them, and often that would not have been allowed in the first place, like Vauban's *Dixme royale*. Even if permission to publish was granted, printing did not increase the "power" of a *mémoire* and the probability of its impact on the final decision, rather the contrary.

Consequently, a comparison is necessarily asymmetrical. While analysis of the French case concentrated on a Richelieu *fil rouge* webbed through the *mémoires*, the corresponding sources for the English are printed pamphlets, books and surveys about trade.[231]

---

[230] On the several steps of development of the French "public sphere(s)" with the central moments 1585–1594, 1614, 1648, cf. Pallier, *Recherches*; Sawyer, *Printed Poison*; Jouhaud, *Mazarinades*; Darnton, *The forbidden best-sellers*; the massive literature on "Revolution/Public Sphere" since the American reception of Habermas after his translation into English in 1989 is not relevant here, but cf. still Calhoun (ed.), *Habermas*.

[231] For the "political economy of empire" concerning the overseas, and mostly Atlantic, colonies cf. Armitage, *Ideological origins*, 146–169.

Three different ideal types of a politico-economic vision of the English/British Empire or of Imperial political economics emerge regarding the Mediterranean. The first was a traditional, legalist and "Mediterraneanist" vision of England's imperial dominion of the sea that proved to be the guiding principle for the Mediterranean. The second type was a concept of Britain's imperial dominion rooted in the colonial settlements, first of all in Northern America, sometimes implicitly or explicitly transferred to the Mediterranean. The third type was a more abstract form of commercial empire based solely on the intrinsic power of economics and spread of values.

1. The legal or legalist concept of the early British empire of the sea was developed by the long renowned anti-Grotian phalanx of Welwood and Selden, in addition to John Davies. The promulgation of the first Act of Navigation in 1651 was soon followed by the publication of the English translation of Selden's *Mare Clausum* (1652), the republication of Welwood's *De Dominio Maris* (1653), and a treatise that had hitherto remained in manuscript form, John Davies' *The question concerning impositions* (1656, new ed. 1659 *Jus Imponendi Vectigalia*). It is fair to interpret this confluence of publications not as mere coincidence but to conceive of them as the theoretical commentaries and legitimatory background for the Act. To clarify the discursive *strata* of legal and theoretical thought involved here, I distinguish between five levels:

1) The concept of the sea as *res communis omnium* in Roman Law (antiquity).
2) The medieval *mos italicus* concept of *dominium* and *iurisdictio* in the Mediterranean as present in the fourteenth century (Bartolus).
3) The late medieval *mos italicus* concept of territorial (mainly Venetian) *dominium* of (parts of) the Mediterranean (Baldus and followers).
--- [humanist epistemic change, *mos gallicus*, historical approach to Law] ---
4) Grotius' modern Natural Law approach, rejecting the *mos italicus* tradition; the transformative point here was the reference back to a universalized form of level one.
5) The English combination of elements from Civil and Common Law, also using an embryonic form of the modern Natural Law approach, creating a nationalized conception of the law of the sea, reviving and transforming the *mos italicus* tradition of level three.

Grotius' *Mare liberum*, published in 1609,[232] is a legal treatise asserting and defending the right of the Dutch to navigate and to conduct

---

[232] Grotius, *Mare liberum* (1609). Cf. van Ittersum, *Profit and principle*; Borschberg, "Hugo Grotius' theory"; Straumann, "Is modern liberty ancient?"

commerce with the East Indies in the Pacific Ocean.[233] That agenda dominates the text which first, as befits any legal argument, states the grounds and basis for its claim, and then, chapter by chapter, refutes all known or thought to be possible rival rights of dominion of the sea of the opponent in question, the Portuguese. Its argumentative framework is Roman property law, the law of things within the usual division of *personae, res, actiones* as in Inst. 2.1. Within property law, as established legal practice held, Grotius had to identify the *sedes materiae* for the larger question of whether the sea could belong to someone and if so, how. And this fragment within the legal texts was by tradition, the Dig. 1.8.2.1. in combination with several other *leges* in the *Corpus iuris civilis* (e.g. Inst. 2.1.1., Cod. 6.46). Dig. 1.8.2.1 is a fragment of the *Institutes* of the Roman school lawyer Marcianus (third century AD) which stated that "some things are, by the law of nature, common [sc. to all: Dig. 1.8.2 pr] which are: the air, the flowing water, the sea and therefore the shores of the sea." All medieval lawyers treated related issues in commentaries to that fragment. Grotius, as a humanist, certainly did not use the traditional form of the *glossa* anymore, but his mind still followed the order prescribed by the Roman law "system." That the *res communes omnium* are such *by law of nature* was, in Roman law, simply an extension from the usual distinction between the spheres of law in *ius civile, ius gentium, ius naturale*. Roman natural law meant merely that there was a sphere of being and communication in the world with possible legal importance which was common to all humans and even animals, like copulation or the familial bonds between parents and children (Dig. 1.1.1.3). As Merio Scattola and others have shown, the merging of this civil law tradition with the different Thomist distinction between *lex divina, lex naturalis, lex humana* had taken place in the sixteenth century in dialogue between Philippist and later Calvinist jurists and theologians on the one hand and the neo-scholastic school of Salamanca on the other.[234] Natural law, which was of little practical legal importance in Roman times, had by then been "sacralized" and lifted to a status of higher importance. In its turn, the theological tradition had been secularized and "juridificated." With that background behind him, Grotius could now, instead of starting with all the technical

---

[233] Perruso, "The development," 90; Muldoon, "Is the sea open or closed?" 121 sees *Mare Liberum* as a "point-by-point rejection" of Alexander VI's *Inter caetera* bull. But only the chapters VI and X of Grotius' book are addressing it.

[234] Scattola, *Das Naturrecht*.

problems traditionally discussed concerning the res communes – how a "sea" is different from other waters, what are "shores," how distant an island can be from a neighboring power, what if two powers are of exact equal distance from an island in the sea – commence with the new platform of establishing legal claims, the early modern law of nature: "the right that we are claiming [for the Dutch and against the Portuguese] ... is rooted in nature [jus autem quod petimus ... a natura enim oritur]." All the energy employed in the first part of the treatise served to strengthen and to solidify that seat of the claim. It was clearly an incipient form of Grotian modern natural law that would be later developed in De jure belli ac pacis libri tres (first ed. 1625), as he was already stressing the fact that natural law was not just operative and binding between Christian powers, but all people on earth. This problem could not be resolved within the framework of constitutional or public law that could figure in debates between Spain and the Dutch about the struggle between claimed and denied independence. Even if there was a lord/subject relationship, the law of nature as root of the claim could not be derogated by a king versus his subjects.[235] Later, this argument led to the formula that the law of nature was valid "etiamsi daremus Deum non esse."[236] The following chapters refute diverse possible claims of entitlement to the dominium maris of the ocean. An entitlement to the dominium is not possible through inventio – one cannot find and appropriate something, where there was no prior owner, Chapter II;[237] not from the pope – the bulls Inter caetera – because even if the pope had dominion over the whole globe – which Grotius refuted through several arguments stemming from the Salamancist conception of the division of spiritual/secular power –, in the very end the correct translatio necessary for the completion of a gift (donatio) was missing (Chapter III, arg. from Dig. 39.5; Inst. 2.7; Cod. 8.53); the entitlement cannot come from the laws of war and conquest, the Indians are free and sui juris following the categories of the Roman law of persons (Dig. 50.6.195.2; Dig. 1.6.4). The standards regarding a gift cannot apply to something that is not traded (following

---

235 Grotius, Mare liberum (1609), f. *5rv.
236 Grotius, De iure belli ac pacis, Prol. 11. Cf. Straumann, "Is modern liberty ancient," 62 n. 28.
237 The inventio was applied in classical Roman law only concerning the finding of treasures (Inst. 2.1.39; Dig. 41.1.63; Cod. 10.15.1.). Grotius argued more logically than in a strictly legal way from Cod. 8.40.13, Dig. 41.3.41 and with Donneau.

Hugues Donneau, Chapters IV, VII). That the entitlement cannot be claimed by *occupatio*[238] took Grotius more efforts to prove in Chapter V. The main point lay in the classification of the ocean as *res communis* and not as *res nullius* in the taxonomy of Roman law[239] and in making a sharp distinction between the infinite ocean and any other form of waters, rivers, or even "internal seas" circumscribed by delineable borders and islands. The force of the argument comes from generalizing and clarifying the meaning of *communis*,[240] and from denying older ideas. He did not mention his opponents directly in the text, but the marginal allegations show that for some points he was doing what a good humanist *mos gallicus* author always would do, refuting the late medieval understanding of the Roman law. That is, he was invoking the legitimacy of allegedly pure Roman law and repudiating the validity of the post-Roman Byzantine law, established by Emperor Leo VI. Grotius furthermore rejected the argument that the sea could be the property of "the Roman Empire,"[241] which was an idea expressed more narrowly by Bartolus and some of his followers. They argued that it was the emperor – thinking of their contemporary medieval emperor – who had to be considered as Lord of the Sea of last resort.[242] He restricted the idea found in Baldus, that there might be lordship over the sea concerning a particular finite zone of the sea, to the question of jurisdiction and protection. Formal agreements between rulers about jurisdictional zones

---

238 On the *occupatio* cf. just Kaser and Knütel, *Römisches Privatrecht*, § 26 I, 138–140.

239 As only a *res nullius* can be object of an *occupatio*, Dig. 41.1.3pr., Inst. 2.1.12. It is wrong that "Grotius believes that by the law of nations the sea is *res nullius*" – if that had been the case, all arguments from the first to the last chapter of *Mare liberum* would not have followed (cf. on such a reading Brito Vieira, "Mare liberum," 370). For misunderstandings that can derive from a wider or nineteenth-century international law terminology of "*terra / res nullius*" and the Roman Law tradition pertaining to Grotius cf. Benton and Straumann, "Acquiring empire by law," 26f.

240 Decisively, he did not refer to the narrower Roman legal concept of the *communio pro indiviso* as in Inst. 3, 25, but formulated a more general "natural law" idea of *communis* based on Cicero and other authors.

241 Grotius, *Mare liberum* (1609), 26f.

242 "Si autem nec alicui regioni nec alicui insule vicina est tunc non possumus dicere quod aliquis in ea iurisditionem habeat nisi imperator qui omnium dominus est [reference to Dig. 14.2.9]" (Bartolus, *Tyberiadis*, 10r – The second part *De insula* of Bartolus' *Tyberiadis* can be considered as a commentary to Dig. 41.1.7.3 and 41.1.7.4; this is received by Cipolla, *De servitutibus*, cap. 26, n. 16, 404. Cf. Barni, "Bartolo da Sassoferrato," 190f.; Cavallar, "River of Law").

would create international law between and for persons, but not property. As for protection, Grotius defined this as the entitlement to engage against pirates *ex communi jure* at sea, therefore also not creating an entitlement to property. In Chapter VII he argued against the *mos italicus* interpretation which had been formulated in the fourteenth century mostly concerning the Thyrennian and the Adriatic Sea. Italian jurists had developed the idea that the Venetians had acquired the dominion over the Adriatic against the Genoese by prescription.[243] For Grotius, there could be no acquisition of property through *longi temporis praescriptio*, meaning through possession, the will of possession and the passage of a long time (Dig. 41.3), as such a way to acquire property was only for things subject to civil law, not for those of the *ius gentium* or *ius naturale*.[244] As a humanist, he simply refuted medieval lawyers whom he believed did not handle the text and ideas of the Roman law properly. The law of nations in the Roman legal sense of *ius gentium* could not entitle the Portuguese to the property of the sea since the law of nations was relatively permissive and guaranteed freedom of commerce, not privileging one nation over another.

Grotius' text is to a great extent the typical work of a humanist jurist who was rejecting, on the basis of a proto-historicist understanding of Roman Law, medieval derivations of its interpretation.

William Welwood's *De dominio maris* (1616) had a precursor in the English *Abridgement of All Sea-Lawes* (1613) which included a response to the fifth chapter of Grotius' *Mare liberum*, but I concentrate here on the Latin treatise which was republished in 1653.[245] Welwood's main

---

[243] See below n. 246 to 247 for Baldus and Barni, "Bartolo da Sassoferrato," 186–188 for some pre-Baldean *glossae* and votes. There is a great deal of literature on the practical political establishment of late medieval Venetian dominion over the sea (cf. only the rich synthesis of Arbel, "Venice's maritime empire"), but much less on that part of the legal justification. Mazzacane, "Lo stato e il dominio" comments on nearly all the fifteenth/sixteenth-century lawyers involved, but only with regard to Venice's lordship over the land (the *terraferma*). Thus, Barni, "Bartolo da Sassoferrato" remains very helpful; Perruso, "The development," 81–84. Borschberg, "Grotius' theory," 34f. sees the importance of the Italian/Venetian reference as opposed to Grotius' arguments, but Grotius could not have known Sarpi (edited 1685) or Pace (published 1619) in 1605/06 while writing *Mare liberum*. Forthcoming is now on these matters Calafat, *Une mer jalousée*.

[244] Cf. only Kaser and Knütel, *Römisches Privatrecht*, § 25, 134–138.

[245] For the Welwood/Grotius discussion on fishing cf. van Ittersum, "*Mare liberum*"; for a precise analysis of the sources used by Welwood cf. Ford, "William Welwod's Treatises."

point was to rebut Grotius regarding the classification of the sea, in particular that the sea was not a *res communis*.[246] Welwood largely did so by reasserting the claim to dominion through *occupatio*. This was a shrewd move as he did not engage at that point with the text of the Digest itself (1.8.2.1 or Inst. 2.1.1). Instead, Welwood moved directly from the general foundation of the Biblical natural law of property to Baldus' commentary on Dig. 1.8.2.1. From Baldus he then took the central idea to negate the attribution of the sea to *res communes* by referring to contrary fragments of the *Corpus iuris* and to conclude that according to the law of nations, "in the sea there are distinct dominions/realms like in the arid soil."[247] While all glossators – Irnerius, Azo, Accursius, Aretinus, even still Bracton – had seen the sea as *res communis* following Marcian,[248] at the end of the fourteenth century Baldus broke from this tradition. He did not do so through a general and explicit rejection, but by building a bridge to the applicability of acquisitive prescription. Interestingly, Baldus took the somewhat hidden final stone of that bridge from canon law: during the Council of Lyon in 1274, Pope Gregory X had issued a decree concerning details of papal

---

[246] "Ex quibus evidenter apparet res inferioris mundi non esse a primordio, sive a natura, ita communes, ut nonnulli mortalibus persuadere nituntur arida sive terras, iure ipso primario per legislatorem primarium, una cum rebus omnibus arido contentis non esse communia." (Welwood, *De Dominio maris* [1615], 4). As proof, he cited Dig. 1.1.5, in which the jurist Hermogenian had postulated that "people are distinguished, kingdoms founded and dominions made distinct," but Welwood omitted that Hermogenian attributed the foundation of that idea to the sphere of *ius gentium*.

[247] "Item in mari est iurisdictio, sicut in terra, nam mare in terra, id est in alveo fundatum est, cum terra sit inferior sphera, ut no.de.ele.c.ubi periculum, li.6. [note to Liber sextus I.6.3.] ergo praescribi potest, ut l. uiros, C. de ier. off.li.12. [Cod. 12.59.8] nam praescriptio aequiparatur, privilegio, de uer. sig. c. super quibusdam, § praeterea [Liber extra V.40.26 § Praetera], & videmus dicitur iure gentium in mari esse regna distincta, sicut in arida terra. ergo & ius civile, in praescriptio illud idem potest operari, & haec praescriptio quandoque aufertur, & applicatur alteri, sed cum applicatur alteri, ita quod alii non aufertur, ista est consuetudo, & sic Venetiani," Baldus, *In primam digesti veteris partem commentaria*, f. 43r – the *Lectura Digesti veteris* is from ca. 1390, cf. Colli, "Le opere di Baldo," 70. Baldus argued also with Dig. 8.4.13pr and Dig. 44.3.7. The reference to the *Liber sextus* is not really clear. Baldus wrote a comment on those decretales (or rather Johannes Andreae's *glossa*) but it was not printed. Cf. the edition Lally, *Baldus*, 88: only two short *postillae* to I.6.3. At least a part of Baldus' textual tradition has "regna" where Welwood put "dominia."

[248] Charbonnel and Morbito, "Les rivages de la mer," 35.

elections (*Liber sextus* I.6.3), on which a commentary tradition had developed. The question was how to act if the pope died on a ship at sea with all the cardinals on board. Addressing that problem, not covered by the 1274 decree, Johannes Monachus (Jean Lemoine) advised that one should not give priority to the Lateran for gathering the cardinals together, due to reasons of ceremonial precedence. It was more important to secure the sacral body of the church which would be without a head for an extended period. Accordingly, a papal election should take place immediately in the city to which that part of the sea belonged: "Because it is said that port cities have a district in the sea like one speaks of the Venetian Sea or the Pisan Sea": This early fourteenth-century verdict therefore provided an early codification of the idea of the divisibility of the sea into districts belonging to adjacent territories.[249] In his comment on the Digest title on dividing things (*de divisione rerum*), Baldus had referred to that canon law discussion and had added the argument that the sea was just water in a hollow of the earth. Under the notion of "earth (*terra*)" one should understand the whole inferior sphere in opposition to the skies, according to the Aristotelian understanding of the world's structure. And so, prescription would be applicable also to the sea.[250] Not only did he refer to the fourteenth-century canon law teaching on the pope dying at sea, but the hidden capstone of that argument was also taken from canon law. He found in *Liber extra* V.40.26 that a right to take duties could be either granted by Imperial or royal concession or by the Lateran Council *or* could be acquired "from old customary law since time immemorial [*vel ex antiqua consuetudine a tempore, cuius non exstat memoria*]" – wherefrom he concluded that granted privileges and prescriptions are equal. In this, Baldus' commentary to the Marcian definition of the sea as *res communis* stated implicitly the opposite to the text of the Digest fragment itself and allowed the application of prescription to it just as to any item subject to quirite law. The usual sixteenth-century editions of Baldus' commentaries already contained an additional note on that passage which clarified that Baldean argument relating to the concrete reality of late medieval shipping and sovereignty over parts of the Mediterranean: "Note that the

---

[249] Johannes monachus, *Glosa aurea*, 246: n. 9–12 to Liber sextus I.6.3. Jean Lemoine's comment to the *Liber sextus* was written after 1304. Guido de Baysio and Johannes Andreae did not yet treat that problem of the pope dying on sea.

[250] Cf. n. 247.

Venetians are Lords of the Adriatic Sea and of its shores, that means, in general, not in particular as each specific shore belongs to the adjacent beaches."[251]

The sentence of Baldus cited by Welwood was a central conclusion within the commentary that legitimated the possibility of the prescription of the sea, as refuted by Grotius. Welwood proceeded to stress that the divisibility of the sea had been already proven "before all the Roman lawyers" and by Aristotle – or rather Pseudo-Aristotle – as he named all the different parts of the Mediterranean according to the peoples residing alongside it.[252] Welwood's next step was to simply state that this divisibility was valid for the ocean as well as for the Mediterranean. This contradicted Grotius, who had made a categorical distinction between the infinite ocean and other seas, while not explicitly excluding the Mediterranean. And again, Welwood continued to support Baldus and Bartolus, "that lamp of the law," and the late medieval lawyer Cipolla, against the literal sense of the Roman fragments of the Digests. Welwood spoke only of adjacent parts of the sea, which usually meant up to one hundred miles away from land. Welwood also remained faithful to Baldus and the Italian tradition of grounding, which is the right to take duties such as anchorage or contributions to finance defense against pirates in the conception of the sea, understood as a *res* very similar to the soil. Baldus referred in that passage to the feudal law definition of *regalia*, which linked the levying of duties with public streets and navigable waters.[253] The roots for British concepts of *protection* and *security*, and for the right to

---

[251] "ADDITIO Venetiani. Adde, quod Veneti sunt domini maris Adriatici, & littorum etiam, scilicet in genere, non in specie, quia littora sunt ciuitatum adiacentium littora," Baldus, *In Sextum Codicis Librum Commentaria*, comment. n. 13, f. 166r. Baldus gives an account of a dispute between Genoese, Paduans and Venetians concerning the concession of a right to take duties at a bridge upon a river. To show the just right of taking duties and to transfer that right to someone else, he alleges the dominion over adjacent parts of the sea in analogy: "nam unum, & idem est territorium, quod eminet super aquas, & quod immergitur aquas, & hoc satis probatur ar. littorum maris. Nam littora, quae sunt sub Imperio alicuius populi, ut Venetorum, vel Ianuensium, sunt illius populi, & ab eo qui praeest, licentia est petenda" (ibid.). The additio also refers to the post-Baldus treatise Cipolla, *De servitutibus rusticorum praediorum* (prior to 1475), cap. 26, n. 7, p. 402 col. a, cf. similarly Bertachini, *Tractatus* (1489), prima pars, quaestio 6, f. 3r, col. b.

[252] Pseudo-Aristotle, *De mundo*, III, p. 7 (transl. Bartolomaeus de Messana), p. 33 (transl. Nicolaus Siculus).

[253] Cf. Baldus' commentary to Cod. 6.46.6/7 (cf. n. 251).

charge subjects and foreigners for those services, became thus deeply entrenched in continental medieval(ist) legal traditions that had been mainly developed regarding Italian territories and city states.[254]

Welwood was himself a professor of civil law, but in Scotland, where the value and applicability of Roman law was more contested than in most other countries in Continental Europe. His reading of Grotius took all alleged *leges* of the *Corpus iuris* at hand. But he gave decisive preference to those medieval commentators and transformers of the Roman law that Grotius had refuted. After 150 years of humanist efforts to historicize, to better and edit the Roman law text, a tradition to which Grotius belonged, Welwood's option for Baldus could not be, in 1615, just a continuation of the *mos italicus*. Instead, it was an intentional and well-chosen revival of Baldus against the humanist lawyer. In this manner, the late medieval Italian concept of dominion over parts of the Mediterranean was received and generalized in Britain and became the cornerstone of the legal response to the Dutch.

While Welwood argued within a civilist framework, another British lawyer started to add common law elements. This was the king's serjeant in Ireland, John Davies (1569–1626).[255] Davies wrote a text in 1625 when Charles I ascended to the throne. He argued therein in favor of the king's prerogatives to legitimately levy duties and impositions such as the much-discussed ship money. Davies had dominated the "Irish legal world" as the "leading figure in the extension and enforcement of the common law" during the "colonization" of Ireland, as Hart and Pawlisch put it. He was central during the establishment of the king's customs offices in nine Irish port cities and legitimated those royal taxation prerogatives by turning to civil law and utilizing arguments from comparative natural law when royal

[254] Welwood, *De dominio maris* (1615), 13 cited in the margins commentaries of Baldus on the *Libri feudorum* I, 1 and II, 53 and 56, from canon law the *Liber sextus* I, 61, Dig. 43.8.2.22 and Dig. 47.9.5. The main legitimatory support he wanted from those citations were (a) the transferability of rules valid for land to the sea, (b) the possibility that the right of impositions could be acquired through long-term customary law (*consuetudo* and *praescriptio*, discussed by Baldus regarding the acquisition of feuds), and (c) that this right of levying impositions belonged to the *regalia*. The citations (especially Libri feud. I, 1 and II, 53) seem wrong or semantically unclear, as it is often the case with early modern prints. For Baldus' commentary cf. only Baldus, *Opus aureum super feudis*, f. vjr col. a, n. 8 (ad I,1); f. 47r col. b (ad II, 53).
[255] Sean Kelsey in ONDB.

claims could not be grounded in common law.[256] The main purpose of this treatise was to legitimate the levy of duties such as ship money. This had also been an objective of Welwood's text in 1615, but it was all the more so for the royalist Davies in support of Charles, newly ascended to the throne. For this purpose, Davies did not start from natural law in a wider sense, but from the law of nations. Here, he continued the revival of the post-glossators. After a definition of the law of nations from Dig. 1.1.9 and 1.1.1.4,[257] the first author Davies cited by name was, once again, Baldus.[258] Davies picked up from where Welwood finished his *De dominio maris*, with the right of impositions (*vectigalia*),[259] and recapitulated that point with Baldus' legitimation and foundation of the law of nations.[260] The following chapters consist of a historical survey, reign by reign, of prior impositions levied by English kings up to the author's present, with some disappointing results for the period from Edward III to Edward VI.[261] After presenting that historical material, Davies then inquired into the "general reasons whereupon this Prerogative is grounded." He formulated his central point of legitimation with "the King of England is Dominus Maris, which floweth about the Island … And he is Lord of the Sea, not only concerning protection and jurisdiction, but also concerning the property on it [*quoad protectionem & jurisdictionem, sed quoad proprietatem*]."[262] Quite obviously, Davies was copying from Welwood without citing him, in those areas which concern citations from civil law, the post-glossators and other ancient sources. He borrowed the central Baldus position from his comment to Dig. 1.8.2.1 on how dominions in the sea are as distinct as on earth, and he also adopted the argument that, in antiquity, the sea was divided into different "national" parts.[263] Likewise, he also used Baldus as a central authority for the legitimation of impositions in general.[264] Probably also copying from Welwood, he imported the fifteenth-century *jus commune* theory

---

[256] Cf. Pawlisch, *Sir John Davies*, 130–141, 161–175; Hart, *King's Serjeants*, 48f.
[257] Cited in extenso: Davies, *The question* (1656), 5.   [258] Ibid., 6.
[259] Welwood, *De dominio maris* (1615), 20–28.
[260] Davies, *The question* (1656), 6. It seems that part of that argumentation was already present in Davies in a pre-Welwood period, during a 1607 lawsuit, cf. Pawlisch, *Sir John Davies*, 131.
[261] Davies, *The question* (1656), 68f.   [262] Davies, *The question* (1656), 87.
[263] Davies, *The question* (1656), 87 (he took that from Strabo at the same place where Welwood was using Pseudo-Aristotle).
[264] Davies, *The question* (1656), 89.

of customs and duties on merchandise, like the *gabella*.[265] Baldus was the only continental legal authority Davies cited here. Moreover Davies, who was significant in strengthening common law "at the expense of Irish customary law,"[266] now also legitimated the king's prerogatives to levy impositions and his status as *Dominus Maris*. Here he used the help of common law sources, an "avowry" case from the times of Edward I given in Fitzherbert's *Grounde Abridgment* which expressed the idea that "the king wants the peace to be protected as well on the sea as on land [*le roy voit que le pease soit cy auxi bien gard en le mere come en le terre*]."[267] From the Yearbook of Richard II he cited a case, where, in a very circumstantial matter dealing with procedural law, the judge Belknap had formulated that "the sea is within the dominion of England as is the crown of England [*la meere est deinz la liegeance Dengleterre come de sa corone Dengleterre*]."[268] This short sentence, also included in Fitzherbert's *Abridgment*, became decontextualized during the sixteenth century, and was embedded in Davies' concept of the king's dominion and lordship over the sea as founded in international law. The third is an allegation from the *Rotuli Scotiae* that probably refers to the times of the early English occupation. The lordship of the English king over the sea and the coast was expressed in those times in every charter appointing a new admiral.[269]

What becomes clear in this nascent paralleling civil law with elements from common law is that Davies wanted to show that the position of the English king as *dominus maris* was already an accepted customary legal

---

265 Cf. for example Bertachini, *Tractatus de gabellis*; Pace, *De dominio maris* (1619), 170–179.

266 Hart, *King's serjeants*, 48.

267 Fitzherbert, *La graunde abridgement* (1577), avowry n. 192, f. 102r. Avowries were a payment by strangers to "buy" protection from a lord and were usually only found in Cheshire and Wales, Stewart-Brown, "The avowries of Cheshire" and Fox, "Exploitation," 529.

268 Year books of Richard II – 6, ed. Thorne, Hager, MacVeagh Thorne and Donahue, Trinity Term 35, pp. 50, 65.

269 The English king always appointed an admiral to be "captain and admiral of the fleet of all our ships from the beach of the river Thames to the Cinque Ports and to all other ports and places on (*per*) the coast of the Sea up to the Western parts [*capitaneum & admirallum flote nostre omnium navium ab ore Aque Thamis' tam Quinque Portuum quam aliorum portuum & locorum per costeram maris versus partes occidentales*]" (*Rotuli Scotiae*, vol. 1, 358, 9 Ed. III membr. 25); cf. Ward, *Medieval Shipmaster*, 28–36. The *Rotuli* were also the place where many royal letters of protection were filed: Harding, "The medieval briefes of protection"; Lacey, "Protection and immunity," 83f.

tradition. He could not cite an English Baldus for this assertion, because there was no English body of theoretical jurisprudential literature directly comparable to the continental *jus commune*. But Davies utilized those arguments to transcend the *jus commune* tradition, and even Welwood at that point, with the specification that the sea belongs to the king not only through "protection," as common law sources supported, or by "jurisdiction," as was the usual late medieval Italian solution, but as "property."[270] Davies showed, unwillingly, in the first part of the treatise, that the practice of levying impositions on merchandise had in fact *not* been very common during the Middle Ages.[271] Therefore, his and Welwood's efforts to legitimate them proved to be a starting point for a new royal self-understanding in those decades, when the first pirate threat appeared on the British coast since 1615/16.

Selden's famous *Mare clausum* marked the final stage of this development. The first version was written in 1619, but the only text we know is the revised version of 1635 which includes a great deal of post-1619 work.[272] What Selden performed was a Grotian refutation of Grotius. He used a method that closely resembled the one that Grotius had developed to generate natural law from historical evidence. The one and only rule or law-like sentence that he produced was that, with all nations in history (book I) and specifically in Britain (book II), there had always been the concept and claim of lordship over a part of the sea. It is worth noting that the first book is almost completely Mediterranean, because the sea in which the people or "nations" of antiquity conducted shipping and trade and whose history Selden was analyzing was the Mediterranean. Likewise, the only medieval seafaring people he discussed, the Byzantines, Venetians and Genoese, were based in the Mediterranean. The Atlantic dimension of the Spanish, Portuguese and Dutch merited two mere pages of discussion that went without major theoretical emphasis.[273] The second book is devoted completely to the Northern Sea. The referential horizon of the whole treatise has nearly no "global" dimension in our sense, even if the

---

[270] Davies, *The question* (1656), 87.   [271] Davies, *The question* (1656), 68f.
[272] Toomer, *John Selden*, vol. 1, 388–437 provides an unsurpassed insight into the sources that Selden cited. The theory of "Natural-Permissive Law" has attracted a significant body of literature, cf. only the recent Somos, "Selden's *Mare Clausum*" and Tierney, *Liberty and Law*, 251–272.
[273] Selden, *Mare clausum* (1635), 74, 115.

concept gained of maritime lordship could be and was applied later to larger dimensions of global trade.

For most ancient peoples, the Romans and Rhodian's aside, Selden could derive a "sea property rule" only from narrative and descriptive texts, and for those exceptions no ancient theoretical legal works could serve as palimpsest for his own "theory." The same is true for many sections of the second book, but now prescriptive common law sources became his major point of reference, including some book-length treatises for more recent times. The somewhat hidden major and central sources of Selden's overall legal concept are therefore once again the Italian, and mainly Venetian, legal treatises. He indicated that shift himself.[274] For the first time in chapter XVI, Selden referred to texts which dealt more or less in the same way as he did with the question of "De dominio maris" like the *consilium* of Angelus de Ubaldis (Baldus' brother, 1328–1407) or like Giulio Pace's *De dominio maris* (1619). Of those, Gentili, Suarez and Angelo Matteacci stressed the pre-Baldus *res communis / ius naturale* tradition and can be therefore seen as precursors of Grotius, relegated by Selden to marginal notes,[275] while the Italian authors that supported the Venetian rights upon the Adriatic were all named by their full name in the text, and introduced by a long citation by Baldus' brother Angelus de Ubaldis.[276] Thus, despite the equalizing manner of presenting each Oriental people of antiquity, the Romans, Byzantines, the Westerners of his own time and the British in the same way, as following the concept of the property and divisibility of the sea, the only prior systematic work on that question that formed the base of Book I, Chapter XVI was Venetian doctrine. Furthermore, if one checks the works Selden cited, one nearly always finds Baldus as the earliest authority for the precise formulation of the Venetian *dominium*

---

274 "Testantur illud & agnoscunt non solum Historici passim & Chorographi [sc. like for all the peoples I have treated before] sed & Iurisconsulti" (Selden, *Mare clausum* [1635], 66).

275 Gentili, *De jure belli libri tres*, ed. Rolfe and Phillipson, vol. 1, lib. 1, cap. 19, pp. 146–149; Suarez, "Consilium de usu maris" (1558), 619–629; Matteacci, *De via, & ratione artificiosa iuris universi libri duo* (1591), lib. 1, cap. 36 ("De iure Venetorum, & iurisdictione maris Adriatici"), f. 70v–72v.

276 A. de Ubaldis, *Consilia seu Responsa* (1532), n. 290, f. 123r–124v; Straccha, "Tractatus de navigatione" (1558), 275–287, n. 8 (p. 278); Peregrino, *De iuribus et privilegiis fisci* (1588), lib. 1, n. 17, 18, p. 7f; Marta, *De iurisdictione tractatus* (1669), vol. 1, lib. 1, cap. 33 n. 25, 26, p. 97; Pace, *De dominio maris* (1619); Bonavides, "Fragment," f. 62v, n. 5; Sarpi (pseud. Franciscus de Ingenuis), *De iurisdictione Venetae in mare Adriaticum* (1619), f. A2v.

maris. Consequently, Welwood, Davies and Selden acted very consciously when citing Baldus expressly so often, not just as the usual "Bartolus and Baldus" eponymy of post-glossatorial wisdom and the *mos italicus*, but as author of an ideological invention, against which the humanist lawyers, not at least Grotius were fighting.

Despite all of Selden's "modernity," what he was doing was reviving and universalizing the fourteenth/fifteenth-century Venetian theory of the dominion of the sea. In terms of Roman Law, Selden was quite prudent. In the chapter in which he explicitly refuted Grotius, he only referred to that author's text and did not precisely discuss the reading of the Roman law fragments in question. His objectives were limited to denying the general assumption of the sea as a *res communis*, admitting only that concerning the "Free sea ... circumscribed by the open Atlantic and Australian Sea ... has something of what the old lawyers [sc. have written] about the common good of the sea."[277] How did Selden understand the Roman concept of dominion over the sea in the classical era? His main strategy in the respective Chapters XIV, XV was to move the focus away from the traditionally central *loci* of the *Corpus iuris* and toward other fragments of the Digest. He also attempted to establish a hierarchy between the Roman lawyers of antiquity by elevating parts of Ulpian's teaching to the rank of his leading authority, always cited directly in the texts, while authors of the other fragments are made invisible or devalued, among them Marcian, author of the notorious Dig. 1.8.2.1 fragment. The same fragment Dig. 47.10.13.7 that Grotius had held as an exception[278] was taken as the cornerstone of the proof that the sea was treated as a *res publica*, the private property of the Roman people, not a *res communis* of all, and that by Ulpian himself. The medieval *loci classici* for the question Inst 2.1pr and Dig. 1.8.2.1 were relegated to a marginal note and downplayed so that the meaning of the *communis* was implicitly restricted or tempered as far as the empire and dominion of the Roman people were concerned.[279]

Welwood and Davies did not dare oppose the great civil law scholar Grotius directly on his most familiar ground, they merely relied on Baldus. Only Selden took up that task, and one sees that Selden and Grotius stressed and universalized two opposing tendencies which

---

[277] Selden, *Mare clausum* (1635), 114. Selden uses the term of *communio* certainly thinking of the *communio pro indiviso*.

[278] Cf. Grotius, *Mare liberum* (1609), 20, 25, 28.

[279] Selden, *Mare clausum* (1635), 59.

were already embedded within dissonant textual fragments written by various Roman lawyers, and then compiled in the *Corpus iuris civilis*. On the one hand, there was the division of the spheres of law into the "sociobiological"[280] *ius naturale* next to the *ius gentium* and *ius civile* which corresponded to the concept of *res communes omnium* in a very general way and which could likewise conform to the argument that in addition to the sea, beaches/coasts (*litora*) were also such free *res communes*. This is what the third-century authors Marcian and Ulpian maintained.[281] Grotius raised those Roman bases to a new level of generality through the framework of modern natural law.[282]

On the other hand, there were fragments from other Roman lawyers in the *Corpus iuris* which consistently presupposed an institutionalized society and which only recognized a distinction between public and private things, refusing to take a more general pre-institutional level of *ius naturale* into account. Corresponding to this was the contention that at least the coastal zones were also *public*. This is the tradition from which Selden drew to support his arguments that the sea and the coast were considered as belonging to the *populus Romanus* (public < *populus*), and so, in a modern sense, national property. The trick was to hold up Ulpian as a main authority by not really citing all of Ulpian's fragments and by obscuring and downplaying the Marcian texts. This tactic was congruent with the *results* of the post-glossators Bartolus and Baldus, but effectuated here in a humanist reading of the Roman law that could conquer Grotius' method.

Finally, Selden far surpassed Davies in the matter of assembling common law sources to prove from precedents the practice of protection and England's lordship over the sea. He did a great deal of historical research for that, using the *rotuli parliamentorum*, but he also worked with

---

[280] I take the term *sociobiological* from Behrends to denote the basic concept of Dig. 1.1.1.3 as natural law being something common to men and animals, often credited with bearing signs of neo-Pythagorean influences (Wieacker, *Rechtsgeschichte*, vol. 1, 644 n. 22 with further literature).

[281] Behrends, "Die allen Lebewesen gemeinsamen Sachen" and Charbonnel and Morbito, "Les rivages."

[282] Miele, "Res publica" charges Grotius with arguing in a contradictory way, but in fact the fragments in the Digest themselves are mutually contradictory. It is not the place to enter here into the discussion whether these differences can be ascribed to solid schools (Sabinians contra Proculians) or even to heritages of different ethnicities (cf. Behrends, "Die allen Lebewesen gemeinsamen Sachen" and Charbonnel and Morbito, "Les rivages" for different views on that, and in general Wieacker, *Römische Rechtsgeschichte*, vol. 2).

archival manuscripts.[283] Using these sources, he derived the status of the English king as lord of the sea, a power he delegated to the admirals as "custos quinque portuum et Maris."[284] Due to the less consistent terminology of common law language, the concept he created was not "unified," but it was still a more general idea of *dominium maris*.

The legal contribution of the troika of Welwood, Davies and Selden to the Mediterraneanist establishment of legitimating ship money, and, in the end, of the horizon of the Navigation Act itself, did not just have an impact on nuanced elements of different text exegesis. At the same time, those different textual interpretations and semantics served to nationalize the concept of sea use and shipping itself. The main subject of Grotian thought was not the nation but the one humankind, the *genus humanum*, as the root and seat of natural law itself. The sea, at least the infinite ocean, allegedly belonged to this general subject of History and of contemporary activity and politics. The revival and generalization of the Baldean and Venetian concept of their lordship over the sea made use precisely of the latest, most territorialized state of development of medieval Roman law, while the older vision of the emperor's supreme terrestrial lordship, as still present with Bartolus, was not of help. With Selden, in the end, this *methodically* Grotian generalization against the Grotian *content* put the subject "people" or "nation" in the forefront of all arguments. Peoples and nations, he maintained, had a *public* relationship of ownership to the sea. There had always been a *plurality* of peoples and nations in the world and so, all of them had, by the law of nations, that public ownership relationship to the sea.[285] Unity versus plurality of the principal subject of History was, on the higher level of thought that only Grotius and Selden reached, the fundamental distinction in approaching the problem. But in terms of technical argumentation, this led the British authors to strongly dig into those sources in the past that had dealt with the delimitation of that unit "nation" in a very practical sense, juridical treatments of border defense and protection and security, those being the everyday forms of defining and defending the "us"

---

[283] For example the "*Ms. Formularum de Rebus Maritimis* in Bibliotheca Cottoniana, 3. Maii, 13 Hen.4," Selden, *Mare clausum* (1635), 198 marg. b.

[284] Cf. for that Toomer, *John Selden*, vol. 1, 412–432.

[285] Already in Welwood and Davies, the reference to Pseudo-Aristotle and Strabo served to stress the root of the dominion over parts of the sea in the vicinity of *gentes*, cf. above n. 252, 263.

against the other, of distinguishing the internal from the external, and, therefore, the fundamental mercantilist distinction. If both sides, the British and Grotius, were humanists in their own way, the British applied the selective historical horizon of source analysis typical for humanist scholarship, but in targeting the *medieval era*, while Grotius was simultaneously a universalist and a humanist legal interpreter who used *antiquity* as his period of reference.

Finally, those texts linked discussions of resource extraction, such as customs, ship money and impositions for the sake of securing the national shipping, to the Navigation Act. The nationalizing of trade therefore has a dimension reaching deeply into the foundations of early modern philosophical and political thought. This was the ceiling and framework for the extremely practical day-to-day decisions of identifying ships, flags, men, of taking legitimate "national" customs or not on goods.

The merchants themselves were not bothered about different concepts of natural law and their deeper roots. However, they still understood that the royalist discourse could work as a somewhat Machiavellian tool to legitimate the extension of impositions. Customs would be:

conceived by some to have its first Original from a safeguard given by those Princes at Sea, to their Subjects and Merchants from all Rovers, Pirats and Enemies, and a Protection for free trading from all such dangers from one Port or City of Trade to another: but we see that in these days the payment of the Duty is still continued, and is daily paid by all Merchants; but the first institution and ground thereof (if so it was) is by many Princes either totally omitted, or at least wise forgotten, and therefore it may now be more properly called a Custom than heretofore.[286]

Lewes Roberts obviously knew how treatises like those of Welwood, Davies, Selden, took arguments from pretended common law traditions and the Mediterraneanist renewal of the Italian *gabella* legitimation, to produce a genealogy of customs such as the one percent duty from levies in order to finance the war against pirates. Roberts even maliciously noted how kings "forget" the legitimation first grounded in specific temporary contexts.[287] Still in 1690, Josiah Child, who had

---

[286] Roberts, *Map of Commerce* (1700), 13.
[287] Cf. for the legitimation of the 1637 ship money via threats presented by "thieves, pirates, and sea-robbers, as well as Turks" *Armitage*, The ideological origins, 116 and literature above n. 18.

used the works of Lewes Roberts,[288] argued against the charge that the *Acts of Navigation* would perhaps, at first glance, be only profitable for maritime merchants and shipowners, while complete free trade might be more reasonable for the majority of British producers and consumers. "[T]his Kingdom being an Island, the defence whereof had alwayes been our Shiping and Sea-men it seems to me absolutely necessary that Profit and Power ought joyntly to be considered."[289] The military defense of the kingdom and the economic prohibitions of the Navigation Act were intrinsically bound together. In many other texts which analyze Mediterranean trade through the lenses of natural law and the merchant law of nations, the granting of the monopoly charters to companies and the establishment of a peace treaty system is understood to be rooted in the prerogative powers of the King.[290] The political economic system was regarded here predominantly from the perspective of protection and was conceived of as a system of security production. The notion of "security" as an explicitly considered general guiding principle of politics was a neologism in the Early Modern era, and it is omnipresent in the texts:[291]

The Security of every Country depends upon the Strength of one Country against another, in case of War between them; and herein Countries are to be considered as they are placed in reference to each other: The Bounds of Inland and Mediterranean Countries, are Rivers, lines, and Forts, which are esteemed sacred; and a Violence done to them, is esteemed a just Cause of War; and so long as these are preserved, the Countries within are secured from foreign Wars.[292]

Until the beginning of the eighteenth century, the organization of navigation and trade, the entire combination of Navigation Acts, and the Act of Frauds,[293] the interplay between armed merchant ships, convoy navy ships, and import/export regulation were all seen as an extension of the defense of the realm's borders. The nationalization of ships was supposed to secure the identification of the moving borders at sea beyond those geographical borders on land. "Defense" was

---

[288] For a different question shown in Letwin, *Origins*, 233.
[289] Child, *Discourse* (1690), 93. [290] Jeffreys, *The argument* (1689), 12.
[291] Whiston, *Decay of trade* (1693), 3; Trevers, An essay (1677); Sheeres, *A discourse touching Tanger* (1680), passim. Cf. Skinner, "Liberty and security" and Zwierlein, "Sicherheitsgeschichte."
[292] Coke, *Detection* (1697), 659.
[293] Also the 1663 Staple Act, less important here.

therefore commercialized, the realm's borders extended into the open space of the sea, and conversely, commerce came under the purview of military defense. This first version of British imperial political economics was, to some extent, conservative, medievalist and Mediterraneanist in both its sources and its application. Using those semantic resources, and by projecting the late medieval Italian as well as common law traditions of maritime dominion and territorial protection into the practice and theory of trade and commerce, all gave a national framework to England's self-understanding of insularity.

2. The second way of conceiving of the empire was to imagine the foundation of a settler colony like in New England. At least one author of importance, Daniel Defoe, imagined North Africa as such a possible colony just at the same time as one finds a new emphasis on French conquest plans. This is certainly due to the political instability in the regencies during the late 1720s and early 1730s and it echoed the second siege of Gibraltar (1727).[294] Defoe did not conceive of the European conquest of Africa as an exclusively British project, but as a (non-crusader) joint venture of all major Christian powers. The short chapter in which Defoe discussed this plan compared antiquity to his contemporary times. The Roman Empire, which Defoe styled as rather disinterested in trade, was correlated with the Ottomans, and the trading people of the Carthaginians was implicitly likened to the European trading nations, foremost the British. Indeed, "Defoe insisted that the Phoenicians were the Englishmen "of that Age." This was an imperial ideology different from the usual Roman classicism that fitted the concepts and realities of a trade empire more exactly.[295] Clearly, Defoe was alluding to a historical narrative drawn mostly from small passages in Strabo, that the Carthaginians/Phoenicians had traded with Britannia in pre-Roman times, mainly in tin, and that they had established colonies there. Therefore, there was not only a comparative but also a genealogical link between Britain and North Africa. Defoe evoked Carthage as a past golden age of well-functioning trade

---

[294] Defoe, *Plan* (1728), part III, Chapter 2, 321–327. For Defoe's earlier concept of empire cf. Dickey, "Power" and Downie, "Defoe"; but this text on the Barbary coast is very rarely considered in scholarship on Defoe's political thought. But cf. Matar, *Turks*, 170–172. For related aspects of Defoe's writing cf. Merrett, *Daniel Defoe*, 160–170; Backscheider, *Daniel Defoe*, 510–515; Novak, "Defoe."

[295] Backscheider, *Daniel Defoe*, 514. For the Roman classicism cf. Levine, *The battle of the books*.

connections between Europe and Africa. The Ottomans and their piratical economy represented a state of depravation to be overcome. The Barbary nations had to be driven out of the coastal cities, Defoe maintained, and each European nation should obtain "separate Allotments of Territory upon the coast" and those territories should be "peopled with a new Nation, or new Nations made rich by Commerce, and the Country adjacent cultivated and peopled after the Manner of Europe." European settlers should cultivate the country, and the Barbary people, driven into the hinterland might even transform the desert into farmland, forced by necessity, and to the advantage of the European coastal region. In Defoe's scheme, this would constitute a successful increase of the "general Product of the Country." By the establishment of a European Africa trading with Europe, using the Barbary hinterland as second subordinated trading partner, the increase of commerce itself would be achieved.

[T]his indeed is the Sum of all Improvement in Trade, namely, the finding out some Market for the Sale or Vent of Merchandize, where there was no Sale or Vent for those Goods before; to find out some Nation, and introduce some Fashions or Customs among them for the Use of our Goods, where there was no Use of such Goods before.[296]

Clearly, the American experience represents the model for Defoe's North African conquest plans in quite precise terms of a British style trade-based "settler colony." Certain elements of this second conception surely had their impact on the Mediterranean, from Tangier to Gibraltar and on what concerns the "colonial culture" performed in the Levant factories in the middle of the eighteenth century. In general, however, it did not have any deeper impact, because it did not fit the Mediterranean circumstances or the Levant Company's traditions and convictions.

3. The third way to conceive of the imperial political economics was concentrating in a far more abstract way on the flow of economic values and the forces of trade itself. In one of the leading merchant handbooks that was widely read among the Levant merchants, Lewes Roberts' the *Merchant's Map* (1638), a "geocommercial" cosmography of all four continents is expressed, providing a prototypical analysis of the invisible imperial power of trade and economic expansion.[297] Roberts was

[296] Defoe, *Plan* (1728), 325.
[297] On Roberts cf. Gauci, *Politics of Trade*, 168 and Zwierlein, "Coexistence and Ignorance."

a member of the Merchant Adventurers, the Levant Company – even with the office of Husband from 1633 to 1641, the East India Company – including a stint as Director in 1639/40, and of the French and the Spanish Company in London, and he had conducted trade in the Levant and lived in Constantinople for at least fourteen years (1611 to 1625):

How had ever the name of the English beene knowne in India, Persia, Moscovia, or in Turky, and in many places else-where, had not the traffike of our Nation discovered and spread abroad the fame of their Soveraigne Potency, and the renowne of that peoples valour and worth? Many parts of the world had, peradventure even to this day, lived in ignorance thereof.[298]

Roberts argued that the small English island, in addition to the small countries of the Portuguese and Dutch, had gained an "over-proportional" renown by means of global commerce. In contrast, he alleged, the "Emperour of Germany, the greatest of our Christian Princes," was hardly known outside Europe, because his nation did not conduct global trade.[299] Roberts concluded that "It is not our conquests, but our Commerce, it is not our swords, but our sayls, that first spred the English name in Barbary, and thence came into Turky, Armenia, Moscovia, Arabia, Persia, India, China."[300] For Roberts in 1641, "the global" started with the North African Barbary Coast, while India and China formed the end of the enumeration. Roberts' general remarks can be understood as an attempt to describe something hard to grasp, the proportion and amount of symbolic capital and functions of global trade for the imperial growth of a nation. Like Italian political authors since Machiavelli who reasoned under the locus *Della reputazione del principe*, Roberts argued that there was a form of power that did not derive from brute military force, but instead from the management of an appearance of splendor and "cultural fashioning," and from the memory of past deeds and wise counsels.[301] It is as if Roberts – who read Italian and possessed books

---

[298] Roberts, *Treasure of traffike* (1641), 91.   [299] Ibid., 94.   [300] Ibid., 92.
[301] Cf. *Il principe* XVIII, 5 (Machiavelli, *Opere politiche*, ed. Vivanti, 166): "A uno principe, adunque, non è necessario avere in fatto tutte le soprascritte qualità, ma è bene necessario parere di averle."; *Discorsi* I, 25, 1 (ibid., 257): "perché lo universale degli uomini si pascono così di quel che pare come di quello che è: anzi, molte volte si muovono più per le cose che paiono che per quelle che sono."; cf. some later examples: "Quanto importi al Prencipe la riputatione per il governo dello Stato: & quello che debba fare per conseruarla" (Frachetta, *Il prencipe* (1648, first ed. 1597), lib. I, Chapter 4, 13–15); Botero, "Della

close to that tradition – transferring those ideas from princes to nations and from military glory to trade, reasoned about the functions of commercial reputation. This is an early original interpretation of how an empire of trade worked while its center was merely a small island off the coast of Northern Europe, so isolated from the resources and markets of the continents. Roberts' rationale shows that this concept of the flag's honor identified the deepest fundament of an empire in the *knowledge* of, and about, the governing nation abroad. Ignorance/knowledge dictated the presence or absence of the imperial power of a nation, long before there might be questions of conquest, settlement and colonies. While Roberts himself belongs more to the traditional Levant and London merchants, his focus on the forces of trade themselves is close to some of the seventeenth-century proponents of free trade. Not every plea for "Free Trade" can be linked to this third type of proto-liberal imperial thought outlined here, as the words themselves did not always correspond to their actual content. In contrast, not every plea for some elements of trade regulation can be understood as safe indicator that a given text would belong to the legal tradition. Nevertheless, this is usually a first indicator. A controversy coterminous with the first Navigation Act was that sparked by the famous Leveller William Walwyn who, in May 1652, addressed a memorandum to the *Committee of the Council of State for Trade and Foreign Affairs*, the first Council of Trade founded in 1651 which only lasted for seventeen months. He strongly advocated for free trade against what he perceived as the monopoly of the Levant Company. He grounded his arguments in quite general notions of the public good and the Commonwealth, pretending that there was "so antient a continuall claymed Right, as freedome to all English men in all Forraine Trade ... an universall freedome in all forraine trades." Allowing every Englishman to be merchant as he asked for and to conduct foreign trade with his means and wherever he wants to go would provide many advantages, among them the "increase of Shipping ... the increase of Marriners ... the increase of Wealth and plenty ... the increase of Merchants." Because he favored the increase of the overall number of merchants and the freedom to use whatever

riputatione del prencipe"; "Quanto importi la riputatione massimamente ne' principii delle cose" (Ammirato, *Discorsi* (1607, first ed. 1594), book XII, Chapter 1, 258–261); Bireley, *The Counter-Reformation prince*, 54–57, 82–84, 171–177, 198–200, 223–225.

means they could afford, he had to admit that the merchant ships that could be used then would hardly be "serviceable ... for Warr or for defence and protection, as those that are built purposely for the uses by the State." He furthermore imagined that, as a result of the freeing of trade, not "so many wealthy men, as [would have been produced] in the same time by Companies, most of them being borne Rich & adding wealth to wealth ... yet it will produce Thousands more of able men to beare publique Charges," thus creating more of a "middle class" type of merchant population than a stratified pyramid of a few rich individuals who could also afford to use extremely costly heavily armed ships versus the many poor who could only serve in subordinate functions.[302] He also tended to distinguish the tasks and the sphere of merchants from that of the state, those simply conducting trade with their modest private means, with the option for the state to provide security by powerful ships, a proto-liberal division of realms instead of its combination. Those points, touched upon only shortly by Walwyn as they were, from his opponents' point of view, his weak points, were taken by those who responded for the Levant Company. The Company started with denying Walwyn's concept of an "original natural state" of universal freedom of all men concerning foreign trade in society, stating that if that had been the case, then "it did precede all Government" and returning to it or the "exercise of it tends to the dissolution of all other Civill Governments whatsoever."[303] This meant that according to the Levant Company, in a constituted civil society no simple complete freedom of trade allowed to all was possible. Instead, the very character of the society to be constituted implied an order according to the laws of rank and specialization.

The principall end of Government in Trade, being to advance the same for the benefit of the publique, by the incouragment and increase of able Merchants, by the exportation of Native Commodities, & importation of forraine, by the promoting of Navigation and increase of Marriners, and in order thereunto, to procure and uphould forraine priviledges, to beare the charges incident to trade, to defend the people and stock of this Nation, and

---

[302] "W Walwins conceptions for a free trade," SP 105/144, f. 36v-39v, edited in Walwyn, *Writings*, ed. McMichael and Taft, 446–452.

[303] "Reasons humbly offered by the Gouernor & Company of Merchants trading into the Levant Seas, to the Council of Trade" (May 21, 1652), SP 105/144, f. 41r–47r, 45r.

to preserve the trade from other Nations: all which may rationally be supposed to be effected rather by a Vnited Society of Persons tutored and bred up to the trade, and injoying the mutuall Councells of each other, then by others trading in a confused and loose way.[304]

The link between the economic and military defense of the people, in addition to the capital of the nation, led then to a rebuttal of Walwyn: "The ships at present employed in the Trade of Turky are such, as a free trade could not produce either so many or so good; and the nature of that trade is such, as cannot be maintained without ships of good defence."[305] The Company remained with the defensive concept of economics whose legitimation was perfected by Welwood, Davies and Selden, pointing to the strength and war-born character of that concept of trade. Aside from the embryonic ideas of an overall increase and the possible global outreach of every Englishman, Walwyn did not develop a complete positive imperial concept that was connected to what he obstinately and repeatedly called "the free trade." Forty years later, such a vision became more discernable in Whiston's 1693 *Discourse of the decay of trade*. Now, the idea of free trade was in fact the leading element for a purely functional economic empire, maintaining that "Strength or Weakness, Wealth or Poverty of this Kingdom wholly depends upon the Good or Ill Management of Foreign Trade."[306] He argued for a more or less complete freeing of commerce, envisioning a global power of Britain in remarkable words. If England could organize its commerce well, the usual opposition between public and private interest would fade away. For that, a military conquest like that of an Alexander or a Caesar would not be necessary, but instead merely the correct management of (free) trade affairs. In so doing, the English would "make our selves in Effect Masters of the Four Quarters of the Earth, and all England become as one City of Trade, and the General Emporie of the World ... the Nation will be abundantly Enriched, and Money being the very Life of War, and Sinews of all Publick Action, we shall be enabled to bring the World into a Dependant Awe."[307] If English commerce flourished, "People from all parts of the Globe, would resort hither to enjoy themselves, and Improve their Stocks: For Trade is the Life-Blood that runs through the Veins of the

---

[304] Ibid., f. 41r.   [305] Ibid., f. 46r.   [306] Whiston, *Discourse* (1693), 2.
[307] Ibid.

Nation, that moves, maintains, and enlivens the whole Body of the People."[308] As an appropriate measure for the government of trade affairs, Whiston recommended the establishment of a Council of Trade whose members should all be merchants. He was essentially making a straightforward plea for functional specialization: "If the business of Salvation be in Debate, we apply our selves to some professing the Ministry: If the Dispute be concerning the Title to an Estate, we desire the Judgment of a Lawyer: If Sick, we Consult a Physitian."[309] So, in questions of commerce, counsel should be provided by merchants specialized in trade. The committee, he proposed, should consist of members of the trading companies – East India, the African and Turkey Company of the merchants trading with Italy, Spain, Portugal – along with merchants from several other colonies and marketplaces from Barbados to Jamaica, New York and New England, in addition to twelve Masters of Ships to be chosen by Trinity House and representatives of the different English places of production with their respective specific goods (Northumberland for coal, Cornwall for tin, Devonshire and Somerset for clothing, etc.).

In 1696, the more stable Board of Trade was finally constituted, where men of politics were, if not by majority, at least in terms of power, still the leading members.[310] The vision and concept behind a text like Whiston's was in fact the strong concentration on the economic power of value flows and growth and on a functionally differentiated society in which economics was more or less self-governed by its own experts. As the idea still started with, and aimed at, the establishment of a powerful empire, the envisioned English Council of Trade became akin to a World Government. In this vision of empire the king was mentioned only briefly; the guiding principle was nearly completely transpersonal. The power that was to drive this empire-building was invisible: not the power of arms and steel, but of economic value acquisition.[311]

As the early opposition between Walwyn and the Levant Company and later quarrels between the East India Company and the Levant

---

[308] Ibid., 3.   [309] Ibid., 4.

[310] On the Board of Trade see Steele, *Politics of Colonial Policy*. On its precursors since 1651 now Leng, *Benjamin Worsley*, 3–79, 153–162, but the Levant Commerce is hardly touched.

[311] Whiston, *Discourse* (1693), 4–11.

Company during the 1670s and 1680s show, although Josiah Child did not advocate a completely free trade like Walwyn and Whiston, but instead a milder version as represented by the joint stock companies, it is clear that the different imperial concepts and forms of economic thought were still in elective affinity with different company structures and trading societies. The regulated trade as conducted by the Levant Company and secured by the security measures of convoy shipping and the defence-born Navigation Acts conformed to the first type.[312] The chartered joint-stock companies, mostly acting on credit stocks, correlated with the second and partially with the third. The Levant Company thought in more bullionist terms, charging that the East India Company always exported bullion and imported only goods, while the Levant Company did just the opposite.[313] Child and others argued against that simplistic vision by grounding the import/export balance in asset values, not in metal, and asserted that more profit was gained through the re-export to places including the Levant of Indian goods.[314] The discussion of the opening of the Levant Company in 1718–1720,[315] and the decay of British trade in the Levant after the 1730s were an effect not only of external competition with the French, but also of internal negotiations over different forms of political and economic empire in Britain itself. Levant traders, at least in general, remained within the older traditions of economic and political thought.

Comparison

Different imperial visions succeeded each other and competed within both countries. Nevertheless, within these national contexts, there were elements which seem to have transcended the divides. The central point is how one can conceive of the relationship between "economics" and "state/nation" in both cases.[316] While reality will always show ambiguous merged forms and hybrids – French British commercialists and British French statists – in an ideal typological juxtaposition these relationships seem to behave as Figure 1.3.

[312] Letwin, *The origins*, 29–36; Anderson, *A consul*, 73; Wood, *Levant Company*, 114f.

[313] *The allegations of the Turky Company* (1681).

[314] Papillon, *East-India-Trade* (1677), 11f.; Philopatris (i.e. Josiah Child), *Treatise* (1681), 6, 12; Philempórios, *Scheme* (1683).

[315] Matterson, *Levant trade*, 226–241.   [316] Cf. the note 4.

Figure 1.3 British and French Imperial Mercantilism: The relationship between Nation, State and Economy.

In the French case, trade was considered as subordinate to the state's interests. It was a hypo- or hyper-tactical relationship. An economic historian might argue that this was a misperception of how economics functioned, but that is another argument and not a question about what the actors thought themselves. This fundamental point did *not* really change during the seventeenth and eighteenth centuries. What changed was the way the state tried to model the trade, from rigid open/ closed decisions (Richelieu/Montchrestien options) to Levant commerce regulated by a populationist state through "general laws" empirically derived from the given realities around 1740. The nation was integral to the state, l'*état* tended to be the global notion. In the British case, in terms of political economy theory, institutional communication and the practice of navigation, the English – or later the British – nation was the general global actor. Trade and state were more connected through parataxis and equal terms, and discussions advocating or disparaging free trade. The development from a legal-defensive concept of merchant empire to a functionally differentiated economic self-government is a question about the shift of dominating parameters on the same level. Only if Defoe's step had been taken in the direction of a settler colony might one have witnessed British populationist regulation in the Mediterranean similar to that evinced by the French. That the French state applied those concepts to what were, in fact, very small communities of some dozens to, at most, two- to three hundred people as in Constantinople shows how the state perception – incorporating the nations in the *échelles* as part of the state abroad – dominated trade from a top-down perspective.

The British model was, from the beginning, more open to a more abstract, invisible form of an empire based on network nodes, possibly even founded in knowledge about itself and perceived by others, as Roberts had put it. The French model conceived of the trade empire as an extension of the French state, itself physically and geographically centered within its hexagon borders. This comparison does not lead to an answer for a classic economic history question, why the French achieved a (fragile) domination of Mediterranean commerce during large parts of the eighteenth century while the British Levant trade declined. The reasons for that are probably only slightly connected to the question of whether one of those models was "superior" to the other. The British acted in the Mediterranean only within the first and oldest of the three imperial concepts outlined above. The Levant trade was structurally conservative. It tended to be excluded by competition within the more and more globalized British Empire, governed within the frames of the second and third models outlined. The French restarted their Levant trade with Colbert after a lengthy hiatus, and they concentrated much more on the Mediterranean as their major "home region." So, the investigation into the differences of general imperial frames of thought is less helpful for explaining what happened in the Mediterranean according to purely economic terms of gain or loss of shares. Nevertheless, those variants are of great importance for analysing the epistemic structures *being formed* in the bottom-up process of specifying norms of nationality and of non-knowledge communication *and also forming* the core and the major perceptional framework of both forms of mercantilist empires, of their specific form to define the distinction between internal and external.

Conclusion: Operative National Non-Knowledge

Ships sailing in the Mediterranean, flags, men on board, corsairs perceiving them, mercantilist forms of trade: all that was seemingly very similar for all Europeans and all others. But the notions of the national and the empire's epistemic structures linked to that and the encounters with each other in the Mediterranean were different, even if the wood of the ships, the flesh of the men and the paper or parchment of the passes had the same physical constitution. Through questioning the national and transforming the unknown into knowns, the French and the English continually processed the very bases of the empires themselves in their different forms. Here, the brilliant insight of Lewes Roberts in 1641 holds true for both: their imperial power was grounded first and before the use of any cannons or swords on *the mere knowledge of their nation* abroad.

The question of how national (non-)knowledge was handled by early modern trade empires relates just to one of the four studied epistemic fields. Those empires and their agents did a great deal beyond asking about and checking nationality, even though this was the crucial point of constantly marking out the difference between internal and external, and it defined the character of their primary "engine" and reason for existence, mercantilist commerce. From the point of view of a history of coping with forms of ignorance, the study of national non-knowledge serves to represent the specificity of *operative* non-knowledge – in its pure forms and in its links and passages over to epistemic, more discursive forms, such as political arithmetic and imperial frames of thought. The autoreferential character of the national proves to be a crucial element here. We are certainly used to, since the 1980s, speaking of the "construction" of nationality, usually referring to discourses of patriotism and similar sources. The path taken here from norms as specifiers of unknowns to the reification of the national and consolidation of discursive formations referring to it, gives way to a different view by putting emphasis on the paradoxical primary element on which everything is built: the logically necessary gap between a normative demand and the impossibility of referring empirically to a reality "just out there" while responding, because the concept encapsulated in the norm is, in itself, artificial. The atoms of nations, so to speak, are specified unknowns, the forces to combine them, are utterances of ignorance.

The processes started by that were powerful and enduring. They quickly transformed all communication in the Mediterranean, starting with the trade itself. And those epistemic movements acquired a shape during the century from the 1650s to roughly 1750, one that needs to be born in mind for an overall view and comparison with the forms of coping with ignorance and the epistemic movements engendered within the other epistemic fields treated in the following chapters: Religion, History, Science.

Using the laws of the 1660s as a starting point provides convenient visibility for the transformation of the national into a specified unknown in comparison to the medieval situation, as shown above. The approximate endpoint in around 1740/50 of the resulting epistemic movement is less firmly connected to similar legislation, apart from the crucial 1740 French-Ottoman capitulations. Inquiries into the national did not stop now, and – despite differentiations between the realms of economy and politics beyond the close mercantilist link between the national and trade – it was not to stop for centuries. This epistemic movement did not have a cyclical form with a starting point, an acme and an endpoint; it was rather an open-ended process. There were, in the 1740s and 1750s, indicators of decisive turning points or shifts in the nature the communication used. This was the high reflexivity about the functioning and the paradoxes created by the processes of identifying the national in the Mediterranean, as evinced by new figures in the field. Reflections such as those on the actual operation of "matching" the lower and the upper part of a Mediterranean Sea Pass betray an awareness of the constructive moment of attributing a national belonging to a ship and its passengers. Considerations like Arnoul's about the functional interdependencies of the two levels of the Mediterranean economy between pirates and Europeans, and the necessity to avoid a "white flag overflow," likewise evince an awareness of the significatory power, but also the rather arbitrary quality, of the national as a brand, a flag, or a color at sea. These reflections do not show that individuals were not taking the attribution of nationalities seriously anymore. However, this reflexivity was an eighteenth-century phenomenon emerging after decades of the system's establishment; it was not present at the start. When paradoxes or dysfunctionalities of hitherto unconsciously performed actions became reflected upon, such as here in thetic or ironic forms, they became germs for a further stage of development. At the same time as those moments of reflexivity, a third point of view materialized in the French case, the paradoxical result of reification of the national. It aggregated the myriad of national (non-)knowledge interactions with *the* nation as object appropriate for populationist reasoning. This was, again, a moment of distancing, of setting off a second-level process. Populationist administrative communication about *the* nation remained associated with fundamental standards like the March 1669 edict, but it also had to be partially detached in order to correspond with new rules of empiricist science driven modeling of that object "nation." The English case is marked, at the same time, by the final decline of the Levant trade itself. On the epistemic level, this was the point when the traditional, legalist and defensive character of the national, as it was "frozen" in the Navigation Act and as the Levant Company was processing it, was surpassed by internal and external developments. This meant that the second level of dealing with the aggregated outcome of national non-knowledge communication that materialized in the French administration, did not apply in the English case for the Mediterranean. The English were far advanced with forms of political arithmetic reasoning and practice relating to other parts and other constitutional circumstances of the empire, but the Mediterranean trade grew increasingly "achronic," conserving older habitus and older corresponding epistemic patterns. These shapes of the epistemic movements – the start of an open-ended process of asking about the national allows, at a certain point, the emergence of a second-level form to deal with the aggregated outcome of the nation – or to leave it with that – is characteristic for this example of operative non-knowledge. This will be compared with the non-knowledge cycles within the epistemic fields of Religion, History and Science.

2 | Religion: Empires Ignoring, Learning, Forgetting Religions

The relationship between empire and religion has typically only been touched upon in the history of oversea missionary enterprises, but recently it has attracted renewed interest as a part of the history of empires in a more narrow systematic sense.[1] For both the British and the French, the religious dimension of their proto-imperial extension into the Mediterranean was linked intrinsically with the complex politico-religious situation in their own countries. In the British case, this was the transition from the Restoration Church to the circumstances of the post-1688 Revolution; for the French it was the increasingly exclusivist opposition between Huguenot and Catholic intellectuals under Louis XIV before and after the edict of Fontainebleau (1685), in addition to the inner-Catholic developments of Jansenist and Port-Royal Catholicism. One has, therefore, to uncover the Mediterranean Imperial dimension of the French debate between Port Royal Catholic intellectuals and Huguenots about the Eucharist and the Mediterranean Imperial dimension of British inner-Anglican church debates about ecclesiology.

Within the wider framework chosen here, the connection between imperial actors, institutions and religion is of interest because it concerns the interplay of demands and moments of awareness of ignorance within the field of religion on the Western side and moments of stimulation and interaction from the East. It involves ignorance and obtained knowledge concerning past theology and contemporary religious groups as well as the faiths and practices of communities in the Mediterranean, insofar as they had bearings upon the French or English contexts. As my more general aim is to discover structural

---

[1] For the neglect of religion in British Empire historiography cf. Armitage, *Ideological origins*, 63–67, 99; for the post-1701 activity of para-imperial institutions like the *Society for the Propagation of the Gospel in Foreign Parts* cf. Strong, *Anglicanism*; Porter, *Religion versus empire?*, 15–38; for recent case studies cf. Trivellato et al. (eds.), *Religion and trade*.

117

congruencies, similarities and different ways of coping with ignorance in proto-imperial discursive fields and administrations, the examples chosen serve to show different religious epistemic forms, functions and movements, and different layers of envisioned "historical depth." While the Greek Orthodox Church was, without doubt, the most important potential ally and Christian dialogue partner for the Western Europeans in the Ottoman lands and thus deserves a prominent position, this chapter does not aim for an exhaustive overview of all Mediterranean religious denominations with which all plural forms of Western confessions and sub-confessions were in exchange at some point. Instead, my focus is on the form of those processes and conjunctures across the different fields by comparing two very different European imperial "machines." First, the changes and developments on the epistemic level itself are described. In the second part, the personal and institutional backgrounds, quite different for the French and the British in the field of religion, are brought into consideration.

Entangling Powers of Non-Knowledge between West and East (Greek Church, Samaritans, Phoenicians)

Orientalizing the Eucharist Debate, Confessionalization of the Greek Church

The Eastern Greek Church, regarded by Rome as schismatic, was certainly not unknown in the West in the strictest sense. Everyone knew about the medieval schism, every ecclesiastical history took it into account. Yet, if one asks what was precisely known about it, what elements of its theology, what periods of theological developments, which texts were known, one recognizes that there was in fact a great deal of ignorance, at least concerning large parts of central dogmatic fields. Eastern theology played no role here. If one referred to "the Greek Church fathers" in Western theology before the seventeenth century, that always meant the early church fathers.[2] Within most sixteenth-century theological debates, discussions about the conception of the Eucharist in Greek Orthodox theology never emerged. The early leading Greek scholar within that debate, Johannes Oekolampad, who had set the standard with his collection of both the Latin and Greek church

---

[2]　Cf. in general Backus (ed.), *Reception*.

fathers' ideas about the *Hoc est corpus meum*, did not take any Byzantine author into consideration.[3] Post–tenth-century developments within the Byzantine Church were not addressed, and often were not even known due to the lack of relevant texts. And the early church fathers who wrote in Greek were not considered as being "Eastern" in a special way within the dogmatic discussion. As one might say in general of medieval theology, it was unshaped regarding many aspects of the empirical worldly roots of the texts – the author's biography, time, context and region. The humanist and Reformation movements had changed much about that during the sixteenth century, but those impulses of historicization and empiricization did not regard Byzantine theology, which remained in the shadow of nescience, with no apparent need for transfer or reception. Two reasons for that are at hand: first the lack of enduring and sustainable contacts between representatives of Eastern and Western churches, second the specific worldview and target time horizon of humanism and Reformation theology.

Besides some individual letters sent and received by Protestant theologians including Melanchthon, the major conduits of perception and exchange between the Eastern Church and Western Protestantism had been texts such as Chytraeus' treatise on the present state of the Greek Church (1563) and the unionist discussion between the mainly Württemberg Lutherans and Patriarch Jeremiah in 1576–1581.[4] If one regards the text closely, it did not contain any citation of a Greek theologian later than Photius, and in the paragraphs on the Eucharist one only finds citations from the Bible. It therefore could not have served as an introduction to a distinct tradition of theology.[5] It was a moment of singular contact and exchange. The next and most crucial period of contact and exchange between East and West was that between the famous Calvinizing Patriarch of Constantinople, Kyrillos

---

[3] Oekolampad, *De genuina verborum Domini expositione liber* (1525); Oekolampad, *Quid de Eucharistia veteres senserint* (1530): Beda and the Berengar conflict is mentioned, also Theophylact (f. 17r–v) but without Greek text. Otherwise only the older Greek church fathers (Cyrill, Cyprian and Chrysostomos). There is virtually no author or text within the long sixteenth century discussion that would have surpassed Oekolampad for that. Cf. Köhler, *Zwingli und Luther*; Bizer, *Studien*; Kaufmann, *Abendmahlstheologie*.

[4] *Acta et scripta theologorum Wirtembergensium* (1584) – English translation: Mastrantonis, *Augsburg and Constantinople*.

[5] Cf. Mastrantonis, *Augsburg and Constantinople*, 49–55, 143f, 188f, 261–263; Wendebourg, *Reformation und Orthodoxie*.

Lukaris (1620–1638, with interruptions), and the Dutch Huguenot ambassador at Constantinople, Antoine Léger, as well as Grotius.[6] The travels to Germany and studies at Balliol College, Oxford, of a follower of Kyrillos, Metrophanes Kritopoulos (1617–1622), with his 1625 Helmstedt Confession, are symbolic of the welcome received by that particular tendency within Greek Orthodoxy on the part of late-Humanist Western Reformed or Irenic Protestantism. But that Confession was only published in 1661, with a Latin translation prefaced by Hermann Conring.[7] The 1629/33 *Confession* of Kyrillos[8] was translated into several languages and was the most widely received of the seventeenth-century Greek confessions – although it was the most heterodox. But neither confession could serve as an introduction into a deeper knowledge of Byzantine theology. Instead it was vice versa, Kyrillos proved receptive to Western Protestant theology and the Western Protestants could mirror themselves in an unexpected Eastern discussion partner. So, in the 1620s and 1630s, there was almost no deeper perception of the historical development of the Eastern Greek theology. Even Grotius, who had known Kyrillos personally, did not cite any Byzantine Greek sources other than Kyrillos.[9] An author like Jeremy Taylor, so important for the English Eucharist discussion, who engaged especially with Kyrillos, did not use any of the other "modern" Greek authors.[10] The texts produced for those particular exchanges were confessions or question-and-answer lists about particular theologoumena where the orthodox theologians themselves rarely gave precise citations of prior doctrines as they were merely stating the current teaching. The Greeks thus perceived the confessional separation in Western Europe and adopted the Western form of communication within that conflict, the genre of confession.[11] However, on the side

---

[6] Roper, "Church of England"; Cuming, "Eastern Liturgies"; Patterson, "Cyrill Lukaris"; still fundamental Hering, *Ökumenisches Patriarchat*.

[7] Karmires, Μητροφάνης ὁ Κριτόπουλος; Davey, *Pioneer for unity*; Davey, "Metrophanes Kritopoulos." The only remaining Ms. of the Confession is Hs. Wolfenbüttel 1048 (= Helmstedt 946); current edition: Karmires, Τὰ δογματικὰ καὶ συμβολικὰ μνημεία, vol. 2, 569–641.

[8] First published in Latin (1629); only in 1633 did Kyrillos authorize a publication in Greek (Genève: de Tournes).

[9] Grotius, *Appendix* (1641); Grotius, *Votum* (1642).

[10] Taylor, *The real presence* (1654).

[11] The concentration on the relationship between state and religion as the central characteristic feature within the 1980s' historiographical debate about confessionalization made us forget sometimes that the text genre of the

of the Western theologians, for a long time, there was no deeper interest in historical developments of that Eastern theology. Only within the field of liturgy studies had medieval Greek Orthodox theology been received to a certain extent since the sixteenth century.[12] The compilatory liturgy publication *Euchologion* by Jacques Goar (1647) was a fruit of that and catalyst for future studies at the same time. Hence it is no wonder that by "the middle of the seventeenth century, all divines who ha[d] an interest in liturgy display[ed] a good knowledge of the eastern rites."[13]

This remained largely the state of affairs for the whole sixteenth and the early seventeenth century. Roberto Bellarmino's *De controversiis christianae fidei*, the most important Catholic compendium of controversial theology, contained in its section on the Eucharist (first published in 1588) a chronological enumeration of patristic testimonies which only went up to the twelfth century, including Theophylact and Euthymius Zigadenus, not citing them in original Greek and providing only the names of some later authors in a more summarizing way.[14] Philippe Duplessis-Mornay, the former Calvinist councilor of Henri de Navarre and head of the Huguenots after Henri's conversion to Catholicism in 1593,[15] published in 1599 the hitherto largest monograph on the Mass and the Eucharist in Western Protestantism.[16] He still cited almost no post-antiquity "Eastern" sources on the question of how the – in his view – false doctrine of "transubstantiation" developed.[17] Against Mornay, the polemical "champion" of French Catholicism during the first half of the seventeenth century, Cardinal

confession itself belongs to the characteristics of the period (cf. Zwierlein, "(Ent) konfessionalisierung").

[12] In France, Morel and Hervet in 1560 edited a Greek Liturgy collection; in England, John Jewel used already in 1560 the fourteenth-century Liturgy manual of Nicolas Kabasilas; John Cosin also used Theodore Balsamon, cf. Cuming, "Eastern Liturgies."

[13] Cuming, "Eastern Liturgies," 235f.

[14] Bellarmino, *De controversiis christianae fidei*, tom. III (1628), "tertia controversia generalis, De Sacramento Eucharistiae, sex libris explicata," 94–195, cap. XXXIV, 138B; cap. XXXVII, 135C; cap. XXXVIII, 135G.

[15] Daussy, *Huguenots*, but Daussy concentrates on Mornay as "homme politique."

[16] Duplessis-Mornay, *De l'Institution* (1599).

[17] Only the second Nicean council and Theophylact are mentioned, the latter by highlighting the Greek terms μεταστοίχομαι / μεταβάλλεσθαι used by him for the transformation of bread and wine: Duplessis-Mornay, *De l'Institution* (1599), 787f.

Jacques du Perron published his *Traité du Saint Sacrément de l'Eucharistie* in 1622, verifying each author cited by Mornay by dedicating a whole chapter to each of them. Besides more precise and longer citations of Theophylact, in the chapter concerning the Council of Florence, he also referred to Samonas, Kabasilas, Nikolaos Methonensis, Markos Ephesios from his *Bibliotheca patrum*,[18] and also the exchange between Patriarch Jeremiah and the Wittenberg theologians from the 1584 *Acta*, in the marginal notes in Greek. These became the standard set of citations for subsequent discussions.[19] One of the most learned reformed authors, Zurich's Johann Heinrich Hottinger, who in 1652 reedited Kyrillos' *Confession* with a long Commentary section revealing an immense knowledge of late antiquity's Greek Church fathers, still cited almost no Greek source beyond Photius (820–886).[20] This was still an "armchair" form of theological discourse ultimately completely focused on Western theology. Paradoxically, the major inner-orthodox transformation that took place between the Ottoman conquest of Constantinople and the end of the seventeenth century, the growth of "Latinizing" tendencies, remained obfuscated for a significant length of time. This process was stimulated by the opposition to Kyrillos, but not only by that. Those developments were happening at the same time as the inner-European Eucharist debate reached its final important peak, and thus it has to be envisioned as a coterminous and dialectically linked process of semantic shifts and changes, but it can be narrated only in a consecutive form.

At its center stood the term "μετουσίωσις (*metusiosis*)," understood as the equivalent to "transubstantiation." It did not exist in classical Greek, where μετουσία denoted participation, and the standard word for transformation concentrated on the form, on the μορφή < μεταμόρφωσις. Today it is established that the term μετουσίωσις was first introduced – after a singular use at a synod in Constantinople in 1277 – into systematic Orthodox theology by a participant in the post-Florentine Council's unionist discussions and later Patriarch of Constantinople, Georgios Gennadios II Scholarios (1403–1472), in his Lent sermon just after the Ottoman conquest (probably in 1453),

[18] Cf. Petitmengin, "Patrologies."
[19] Du Perron, *Traitté du Sainct Sacrement* (1633).
[20] Hottinger, *Analecta* (1652), 398–567 (exception: p. 444 Balsamon [1140–1195]).

borrowing much from Thomas Aquinas.[21] Gennadios' early use of the term had not been accepted within the Orthodox Church as canonical. The particular text in question was not known by Bellarmino, Mornay, du Perron, Aubertin or by anyone else in the West before the edition by Eusèbe Renaudot in 1709 from the papers that Nointel had gathered from the Greek dragoman at the Porte Panagiotis in the 1670s. At that time, Renaudot also published the seventeenth-century polemical texts of Meletios Syrigos, which strongly defended the μετουσίωσις against Kyrillos.[22] Of high importance were three Greek Catholic converts, Allacci, Karyophylles[23] and Arkudios.[24] Allacci responded to the 1652 reedition of Kyrillos' Confession by Hottinger, and in this treatise, we find a huge amount of citations from Byzantine theologians.[25] In his reply to Allacci, Hottinger still did not make use of almost any medieval Byzantine text.[26] The semiotic and semantic changes resulting from that theologico-confessional contact between Greece and Latinizing Greek exile writers were sometimes noted but not fully understood and contextualized by Western observers. Grotius had noted a fluidity of Eastern Greek terminology (μεταβολή, μεταποίησις, μεταστοιχείωσις), Casaubon even thought erroneously that those words were inherited from the "Greek fathers of the first [sc. Christian] centuries [*Patres primorum seculorum*]"[27] and in 1654, Taylor still noted, disparagingly and even ironically, the inconsistency of terminology and that there

---

[21] Steitz, "Abendmahlslehre"; Jugie, "transsubstantiation"; Tzirakes, Ἡ περὶ μετουσιώσεως εὐχαριστικὴ ἔρις, 35–48 (exact dating of the sermon is not evident, ibid., 35f.)

[22] Gennadios, *Homiliae* (1709). The *praefatio* (ibid., i–xxxi) gives some information about the process of manuscript gathering and editing from 1672 until the publishing date. Another work of Scholarios, his dialogue with a Muslim in defense of Christian faith (today edited as *Confession* / Ὁμολογία, in Early Modern Times normally known as Περὶ τῆς ὁδοῦ τῆς σωτηρίας τῶν ἀνθρώπων / *De via salutis humanae*), was already printed in 1530, but it did not contain any passage on transubstantiation doctrine, cf. Karmires, Τὰ δογματικὰ καὶ συμβολικὰ μνημεία, vol. 1, 429–436.

[23] Johannes Matthaios Karyophylles, not to be confused with the follower of Kyrillos, Johannes Karyophyllos Byzantios.

[24] For literature on these cf. Podskalsky, *Griechische Theologie*, 156–160, 181–183, 213–219; Cerbu, *Leone Allacci*; Sojer, "Il manoscritto autografo."

[25] Allacci, *Hottingerus* (1661).

[26] Hottinger, *Enneas dissertationum* (1662), 179–212. He only refers to the narrative accounts of the Basel (Sylvester Sguropulos) and of the Florentine council.

[27] Grotius, *Appendix* (1641), 50; Casaubon, *Ad Epistolam Responsio* (1612), 37.

was now coming "another production from Africa, a transaccidenti-substantiation, a μεθυφισταμενομετουσία."[28]

Between 1650 and 1730 this rather superficial form of perception changed, the disproportional attention given to Kyrillos ceased, and a more thorough awareness of ignorance and eventually a need for knowledge about the Greek Church's recent past and its present theology became the engine of a large epistemic movement entangling Western and Eastern fields of discussion. An "Orientalization of the Eucharist debate" took place at the same time as the Greek Church was conceived as something like a confessional church equivalent to the Western Christian confessional churches after the Reformation. The reason is again to be found on two levels:

(1) The infrastructure of the trade empires was used now for supplying participants in Western discussions with necessary information. The "reality" of the Greek Church was thus entering into the descriptions and narratives instead of the more pre- and praeter-empirical topical forms to conceive of the "ecclesia orientalis." This contact and exchange had its impact on Western discourse as well as on the Eastern. (2) Second, the concept of time and history active in the relevant circles and their hermeneutical approach to scripture and theological tradition had changed, mostly in France, but also in Britain.

The European Controversy on the Eucharist and the Greek Church
After the wars of religion, the inter-confessional debate in France reached a new degree of intensity as the war with swords also became a war of words. After the conversion of Henri IV in 1593, there were 166 conferences between Catholics and Protestants, with a further 72 not definitely attested to throughout the seventeenth century.[29] The amount of specialized polemical literature exploded.[30] France was for that the most important place in Europe, as the monarchy was no longer ravaged by wars – if one excludes the La Rochelle war of 1629/30 and some aspects of the Fronde – as was the case in Britain, the Netherlands and Germany until 1648/50. The central debate of the Eucharist had been shifted to a new level of scrutiny and sophistication from Mornay to the Port-Royal theology.[31] In 1654, Edme

---

[28] Taylor, *Real presence* (1654), 153.
[29] Kappler, *Les conférences théologiques*.   [30] Desgraves, *Répertoire*.
[31] The Oriental dimension of the French Eucharist debate, which was the central and starting point in the perception of contemporaries, is treated quite

Aubertin, the learned Geneva Calvinist theologian, published a second Latin edition of a treatise on the Eucharist which originally had been written against du Perron in French in 1633; now this work had developed into a learned compendium of the church fathers' teaching on the Eucharist.[32] Aubertin took the lead over Mornay and Perron regarding the length of the work and the Greek Orthodox authors included. Aubertin's main point was to show that Kyrillos' Calvinizing confession was to be seen as a return of the present Greek Church to their own ancient teaching. For the sake of that argument, Aubertin necessarily adopted a historical perspective and he had to prove by citations that the older Byzantine church fathers had been close to Kyrillos – in the same way as Protestant theologians liked to compare their teaching with Hus, Wyclif and the old church fathers to construct a small and thin but existing continuity of correct teaching since the earliest times. In response to that, and stimulated by a conflict between the Calvinist minister at Charenton, Jean Claude, and Pierre Nicole during an attempt to convert the Huguenot maréchal Turenne,[33] the Port Royal authors Antoine Arnauld and Nicole published their work in different states and editions under the title *De la perpétuité de la foy de l'église* (*On the eternity of the faith of the church*), perhaps the largest work ever written on the Eucharist in church history. It grew out from a small first version by Nicole in 1659 into a more important 1664 rendition.[34] Claude replied with a *Défense* in 1665 which went through many editions and which upheld still more strongly the argument that the Greek Orthodox Church – a church without a pope and presumably without a transubstantiation

marginally by recent research. It receives still the most attention in Snoeks, *L'argument*. In his magisterial studies of the erudite Port-Royal theology, Bruno Neveu never touched upon their interest for Greek Orthodox sources, cf. Neveu, *Erudition*; cf. Quantin, *Catholicisme*, 321–356; Quantin, *Church*. In the historiography on the Greek orthodox church itself, the period between 1650 and 1730 remains far less studied than the decades of the Kyrillos controversy, cf. Tzirakes, Ἡ περὶ μετουσιώσεως εὐχαριστικὴ ἔρις, 235–246; Ware, "Orthodox and Catholics"; Kitromilides, "Orthodoxy and the west," 201.

[32] Aubertin, *De Eucharistiae Sacramento* (1654). For du Perron's part in the controversy cf. Snoeks, *L'argument*, 84–128 and Kappler, *Les conférences théologiques*, passim.

[33] Nicole had in 1661 given the treatise in a first handwritten version to the maréchal as an argument for conversion; Claude had replied with another manuscript response to that at the request of Turenne's wife.

[34] Nicole, *La perpétuité de la foy* (1664).

doctrine – was in accordance with the Calvinist teaching and thus a living tradition and argument against the "papists." Claude's seventh edition of 1668 also incorporated direct Greek citations from the medieval Greek theologians – while the work of Arnauld and Nicole was completely in French.[35] Arnauld and Nicole responded with a three-volume version in 1669, which was subsequently enlarged throughout the century, ending with six-volume editions in 1713, 1781/82, 1841.[36] During this debate in France and with the *Perpétuité*, Greek theology became integrated to the highest degree possible at that time into the French Eucharist debate.

What was entirely different from the sixteenth-century forms of the Eucharist discussion was the use of the infrastructure of the trade empires for the gathering of information and for direct contact with the Eastern churches. It was thus not just a form of purely academic reception, it was embedded in the imperial communication network and became part of politico-religious exchange in the Levant itself.

Because of the conflict in France, the Jansenists asked the French ambassador in Constantinople, Charles-François Olier, Marquis de Nointel, for help in obtaining direct information about current Greek theology. The conflict was also a major catalyst for research and the sending of relevant Greek manuscripts to Europe and for their editing. Nointel was supported for those purposes by the Orientalist and author of the *1001 nights*, Antoine Galland.[37] Later, others continued to gather manuscripts, such as the French consul at Aleppo (1652–1661) and future Bishop of Babylon François Picquet, who collaborated with Hugo Jannon, priest of the collegiate church St. Juste at Lyon.[38] Other European representatives, merchants and dwellers in Constantinople also became involved, including the German Calvinist Pestalozzi and the British Thomas Smith and John Covel. Because the Europeans insisted on it, the Patriarch of Constantinople, Dionysios IV, sent out questions throughout the Mediterranean about how the issue of transubstantiation was taught in the Greek churches, and he

---

[35] Claude, *Réponse* (1668).
[36] Arnauld and Nicole, *La perpétuité de la foy* (1669).
[37] Galland, *Journal*, 34, 37, 44, 49, 55–56, 65, 104; Vandal, *L'Odyssée*, 48, 101–102; Abdel-Halim, *Galland*, 38f.; Hamilton, "From East to West," 85–92 has made very fruitful use of BNF Ms. Arménien 145.
[38] Gennadios, *Homiliae* (1709), f. ij. Cf. Arnauld and Nicole, *La perpétuité de la foy* (1669), vol. 2, 147–150; vol. 3, 628–630.

assembled in January 1672 a council of Greek patriarchs and churchmen in the city to resolve the question of the Eucharist, church hier-
archies, idolatry and the number of canonical Biblical books. That
declaration was transmitted by Covel to England, but later printed in
Paris together with the acts of the Greek synod of Jerusalem which was
held just two months later in March. The French conflict had forced the
Greeks to formulate a doctrinal position in two large assemblies,
adopting it more officially and *uni sono* than had ever been the case
before the doctrine of transubstantiation in its current Greek terminology.[39] The controversy did not stop there however, even if it might
seem that its main points had been settled. The French Calvinists, who
did not possess an appropriate information network, used the British
communication channels for their purposes. On August 22, 1674, John
Crawford, chaplain to the British ambassador in Venice, Thomas
Higgins, wrote to his colleague in Constantinople, Covel, providing a
short account of the French Eucharist controversy. As Jean Claude was
"destitute of all such helps," that is, of ambassadorial support like
Arnauld, he sent a memorandum for precise inquiry to Crawford and
the latter sent it further to Covel, who was asked to inquire among the
Greeks about their true beliefs – during which he should make a
prudent distinction "betwixt those who are purely of the sentiments
of those churches, and these who have anywise drunk in the Roman
principles by their frequent communication with these emissaries."[40]

Claude's memorandum consisted of fifteen points which one can
interpret as a detailed specification of ignorance, a concrete formulation of non-knowledge. Perhaps surprisingly for the realm of theological research, he used a method which resembled the queries with which
natural historians and voyagers were sent out by the Royal Society, as

---

[39] *Synodus bethlehemitica* (1676); Georgi, *Confessio Dosithei*, 24–30 for the
context and 72f. for the confession's article on the Eucharist with its "massive
realism" expressed with the words μεταβάλλεσθαι, μετουσιοῦσθαι, μεταποιεῖσθαι,
μεταρρυθμίζεσθαι). John Covel gave an account of how he returned from the
Levant with the text of the synod, first staying in Holland, then in Paris, always
being prevented from printing it, as King James was on the throne and "absolute
Popery hang'd over our heads." While the text had been translated and printed
in Paris, he still proceeded with the project of his own edition, Allix was
commissioned with the translation. In the end, all that work remained in his
"study at Cambridge" (John Covel to George Wheeler, May 28, 1717, BL Add
Ms. 22911, f. 218).
[40] John Crawford to John Covel, Venice, August 22, 1674, BL Add. Ms. 22910,
f. 76–77.

will be touched upon in the last chapter on "science."[41] The central first point was (1) if the "dogma of transubstantiation, of μετουσίωσις, ... as it was recently defined by the Romans in their Tridentine council ... was held as an article of faith."[42] The other points are consequences of the first: (2) if the Greek believed in the idea of Christ's ubiquity ("multiplicatio praesentiae suae in pluribus locis"); (3) if they believed that the body of Christ with its distinct members such as a head, neck, arms and feet was communicated orally to the believers being then "in the stomach and belly" of the communicant; (4) also if infidels or dogs eating the consecrated bread by accident, would eat Christ. Several other questions concern the consequences of the supposed physical transformation – points five to seven – and the adoration of the bread or the body of Christ. Claude wanted to know – questions number 12 to 14 – if the doctrine of transubstantiation had been accepted by any Greek synod and how they would deal with the memory of Kyrillos Lucaris who had rejected the μετουσίωσις in his Confession, with the anti-Kyrillos synods of 1639 and 1642, and with the confession of Petrus Mogilas and Meletios Syrigos (1642). Finally he provided a list of the Greek patriarchs and bishops who had signed the 1672 synod organized by Dositheos. The English chaplains were instructed to investigate their exact doctrine and faith.[43] Unfortunately, not much of the correspondence that must have been conducted by the chaplains on this matter has survived.[44]

Covel returned in 1688 to Cambridge, becoming the university's vice-chancellor two years later, but he worked for decades, struggling with the subject, keeping contact with all possible informants in the Mediterranean, continuously inquiring and receiving further

---

[41] Jean Claude Ecclesiae reformatae Parisiensis Minister to the "Reverendos et clarissimos pastores Anglos in partibus Orientalibus degentes et pro suo in Religionem reformatam zelo" (BL Ms. Add. 22910, f. 83r–v).

[42] "An apud eos dogma Transsubstantiationis μετουσιώσεως, hoc est conversionis realis et physicae totius substantiae panis in eandem numero substantiam corporis Christi ... nuper definitum a Romanis in suo concilio Tridentino habeatur pro articulo fidei" (BL Ms. Add. 22910, f. 83r).

[43] Ibid.

[44] Without claim of exhaustiveness, I have inspected important parts of the Levantine correspondence of Pococke, Thomas Smith, John Covel, Thomas Shaw, Edward Williams, Robert Huntington, William Hallifax, John Luke and Edward Smyth, where the interest in the Greek theology is a recurring issue, but the inner-Levantine correspondence or contacts with representatives of the Greek Church is not documented.

information and gathering his notes[45] to finally produce his 1722 book on the Greek Church.[46] Likewise, in France, Eusèbe Renaudot studied "during the longest parts of my life the religion and the discipline of the Oriental churches,"[47] collected information and manuscripts and drafted many notes and treatises on the subject of which only a minor part found its way into publication in his *Défense* of the *Perpétuité*, his later publications or that was integrated in the successive enlarged versions of the *Perpétuité de l'Eglise*.[48] A large sequence of treatises, some in French and many in Latin on the Greek Church's doctrines, mostly concerning the Eucharist, was the fruit of his lifelong endeavors. His lifetime opponents, constantly recurring in his writings, were Claude, Aymon, Smith, Hottinger, Allix.

A similar revival of interest in the Greek Church happened at the same time in Britain as the traditionally more famous French Eucharist debate. No significant printed work bears testimony for that apart from the short books of Thomas Smith (1678) and John Covel (1722). In fact, intellectual correspondence of the time show that the Anglicans had already started to reconstruct the Greek Church's theology in a historical way, at exactly the same time as the 1669 edition of *La Perpétuité de l'église* codified the matter within French intellectual circles.

The best source to show this is a hitherto unused manuscript treatise of "the leading light in the famous Nonjuror colony,"[49] by Henry Dodwell, written for and sent to Edward Bernard of Trinity College,

---

[45] "The materialls for my travayl in the East and my minutes about the Greek Church ... lay all middled up in my study at Cambridge, yet I was busy all my stay in Holland ... being now an old Octagintarian, I desire to liue the seueral remeinder of my evil days in peace and diete ... [what I have written is of no offense] to our own Apostolick and Catholick Church ... and therefore as I told you, if I doe ever print it, I shall dedicate it to the Turky Company whose bread I did eat" (Covel to George Wheeler, May 28, 1717, BL Ms. Add. 22910, f. 218r). Important for the information about the Oriental churches are several letters from Wanley, original confessions from Nestorians, Jacobites, Maronites, Armenians, several Greek patriarchs, in BL Ms. Add. 22910, 22911.

[46] Covel, *Some account* (1722).   [47] BNF NAF 7468, p. 368.

[48] The complex flow of manuscripts between the Orient and Paris in those years from 1670 to roughly 1730 has not yet been reconstructed and correlated with the printed publications in all details. The material of the Renaudot papers (BNF NAF 7456–7478, mostly 7464–7466) is a starting point, the introduction in Malvy and Viller (eds.), *Pierre Moghila*, LXV–LXXXI has probably made the most prolific use of it in research until now.

[49] Rose, "The origins and ideals," 176.

Dublin, on June 2, 1670. These dense five folio pages are almost completely taken up with Eastern authors with full Greek citation and the whole argumentative structure for which *La perpétuité de la foy de l'église* needed several hundred pages – the important difference aside in its Anglican conception of the Eucharist –, and apparently without direct knowledge of the work by Arnauld and Nicole. It shows how a rigid Anglican exegetical approach could lead to something astonishingly similar to the Jansenist Catholic interpretation. While Dodwell has received quite a lot of scholarly attention in recent times, his interest in the Byzantine Greek Church has not been appreciated.[50] Apart from that treatise, one can reconstruct parts of his Byzantine readings from five of his preserved notebooks, which show that he was using the best sources available, namely the Barroci Manuscripts that the Bodleian had acquired in 1629,[51] the Laudian manuscripts and those of Kenelm Digby, transcripts from the Vatican and the Florentine Medici library, as well as texts from the Alexandrine library once collected by Kyrillos Lucaris.[52] He also received many copies from Edward Bernard.[53]

Dodwell's letter responded to the question of Edward Bernard about "the sense of the modern Greeks since Photius's time, touching Transubstantiation." This seems to have belonged to the larger enterprise at that time conducted by Anglican theologians. Dodwell mentions that they were waiting for "our friend Mr Smith,"[54] meaning Thomas Smith, who was a chaplain in Constantinople between 1668 and 1670.[55] Dodwell was well aware that the seventeenth-century Anglicans were doing something new by looking into the Greek Church's teaching concerning the Eucharist, as he explicitly recalled the reformed Peter Martyr Vermigli and the Lutheran Martin

---

[50] Cornwall, "Divine right monarchy"; Leighton, "Ancienneté"; Quantin, "Anglican scholarship"; Quantin, *Church*, 366–395.

[51] Humfrey Wanley judged in a letter to John Covel from January 14, 1715/16: "the Baroccian MSS now at Oxford, being the most valuable [parcel?] that ever came into England, consisteth of 264 Greek books, beside some few others," BL Ms. Add. 22911, f. 185v. These are the manuscripts of the Venetians Francesco and Iacopo Barozzi, brought by Henry Featherstone to England.

[52] The Epistle of Clement to the Corinthians was collated with an Alexandrine manuscript by Kyrillos Lucaris, cf. BL Ms. Add. 21081, f. 114–116.

[53] BL Ms. Add. 21078–21082.   [54] Bodleian, Ms. Smith 45, f. 5v.

[55] Pippidi, *Knowledge*, 241 is the only one who briefly mentions that treatise as far as I know.

Chemnitz as the "onely ancient Reformers, that I know of, that have as much as payd any claim to the Graecians in this question."[56] Eusèbe Renaudot also noted around 1700 in retrospect that both Catholics and "Protestantes pauci de illis scripserunt."[57] The recognition of ignorance concerning the Greek Church by the Western men of erudition was thus shared.

Dodwell wanted to write "onely of them who still maintain the warrantableness of their first separation, wherein a man had need be cautious what authors he trusts, For both in Italy and most of the Mediterranean Isles, and the Maritime coasts of Greece itself, especially the dominions of the Venetians, either by the Fflorentine [sic] union or the influence of their Latine masters, and the naturall servility and flattery and barbarousness of the nation, they are more or less Latinized."[58] He claimed that what would be said applied only to the Constantinople branch that upheld the schism. This is an approach very different from the medieval and sixteenth-century practice, where the "ecclesia orientalis" was rather a vague denominator without concrete empirical roots.

After a short sketch of what he understood as their current teaching, Dodwell then analyzed Kyrillos' Confession, always precisely citing its Greek text, and he affirmed that his understanding of the Lord's supper is in fact "in our [sc. Anglican] sense,"[59] as it stated real presence

---

56 Bodleian, Ms. Smith 45, f. 5r. Vermigli had already made use of Theophylact in his 1562 treatment of the Eucharist opposing Stephen Gardiner. This was incorporated into Vermigli's *loci* (Vermigli, *Locorum Communium*, tom. I (1580), col. 1603f.) and thereby widely received, for example by Bellarmino, *De controversiis*, tom. III (1628), 138G. For Chemnitz, one of the Lutheran co-authors of the Formula of Concord and therefore in close contact with the exchange of the Tübingen theologians with Patriarch Jeremiah cf. Mastrantonis, *Augsburg and Constantinople*, 9, 13.

57 "Protestantes pauci de illis scripserunt: Cum tamen adversus Catholicam de Eucharistia fidem declarare Claudius Carentoniensis coepit, inquisitum est diligentius in Orientales Liturgias" (BNF NAF 7459, 160v); "On connoissoit peu l'estat des Eglises d'Orient" (BNF NAF 7459, p. 118).

58 Bodleian, Ms. Smith 45, f. 1r.

59 Bodleian, Ms. Smith 45, f. 1v. Dodwell did not provide his own understanding of the original Greek of Kyrillos, but simply stated that the phrase "τὴν ἀληθῆ καὶ βεβαίαν παρουσίαν τοῦ κυρίου ἡμῶν Ἰησοῦ Χριστοῦ ὁμολογοῦμεν καὶ πιστεύομεν. πλὴν ἦν ἡ πίστις ἡμῖν παρίστησι καὶ προφέρει, οὐχ' <ἦν> ἡ ἐφευρεθεῖσα εἰκῇ διδάσκει μετουσίωσις" would be "in our sense": "[. . . in which institution] we confess and believe the true and firm presence of our Lord Jesus; but that which the faith presents and produces for us, not that which teaches us the strangely

(„ἀληθή[ς] καὶ βεβαί[α] παρουσία τοῦ κυρίου"), but explicitly denied transubstantiation („οὐχ' ἦν ἡ ἐφευρεθεῖσα εἰκῆ διδάσκει μετουσίωσις"). Then, Dodwell skeptically pointed out that Kyrillos' position was an historical exception and he argued this by investigating the Greek Church's history, showing how the Metropolitan of Philadelphia, Venice Gabriel Severos (1577–1616), had taught that the holy spirit would transform and transubstantiate the bread into the body of Christ, already using the terms μετουσιάζειν and μετουσίωσις. Dodwell knew the writings of the Latinizing Allacci, Arkudios and Karyophyllos; he cited the Tübingen exchange of 1576–1581. He then moved back in time, step for step, to the Council of Florence (1438–1445) and the anti-unionist writings of Markos Eugenikos, Archbishop of Ephesus (1392–1444), which would already express the meaning of transubstantiation;[60] to Nikolaos Methonensis, Nikolaos Kabasilas (1319–1391), Euthymius Zigadenus (+ after 1118) and Theophylact of Bulgaria (1055–1107), to Samonas, the Melkite bishop of Gaza (11th c.), and beyond Photios (810–891) to Germanos of Constantinople (650/60–730), John Damascenus (676–749), and to Anastasius Sinaita (630–700) and to the decrees of the second Nicene Council (787), where he evinced a very old proto-transubstantial conviction present within the Eastern Church as a by-product of the discussion on idolatry: "as the assertion of Images first got footing in the Oriental Church, so also theirs [sic] are the most ancient Testimonies for Transubstantiation."[61] We see that Dodwell had a very concise

invented *metousíosis.*" Cf. for a modern print of Kyrillos' confession Kimmel (ed.), *Monumenta fidei* (1850), pars I, 24–44, 36.

[60] He had to argue here in a quite sophisticated manner because the terminology used by Markos Ephesios (transformation of ἀντίτυπα into πρωτότυπα) is the old Greek fathers' diction, independent from the Latin medieval tradition: "that afterwards they are so changed into the πρωτότυπα that they cease to be ἀντίτυπα, so that the change is not onely perfective but corruptive; and that they have not ceased to be what they were formerly as to the sense, it must follow that the cessation is in the substance of the elements, and consequently that the change is proportionally into substance of the body and bloud of our Lord." (Bodleian, Ms. Smith 45, f. 3r).

[61] Bodleian, Ms. Smith 45, f. 4v: while the Iconoclasts asserted that the institution of the Eucharist was meant by Christ to be the only and sufficient representation (εἰκών) of his humanity "and consequently left no necessity of other resemblances," the adversaries – whom Dodwell takes to be precursors of the "Eastern" Church – argued that due to the transformation of bread and wine into body and blood during the Eucharist, the signs were no longer signs, but real body and blood, and consequently, there was still need for pictures and representations.

concept of the chronology, geography and history of the Byzantine church. These texts are not nodes in a fluid web without order and historical depth, there is also a nuanced understanding of their contexts. Dodwell gained that knowledge from several Byzantine chronicles and patriarchal lists,[62] and he also possessed copies of many important testimonies usually favored by the English church authors. From the early times he had the letter of Polycarp to the Philippians,[63] the *Acta S. Ignatii*[64] and various editions of the Ignatian letters[65] and from the medieval times he utilized the letter of the Patriarch of Constantinople Michael I Keroularis to the Patriarch of Antioch, Peter Butrus III, from the time of the Schism (1054),[66] a short treatise *De haeresibus*,[67] and copies of letters of the non-juring bishop William Lloyd to Antonio Pagi from 1686/87 with theological content concerning the Ignatian controversy as well as Byzantine church history.[68]

In both the French and the English cases, the successive cognitive shocks of the Kyrillos scandal, as well as of the renewed debate on the Eucharist in Europe led to an awareness of ignorance and set off a process of empirical research completely unfamiliar to the sixteenth-

---

[62] Cf. "Ἱππολύτου Θηβαίου παπ. Ῥώμης ἐκ τῆς Χρονικῆς αὐτοῦ Ἱστορίας περὶ τῆς τοῦ Σωτῆρος ἡμῶν οἰκονομίας καὶ περὶ τῆς ἁγίας Θεοτόκου πόσα ἔτη ἔζησεν εἰς ἀκρίβειαν ἔτι δὲ καὶ περὶ τοῦ Βαπτιστοῦ Ἰωάννου καὶ περὶ τοῦ Θεολόγου Ἰωάννου καὶ τοῦ Ζεβεδαίου δὲ καὶ τῶν ἡμῶν αὐτοῦ καὶ περὶ τοῦ ἀδελφοδέου Ἰακώβου καὶ περὶ τῶν μαθητῶν τοῦ Χριστοῦ," BL Ms. Add. 21078, f. 100r–104r (pp. 193–201); "Νικηφόρου Καλλίστου τοῦ Ξανθοπούλλου εἴδησις ἀκριβεστάτη περὶ πάντων τῶν ἐν τῇ Κωνσταντινουπόλει ἐπισκόπων καὶ πατριαρχῶν," BL Ms. Add. 21078, f. 105r–115r (pp. 203–223), ibid., f. 105r–115r (pp. 203–223) and the "Fragmenta Historiae Graecae de tempore Theodosii junioris, and Concilij Caesariensis tempore Victoris sub Theophilo Fragmenta," ibid., f. 118f (pp. 229–231); a list of 75 patriarchs and rulers "Χρονογραφικόν διὰ στίχων καλλίστου τοῦ Ξανθοπούλου Κυρίου Νικηφόρου" accompanied by a (mutilated) chronicle from late antiquity to the 13th century (BL Ms. Add. 21080, pp. 338–364); a "Chronologia Patriarcharum e Demetrio," BL Ms. Add. 21081, f. 81r–85v.

[63] From the Florentine codex that had used Vossius for his edition, BL Ms. Add. 21078, f. 62r–67r.

[64] From the Laudian codex C. 93 the dialogue between Ignatius and Trajan, BL Ms. Add. 21079, pp. 245–275.

[65] From a Florentine codex, transmitted by Edward Bernard, BL Ms. Add. 21081, f. 108v–113.

[66] A copy from a Vienna Ms. that had been taken by St. George Ashe, followed by Abednego Sellerus, then Dodwell – BL Ms. Add. 21079, pp. 281–306.

[67] "De haeresibus fragmenta quaedam," from the Ms. Barocc. 76 about the Melkite Church, BL Ms. Add. 21079, pp. 204–206.

[68] BL Ms. Add. 21082, f. 1r–63r, on f. 23v–28v: "De vero tempore Martyrii Ignatiani."

century debates. Neither Zwingli, Bucer, Karlstadt, Oekolampad or Luther in the 1520s, nor Calvin, Westphal and Bullinger in the 1550s, sent out emissaries to ask for information about current and medieval Greek teachings on the Eucharist. Those were still worlds apart.

Time Horizons, Reflexivity

Beyond the question of particular events and texts stimulating such an epistemic movement, a major important reason for the ignorance to become "visible" is the change of the theological time horizon that provided the frame of reasoning. While the ideas about when the church had been in a (nearly) perfect state was common to Protestants and Catholics,[69] it was nevertheless a more exclusive, quasi-cyclical vision in the Protestant case. Protestants devalued medieval times by preselecting and emphasizing the early and ancient eras, sometimes nevertheless still reserving a special place of dignity for the time of the first Christian emperors. This selective hermeneutics of target and non-target time horizons explains on a second level why the Greek Orthodox Church and its eleventh- to sixteenth-century history lay out of range, occupying a blind spot in early Reformation theology. They thus belonged to the realm of ignorance, a largely unconscious, disinterested ignorance, a field of non-specified nescience. What changed in the seventeenth century?[70]

The first 1664 edition of *La perpétuité de la foy de l'église* culminated in a brief final chapter (III, 8): "that all the sects separated from the Roman Church do agree with it concerning the issue of transubstantiation, and most of all the Greeks."[71] In the 1669 edition, which now competed with both Aubertin and Jean Claude, and which was solemnly approved through distinct letters prefixed to the print by nearly all French bishops, the concept of the Greek Church as the most important ancient schismatic church became a cornerstone of the whole work and the dispute with the Calvinists. The whole second, third and fourth book now consisted of an historical-exegetical

---

[69] Cf. Quantin, *Catholicisme*, passim and 97–124.
[70] For the concept of history in the English Church cf. for this period Cornwall, "Search"; Bennett, "Patristic Tradition"; for a slightly later period Doll, "Idea"; Quantin, *Church*; for the early reformation cf. as a case study Zwierlein, "Reformation."
[71] Nicole, *La perpétuité de la foy* (1664), 464–475.

argument that, at all times, the Greeks had agreed with the Romans about transubstantiation and the real presence. The "eternity of the church" was proven for the eleventh century – Theophylact, during the crusades –, for the twelfth century – Euthymius Zigadenus, Nikolaos Methonensis, Zonoras, Niketas Choniates –, for the thirteenth century and so on right until the present date and the recent events after Kyrillos Lucaris' death, with each chapter dedicated to the presentation and exegesis of a given conflict or author.[72] Arnauld and Nicole thus elaborated into hundreds of pages what had been Bellarmino's conclusion in 1588: "Hence we see the continuity of the catholic doctrine throughout all eras of the church, not only in the Occident, but also in the Orient (*Habemus igitur continuationem Catholicae doctrinae per omnes aetates Ecclesiae, non solum in Occidente, sed etiam in Oriente*)."[73] – This provided the title for the book (*La perpétuité de la foy de l'église*). All of this specifically dealt with "the Greek Church," which was now treated as something very distinct. The manageable and coherent chronological enumeration of church fathers and authors, as had been the case in the *Bibliothecae patrum* and then in the polemical literature – e.g. for the Eucharist – in Peter Martyr Vermigli and in Roberto Bellarmino's large treatises, had been changed utterly. Hundreds of pages, nearly more space than for the Western church fathers, were now dedicated to the theological treatment of the Greek Church.

French baroque theology and patristic studies became grounded in a new understanding of history. The "perpetuity" of the transubstantiation dogma was brandished most strongly in contrast to the historicized perception of the development of human affairs.[74] This was a newly sharpened bipartite historical conception of "eternity *plus* human historicity" within Catholic France. It was solidified through opposition to the already briefly mentioned Protestant and foremost Calvinist conception of History of a cyclical movement and "return-to" with the target time horizon of the early church, in so doing bypassing the "dark Middle Ages." A second generation of reformers,

---

[72] Arnauld and Nicole, *La perpétuité de la foy* (1669), the approbations in vol. 1, ***6^V-*****4^V, the books II, III, IV in vol. 1, 157–620; still devoted to the Greek Church are book VII, vol. 2, 280–455, and book XII, vol. 3, 411–630).

[73] Bellarmino, *De controversiis*, tom. 3, lib. II, cap. XXXVIII, 139G.

[74] For remarks on the approach of history of French Catholicism at that time Quantin, "Fathers," 963; Snoeks, *L'argument*, 522–542.

that around 1600 included both Catholics and Protestants, had started to rediscover those dark ages, but differently. Protestants could only try to find "some small glimmering and shining lights of truth" after the times of the early church and maybe still the Christian emperors, in apocalyptically-framed conceptions of histories, since the High Middle Ages at least, the light of Waldensians, Hussites, Wycliffists was allegedly miniscule compared to the conquest of church by the Antichrist, but it guaranteed its persistence until the times of revelation through the prophetic word of Luther (II Thess 2, 8). The major shape of that historical world-view remained cyclical: Christ on earth, early church, decline through Antichrist's activities, revelation and return to the truth through Reformation.[75] The Catholic side stressed and hardened a conception of continuity. Both visions of history were challenged by rising secular concepts of history which, themselves, tended to move from cyclical conceptions to more linear-teleological ones.

The inclusion of the Greek Church into the Western inter-confessional theological controversy about the Lord's Supper followed that scheme. Aubertin could interpret Kyrillos as the Eastern Luther or Calvin who returned the teaching of his church to its early beginnings, and he was eager to show that in doing so he did not just adopt Western Protestant ideas, but that he returned to the older roots of early Eastern Greek teachings which were absent during the Middle Ages. Catholic authors tried to show that – while there was dissent over the *filioque* and other issues with the Eastern church – for the central dogma of transubstantiation and real presence, they belonged to the unchanged and unchangeable eternal part of the church. In Renaudot's papers, we find a constantly recurring, even obstinate, repetition of the *eternity* of all Oriental churches' doctrines against the false idea of the *changement* (*mutatio*) proposed by the Calvinists.[76] The French authors reflected

---

[75] Cf. Seifert, *Rückzug*; Firth, *The apocalyptic tradition*; Ward, *Early Evangelicalism*, 85–98; Starkie, *Church*, 103–125.

[76] "Nullam ab haereseon vel schismatum initiis inter Orientales fuisse de Eucharistiae Sacramento controversiam, aut inter se aut cum Ecclesia Catholica, ita certum habitum est usque ad eam quae cum Calvinistis nostris suscepta est ante annos aliquot disputationem" (BNF NAF 7468, p. 266); "On fait voir que le changement que supposent les Calvinistes n'est animé dans aucunes Églises Orientales: ny en particulier dans l'Église Grecque" (BNF NAF 7459, p. 123f.); "Cum certissimis argumentis huc usque probatum fuerit Orientales Christianos fidem de reali corporis et sanguinis Christi praesentia Eucharistiae Sacramento *semper habuisse* … Nulla mutatio introducta esse probatur … Mutatio aliqua

how, more than 150 years after the Reformation, the whole inter-confessional struggle of the sixteenth and seventeenth centuries relied on different and competing visions of time and history.[77] This reflexive, historical attitude toward visions of history forming the backbone of an argumentative structure is perhaps the best sign that the whole heated debate about the Eucharist, going for decades, and forming the noisy surface of Ludovician French religious discourse, was, actually, only feeding that increasingly proto-historicist and positivist approach to the dissolution of the confessional itself. Once the Huguenots were expelled and the importance of confessional conflict decreased, the Enlightened age might struggle more with the relativist implications of historicization for religion in general.

Dodwell's point of view – which was the most rigid among the English Church divines – reconstructed the historical (non-)development of the Greek Church almost exactly like the Catholic Jansenists, but from the dogmatic perspective of the anti-transubstantiation real presence doctrine of the Anglicans. However, he also stressed the continuity and unchanging position on this matter while at the same time applying a "modern" empirical historical sense of the development of the church, its branches and its regional diversity. So he handled the framework "Eternity *plus* human History," although his own historical time horizon of the Anglican Church, regarding Mediterranean sources and history, was a reinforced cyclical one as his interest in the Ignatian letters and early Christianity shows. It was a

fieri non potuit ... Nullum mutationis in doctrina aut disciplina vestigium ... Mutatio doctrinae circa Eucharistiam, quam Protestantes supponunt esse impossibilem, ostenditur ex historia" (chapter "An aliqua circa fidem Eucharistiae mutatio in Orientalibus Ecclesiis supponi queat per quam doctrina realis praesentiae introducta fuerit," BNF NAF 7465, f. 152r, 160v, 162r, 162v, 164r).

[77] "Scilicet ubi scribere pro fide catholica adversus haereticos volunt, vindicare antiquam doctrinam id unum spectare et opin\<ari\> debent, ut quidquid antiquarum est originum, exponant accurate, neque ita extra traditionis limites sibi vagandum existiment, ut humanis documentis, non divinis Ecclesiam conentur demonstrare ... Cum superiori saeculo novae haereses exortae sunt eo praetextu, quod Ecclesia Romana defecisset ab antiqua fide, atque adeo reformari in integrum postularet, erat tunc temporis apud Catholicos ignorantia rerum Ecclesiasticarum supinissima ... At erant novi haeretici, a doctrina Ecclesiastica paratissimi, nihil nisi antiquitatem spirabant" ("Disputatio prooemialis ad Opus de fide et institutis Ecclesiarum Orientalium," BNF NAF 7468, p. 638).

longing for the revival of a lost past, also to be found in the heritages and traditions of the Mediterranean shores.

The orientalizing triangulation of the Eucharist debate was a dimension which distinguished the seventeenth-century controversy sharply from its sixteenth-century counterparts. Western Protestants and Catholics now elaborated a distinct history of the Greek Church in their writings. They distinguished the Greek theological tradition from their own. The Greek Church thus became a distinct actor in History. If Western theology had been in a state of nescience about the contents of Eastern Greek theology, the process of specifying and surpassing that ignorance through direct interaction with the church's contemporary leaders, through the collection of manuscripts, through integrating that knowledge into their own theological structures, all led to the construction of a new European confessional system. While the Western churches orientalized their own Eucharist discussion, the Greek Church under Ottoman rule was westernized as a confessional church.

Owing to these developments, direct interest in the Greek Church faded out in the West as far as a belief in an effective merging and alliance might have been concerned. The French interest in Greek theology and in unknown texts or better manuscripts remained strong.[78] But concerning the pressing question of the contemporary teaching and confession of the Greek Church, the situation was already settled in 1672. On the British side, the non-jurors still kept hoping to gain the Greek as well as other smaller Oriental churches as potential allies or sister churches. The Anglicans were in close contact with the early Lutheran Pietists, with the brothers Hiob and Heinrich Wilhelm Ludolf playing a leading part. It is significant that those were minorities in defense in their homeland, as were the non-jurors, or, with the Lutherans, a confession that was strong only in Central Europe, but exceptionally weak in terms of extra-European outreach until that time. Hiob Ludolf, known for his Ethiopian voyage and studies, still imagined around 1700, like the British, the possibility of a (re)union of

[78] For the collaboration between the French (ambassador, school of the enfants de langue . . .) and the Greek ecumenical patriarch concerning those matters, as for instance the purchase of the library of the Prince of Valachie Stefan Cantacuzino cf. letter of Abbé Sévin, Constantinople, December 12, 1728, AN AE B III 24, list of the Greek manuscripts (March 28, 1733, AN AE B I 407). Cf. Pippidi, *Tradiţia*, 340f.; Bacqué-Grammont et al., *Représentants permanents*, 31f.; Hitzel (ed.), *Istanbul*.

the Protestant churches with the Christian Levant churches. In 1691, he published the *Fasti Sacri Ecclesiae Alexandrinae* and sent it to Louis Piques (1637–1699), a work containing an irenic-unionist dialogue.[79] Piques' response of December 1698 shows well how the orientalizing debate about the Eucharist had refined the French understanding of the processes of specifying ignorance through queries, and interrogations of the different Christian groups in the Levant about their confessions; it had produced cognitive skepticism. In reaction to Ludolf's account of the questioning of an Armenian Murat who was, in Ludolf's rendering, likely to believe a Protestant-like version of the Eucharist, Piques began with something akin to a methodology of religious ethnology and query-guided field studies. The student of religion should employ, Piques maintained, the *critical nose* that Vossius had recommended for philological studies. He should not just indifferently ask the first person he encountered, but someone "professional" – so, a theologian –, and not only one, but many. Moreover he should not question them through elaborate interrogations in the way of sophists. Rather, he argued, one had to ask them about their practices through a narrative. If they had books such as catechisms and liturgies, one should endeavor to acquire samples. This was all because an "artificial judge can render every innocent man criminal and worthy of capital punishment."[80] This went beyond the usual recommendations of *apodemica* literature and even the contemporary queries published by the *Royal Society* between 1666 and 1690, more concerning scientific objects than religion, and it did not contain similar reflections on the malleability of the answers caused by the want and needs of the one who asks. It is one of those early hermeneutical reflections about what we might call the

---

[79] Ludolf, *Fasti Sacri* (1691). Cf. UB Frankfurt/M Nachlass Ludolf Nr. 592. On Piques cf. Richard, "Un érudit."

[80] "[If one wants] employer *Nasum Criticum* pour entrer en connoissance de la Religion d'un pays j'estime qu'il ne la faut pas tirer d'un entretien avec le premier venu, mais qu'il faut s'addresser à un homme de la profession, et mesmes à plusieurs consecutivement, ne pas mesme les surprendre par des interrogations estudiées à la mode des sophistes. Il faut leur demander leurs pratiques par forme de récit, et s'ils ont des livres comme Cathechismes, liturgies, qu'ils ayent à nous en instruire simplement, comme si nous voulions estre prests de suivre leurs créance et leurs pratiques. Cette conduitte me semble innocente, et nullement insidieuse car vous sçavés qu'un juge artificieux peut rendre l'homme le plus innocent, on le peut rendre dis-je criminel et digne de mort." (Louis Piques to Hiob Ludolf, Paris, November 12, 1698, pres. November 10, UB Frankfurt/M Nachlass Ludolf Nr. 592).

Heisenberg problem of ethnology or cultural contact, that the act of observing and the observer with his motives changes the observed – on both sides. Piques reflected on both of those problems. The process of observation was biased because of the researchers' aims to use the Levantine Christians' confessions within Western European debates. Piques gives the example of the famous German traveler Wansleben, who first gave an account of the religion in Egypt from a Protestant point of view and then, after his conversion and entering into the service of Colbert, from a Catholic point of view. Piques was also thinking of the other problem, much discussed during the Western Eucharist debate, that the Levantine Christians, after having had contact with missionaries, had already changed their own beliefs and practices. This was an obvious fact for some Latinizing representatives of the Greek Church, but it also applied to all other churches. Through a critical philological scrutiny of Ludolf's accounts of interviewed Levant Christians, Piques attempted to show that some specific Latin words were highly unlikely to have been said spontaneously by those interrogated. This was especially the case, he argued, when the interrogated persons emphasized the uninterrupted continuity between their belief and religious practice and that of the most ancient times, or when precise Western terminology was applied which made use of minute and intricate theological distinctions, such as the differences between transubstantiation, consubstantiation and impanation. How likely was it that an ignorant – as they represented the interrogated – would have used that sophisticated language? Rather, the succession of (dis)informing communication nodes – the use of different languages, use of dragomen, the indirect transmission of what was said through narratives by a traveler to a Western theologian who, in turn published it in Latin – was prone to many errors.[81] He also demonstrated that some of Ludolf's translations from the Levant languages were highly biased by Ludolf's purposes. Ludolf, he argued, had made a perfect Calvinist of his Gregoire and of his Murat by suggesting that their

---

[81] "Vous me renvoyés à vos remarques sur les responses du Chogia Murath et vous me reiterés qu'il estoit un parfait ignorant ne sçachant ni lire ni escrire. Mais la teneur de ses responses ne marque point un homme ignorant: Paul de Roo luy a proposé vos questions. *Responsiones diligentissime excepit* (qua lingua, quo interprete ?) *belgice vertit, latine hic traditae leguntur.* Surquoy faittes vous tomber son ignorance? Je croirois pour moy que Paul de Roo, son truchemant et d'autres ont beaucoup profitté de son Ignorance" (ibid.).

answers conformed to a symbolic understanding of the consecration enunciation (*Hoc est corpus meum*). Piques finally invented a fictitious dialogue in Paris about Ludolf's works in which one of the participants remarked that, if Ludolf had conducted his investigations on religious matters not in Ethiopia, but in the Russian port city of Archangelsk, his results would have been the same: all would profess that which he wanted them to profess.

So, not only had the confessional situation in France stabilized by 1700 and interest in the Greek Church had transformed from searching for a tool within the inner-French debate, into an object of purely academic interest. The high tension of the debate under the new conditions of pre- or early Enlightenment scholarship had led to a hermeneutical skepticism even at the Sorbonne. Believing that the Levant churches "were just like ours" was an act of wishful thinking. Like the officially Lutheran Ludolf, the British still aired such projects for some time. One prominent example is the attempt, started in 1716, by the three to five[82] non-juror bishops Jeremy Collier, Archibald Campbell, Jacob Gadderer, John Griffin and Thomas Brett to submit the non-juror English Church to the Patriarch of Jerusalem. It seems that the non-jurors asked again for responses to their dogmatic questions not only in Constantinople, but, as in 1672, all over the Mediterranean.[83] The attempt failed and showed finally that Dodwell had been right. The Greeks answered by copying and repeating literally the confession text of the 1672 Bethlehem synod, stressing their doctrine of the μετουσίωσις and explicitly citing for that the strong opponent of Kyrillos Lukaris, Meletios Syrigos. While it is surprising how willfully the non-jurors tried to submit themselves to the faraway

---

[82]　Only Collier, Campbell and Griffin continued the enterprise until 1725, also involving the Russian Church and Peter the Great.

[83]　John Covel asked Humfrey Wanley to inquire again among the Greek clergy about transubstantiation, now referring to the early "canonical" use by Gennadios Scholarios recently edited by Eusèbe Renaudot. Arsenios, Metropolitan of Thebes, answered they did not know the term μετουσίωσις: "They said that in the Blessed Sacrament of the Eucharist, they receive Christ's Body πνευματικῶς, and declared that they would give me a Certificate of such their Belief when I should require it" – but they handed him over only citations from Chrysostome and Basil (Wanley to Covel, December 21, 1714, BL Ms. Add. 22911, f. 163f.). Cf. also a reference to a manuscript containing the Ἀποκρισείς ... πρὸς τὰς ἀπὸ Βρετάννιας ἀποσταλείσας of the metropolit Hierotheos of Dristra (not dated, but after 1709, Library of the Rumenian Academy of Sciences, Pippidi, *Byzantins*, 258).

Greek jurisdiction in Jerusalem, conceding several important points, they had to remain adamant on the crucial article on the Lord's Supper: real presence, but no transubstantiation. No compromise could be possible on that matter.[84] For more than a century, no real new attempt at union or alliance took place, while both sides considered that they "knew" each other now.

No Unknowns: The French St. Louis Cult on Mediterranean Shores

The history of how the French and the British coped with ignorance regarding Mediterranean theologies and religious practices has some further *strata* and dimensions beyond the very concrete exchange with the Greek Church and its very realistic political dimensions. But those matters to which I will return shortly were of higher importance for the British than for the French. It was already mostly the reformed Catholic wing of the Port-Royal theologians and Catholics close to Jansenism, not the orthodox center that was using the French imperial infrastructure. I do not want to establish here a general "epistemic rule" that it is those who are "discursive minorities" in an (informal) imperial center who use the empire's contacts and periphery to strengthen their position in the heartland. Nevertheless, this seems to be at least one of the typical situations that starts the engine of such a non-knowledge-cycle. In the following examples it is likewise a demand created by a lack, an awareness of apparently needing something, which is the starting point: the partial lack of religious, ecclesiastical and also national foundations. As this was more the case with the British, one has to ask why it was not so for the French. By drawing a short illustrative sketch of the French imposition of royal Catholicity on Mediterranean shores, I will try to answer that systematical question by showing how the French were partially immune or saturated in that regard. This does

[84] Mansi 37, 387f. for the consubstantial understanding of real presence in the propositions offered to the Greek in 1716; the Greeks' response, being a copy of the declaration of Dionysios IV given to Nointel and Covel in 1672, has the explicit denial of Kyrillos' teaching on col. 401f., the transubstantiation doctrine on col. 443f., but with μεταβάλλεσθαι as the term with reference to Syrigos; the second synodal declaration of 1691 under Patriarch Kallinikos of Constantinople, which the Greeks also sent to the non-jurors, now had the firm μετουσίωσις, ibid., col. 465, 467, the major issue of the non-jurors' 1722 response was the Lord's Supper, ibid., col. 471–492.

not mean that they did not use the network of consuls, merchants and missionary friars for purposes of manuscript collection and proto-Orientalist studies in a very competitive if not superior form to that of the English. Nonetheless, in terms of the inner-French functions of that work of erudite collection, it was far more detached from effects of cohesion, identity building and similar forms than in the English case.

Concentrating on aspects of the largely official early modern French cult and veneration of the king St. Louis, who had died in 1270 during the Seventh Crusade on the coast of Tunis, serves to exemplify something like a cognitive closure and why there was no or less awareness of ignorance concerning the different forms of Oriental religion in the Mediterranean that had direct bearing on France's own demands and needs. In nineteenth-century Maghreb, it is known how a renewal of the cult of St. Louis took place after the French colonization of Algeria – the building of a *chapelle de Saint-Louis* in 1841, archaeological research on the crusade, the institution of memorial days and celebrations – in the context of French nationalist ideology and even the racist construction of a "Latin Mediterranean."[85] One may add that the major forms and structures of that veneration had already been created in the seventeenth century as an integral part of Ludovician cultural politics.[86] The greater number, if not all, of the consular chapels in the Levant seem to have been devoted to St. Louis.[87] This was quite logical, as Louis was the only saint king of France and all churches linked with France as a proto-nation were usually consecrated to that patron such as, most prominently, the French nation's church in Rome, San Luigi, home church of the cardinal *prottetore della nazione francese*.[88] During the reign of Louis XIV however, the politics of "Ludovician branding" discovered the chapels as a convenient place to blend the *gloire* of the current monarch with the holiness of the former. A very active consul in the field of religo-cultural politics like Claude Lemaire

---

[85] Delattre, *Souvenirs*; Dupront, *Le mythe*, vol. 2, 903–929.

[86] Cf. for the reign of Louis XIV Dupront, *Le mythe*, vol. 1, 502–518 and 399–430; vol. 2, 817f. Sometimes Dupront's work has to be read with care as there is a tendency to count every use of epithets like "saint," "sacred" in relationship to war and military enterprises as a revival of crusader myths.

[87] At least I have not encountered a different saint's name for a consular chapel in the consular correspondence.

[88] The strategic choice of the name becomes evident there where the consul's chapel was somehow in pastoral competition with churches and chapels under more direct rule of the *Congregatio de propaganda fide*, cf. Ladjili, "Paroisse."

worked for the renovation and partial rebuilding of the chapels of all the consulates of his career, such as at Tripoli de Barbarie and in Cairo.[89] The revitalization of the St. Louis cult and the new ornamentation of the churches served as the exposition of a specific French version of Catholicism on the Mediterranean shores – also in mutual competition with the other nations.[90] The day of St. Louis was celebrated every year in every consular chapel of the Mediterranean and the merchant's nation of the localities reserved a certain amount of money for its support.[91] Occasions for further blendings of the memory of the medieval St. Louis with the glorious Louis le Grand were typically births and deaths within the royal family. For example, when the Duke of Brittany, son of Louis, Duke of Burgundy, and Marie-Adélaide of Savoy, therefore the great-grandson of the reigning Louis XIV, was born in 1704 (June 25), Seignelay's successor as secrétaire d'Etat de la Marine, Pontchartrain (1643–1727), issued an order to all consuls and vice-consuls to hold official celebrations in all their locations. The very diligent consul of Seyde, Jean Baptiste I Estelle,[92] for example, used that order to prepare a huge event in 1705, preferring the spring time to the winter and therefore having some months to prepare it. The celebration was held on April 29, 1705, ironically sixteen days after the baby had already died at Versailles, but news of that tragic death did not arrive in Seyde before June, and as a manifestation of French glory, the event was nevertheless successful. The consul negotiated all necessary permissions with the local Ottoman governor, Mehmet Pasha and the Divan. It was a huge spectacle with specially-built triumphal arches, processions, portraits of the royal family

---

[89]   Cf. for that AN B I 317, f. 147v (Lemaire to Pontchartrain, Cairo, October 10, 1713), f. 189–194 (March 7, 1714: letter with list of costs for the renovation of the chapel that was paid for first by Lemaire himself).

[90]   For Lemaire's care for the chapel cf. AN B I 317, f. 199v, 245r; AN B I 318 (February 26, 1719); AN B I 319 (February 18, 1720); for Tripoli cf. AN B I 1089 (1704) until 1090 (1707, recurring theme in many letters: building of a St. Louis hospital), AN B I 1093 (consul Raimondis, December 6, 1729); for a St. Louis chapel in Oran cf. AN B I 928, f. 116; for disputes between the Minorite friars and the French Capuchins about the building of a new St. Louis church in Pera/Constantinople, cf. AN B VII 466, f. 130v. Many decisions of the secrétaire d'état concerned the consular chapels, cf. for that the minutes in AN B III 1 and 2 for the years 1696 to 1736.

[91]   Cf. the decision letter in AN B III 2, p. 86 (November 30, 1735): the contribution of the nation for the St. Louis day in Tripoli, Barbary is reduced to fifteen piastres.

[92]   He had been consul in Morocco before, cf. Caillé, "Jean-Baptiste Estelle."

erected on pedestals representing royalty abroad, Latin inscriptions on flags and banners, *fleurs de lys* all over the place, fireworks around a royal obelisk lit up by 400 lights, and, certainly, with discourses and sermons.[93] The consul described the field framed by three buildings on the shore – among them the consul's house and the hospital of the fathers of the Holy Land – as resembling a ship coming from the Christian shores controlled by a fortress said to be built by St. Louis, so the holy king was virtually watching all over the ceremony.[94] The central sermon by the Jesuit Grosset – a copy was sent with the consular dispatches to Versailles – used all the ingredients of Ludovician panegyrics:[95] the whole celebration and the prayers at Seyde were envisioned as uniting the French nation on the Mediterranean shore of "Asia" where they were, across the sea with all the other citizens in France. Grosset underlined, as usual in those times of the War of Spanish Succession, how much "heaven loved" the king. The sermon, given in a small fortress in Seyde said to have been built by St. Louis in the Middle Ages, did not hesitate to adopt an aggressive militarist tone: the God to whom they prayed here in 1705 for the birth of the heir to the throne was the "God of the armies" who protected Louis le Grand. Praise of the worldly deeds of the king culminated in the idea of a possible unification of the "two Empires" of Europe after the extinction of the Holy Roman Empire, the foundation of the "universal monarchy." The sermon went on to add aspects of piety and even holiness to the king's person and aura. By the extermination and expulsion of the Huguenots, the king allegedly has shown himself worthy of his glorious title of "Roy très Chretien," his military successes demonstrated that heaven had "destined" him and had made "great designs" for him to something great. The pious king would reform manners, rebuild the temples of God after the expulsion of

---

93 Cf. the report letter to Pontchartrain by the consul Estelle, Seyde, May 22, 1705, AN B I 1018, p. 86f. and the enclosed "Relation des Réjouissances faites par la nation françoise establie à Seyde à l'honneur de la naissance de Monseigneur le duc de Bretagne le Mercredy 29 avril 1705," ibid., pp. 113–126 – the extraordinary celebration was so expensive that the day of Jeanne d'Arc had to be cancelled.

94 AN AE B I 1018, p. 116.

95 "Discours Prononcé sur la Naissance de Monseigneur le Duc de Bretagne par le Reverand père Grosset de la Compagnie de Jesus en présance de Monsieur Estelle Consul de France et de toute la Nation à Seyde le 29e avril 1705," ibid., pp. 88–112.

heretics, a monarch greater than Solomon and Titus. Grosset, across the Mediterranean Sea, reminded the prince that "there is no true greatness than in holiness, that all other greatness outside of holiness has only the name and the external appearance of greatness,"[96] and so his sermon culminated in the invocation of holiness as an aim of royal activity, recalling St. Louis as the prince's precursor: "Must I give you an example, my prince? An example which to imitate you will not be ashamed is Saint Louis," St. Louis as a king who understood the matters of war better than any other hero of his time, and who remained in the "memory of mankind" not only because of his military actions – that would be a glory which he had "in common with so many other kings," but because he has ascended to a "holiness" that only a few kings share with him. He therefore wished that the newborn prince might unify "the holiness of St. Louis, the significance of Charles Martel, the successes of Charlemagne, the courage of Henri IV, the piety of Louis the Just and the felicity of Louis le Grand ... to live forever in the memory of men on this world and to reign in eternity in the heavenly Jerusalem with all the saintly kings."[97] It is remarkable that in this way, under Ottoman rule, under the eyes and with the permission of an Ottoman Pasha, and even with the participation of Turks interested in the spectacle, the French could openly refer to their crusader heritage,[98] blending their current Louis with the Christian precursor of the Pasha as governor of the city and celebrating Christianity, in a quite martial and even imperial manner, as a part of the monarch's rights to universal monarchy – in competition with the other European powers, but implicitly by that also in the Mediterranean.[99] It gives us a sense of how mightily the French played that card of baroque royal cultural politics through the medium of their

[96] "Mais vous vous souviendrés, mon Prince, quil n'y a de véritable grandeur que dans la sainteté, et que toute autre Grandeur hors de la sainteté n'a de la grandeur que le nom et l'aparence" (ibid., p. 109).

[97] AN AE B I 1018, p. 112.

[98] For crusader reminiscences in the context of Louis' participation in the military campaign against the Ottomans on the side of Emperor Leopold in 1663 cf. Rousset, *Louvois*, t. I, 35–57.

[99] Cf. for some other examples AN AE B I 317, f. 57–63 ("Relation des obsèques celebrés au Caire dans la Chapelle consulaire le 29e octobre 1712 pour feus Monseigneur le Dauphin et Madame la Dauphine par les soins de Mr Lemaire"); f. 410–415 (letter indicating the reception of the news of Louis XIV' death); AN AE B I 1116, f. 294ff (letter of August 21, 1729 with report about the celebration of the Dauphin's birth).

consular network, chaplains and French *religieux* in the Mediterranean. The importance of the king's "self-fashioning" has long been recognized for the internal affairs of the kingdom,[100] but it was also part of a well-orchestrated foreign cultural politics. Moreover, this card could not have been, nor was it ever, played in that way by the other European powers, whose bilateral capitulations with the Ottomans did not contain the protection of Christian faith in the same way as a part of the French state's presence in the Mediterranean. To my knowledge, no English consul or Anglican chaplain referred to Richard Lionheart in such a way, nor could or did the Dutch or the Swedish depict their Republican or royal sovereignty in a similar sacralized manner as the French regularly did, always using that aspect of Ludovician holiness.

This can only be an illustrative sketch, but it serves to explain why the medievalist reinvention of a sacral Ludovician continuity within a more or less unchanged Catholicism – "perpetual" as the title of Nicole and Arnauld's work – meant a certain self-sufficiency in terms of religious identification. Yes: Contact with and research into the Oriental churches for missionary purposes, but there was far less of a sense of lacking knowledge and no impulse for gathering new knowledge within royal Catholicity. A history of ignorance has also to take seriously situations and moments of cognitive closure, of negative knowledge or of non-perception of ignorance.

Meet the Unknown Past in the Present: The Samaritans

The non-knowledge cycle observed in the case of the Eastern Greek Church was reinforced by a two-sided inner-European demand relating to the Eucharist debate and by the possibility of finding an ally in the most important Christian church in the Orient. This might have been an important factor, especially for the Anglican non-jurors.

For the many other smaller communities in the East, including the Copts,[101] the Melkite and the Maronite churches or denominations,[102] the hope of confessional alliance could not play a serious political role. Their significance was different. Either as object of missionary work and "reunion with the church" – the Catholic version –, or, and systematically prior, as an object of investigation following the framework of

---

[100] Burke, *Fabrication*.   [101] Hamilton, *Copts*.
[102] Heyberger, *Les chrétiens*.

each confession's own theologico-historical conception. I remain here with the English case as the "new-comers" in the Mediterranean – meaning the relatively late arrival of the English merchants, and appointment of the English ambassador at Constantinople in 1578/80. Protestants sometimes had very different interests than the French Catholics. The "hunt" for the best manuscripts of the Ignatian letters by Levant Company chaplains, for instance, was linked to the English debate between Presbyterians and Episcopalists. Their so-called middle recension had been edited in 1640/45 by Ussher and served for the Anglican Church during the Civil Wars and the Restoration as proof of the very old roots of a strong episcopal hierarchy in the early Christian church without and before the absolute supremacy of a pope.[103]

Also beyond the times of the already-established early Christian church, the refined seventeenth-century framework of the Anglicans' cyclical concept of history served to sharpen and crystallize the borders between knowns and unknowns, to specify the ignored. The fascination that the Samaritans exerted on the Europeans, and in the late seventeenth century mostly on the Anglicans, is a good example to show another type of religious non-knowledge cycle. This one was motivated by a general desire to discover the earliest times and the most authentic transmission of God's word, and by that the very borders of the overall historical framework. Its outcome also served to question the foundations of that historical concept itself.[104]

Interest in the Samaritans, mentioned several times in the Old and New Testaments as a religious community different from the Jewish orthodoxy, focused firstly on their language.[105] In the general quest to determine the number, similarities and parentage of the different human languages following the punishment inflicted for the Tower of Babel, the Samaritan alphabet was of major importance because it was considered to be very close to "palaeo-Hebrew." But beyond the question of language, the Samaritan community as a whole, and its specific

---

[103] Cf. Smith (ed.), *Roberti Huntingtoni epistolae* (1704), 20, 24, 25, 32, 34. On the importance of the Ignatian letters cf. de Quehen, "Politics and Scholarship"; Thomas Smith was working on a new edition of the letters which strongly occupied him after July 1708, cf. Harmsen, "Antiquarian Publishing," 9; Smith, *Ignatii epistolae* (1709); Quantin, *Church*, 166, 267, 284, 321, 340f.

[104] For the first generation of European scholars who were interested in the Samaritans, cf. Miller, "A philologist."

[105] Edward Bernard to John Lightfoot, St. John's College, Oxford, March 5, 1673/4, in: Lightfood, *Whole works*, vol. 13 (1824), n. 74, p. 452f.

kind of worship and veneration of the law since the times of the Israelite establishment in Canaan after the Egyptian captivity (Exodus 19ff), was of great interest because it might have served to increase the existing knowledge about the general starting points of Judaeo-Christian religion in its forms constituted by God's law as revealed to Moses. The idea, contested, but still maintained during the eighteenth century, was that the Samaritan Pentateuch could have been the earliest text version of Moses' books not written in Greek by diaspora Jews like the *Septuaginta* but close to Moses' lifetime, and perhaps closer to the original Hebrew text than the current Hebrew vulgate version.[106]

Unfortunately, not much was known about the Samaritans as a distinct religious community, which could have lent insight into Jewish culture prior to the Babylonian captivity. It was therefore a little sensation when, around 1630, Pietro della Valle (1586–1652) first gave an account of his (re-)discovery of a still-existing group of Samaritans that lived at Sichem near Mount Gerizim. After the 1670s, the Anglicans and the Pietist Lutherans became interested once again in the issue, commencing first of all with Robert Huntington[107] and subsequently with his successor as chaplain in Aleppo, William Hallifax.

Huntington traveled to Sichem in 1672. He estimated the group to consist of just thirty families.[108] According to Scaliger and Pietro della Valle, they had previously also lived in Damascus and in Cairo.[109] He obtained a letter from those, "who live in Sichem [addressed] to their brothers in England."[110] As he later wrote to Ludolf, this letter was written by Merchib Ibn Yacob, called Mopherrege, their leader, and he

---

[106] Morin, *Exercitationes* (1631); Hottinger, *Exercitationes* (1644); and still Wells, *Specimen* (1720); Boberg and Ericander, `Al-leson we-torat has-somronim* (1734).

[107] Toomer, *Eastern Wisedome*, 284f.; de Sacy (ed.), *Correspondance*.

[108] Bernard to Lightfoot, March 5, 1673/74, Lightfood: *Whole works*, vol. 13, p. 453.

[109] Smith (ed.), *Roberti Huntingtoni epistolae* (1704), 47.

[110] Cf. a copy of Edward Bernard's Latin translation in the Bodleian, Ms. Smith 28, f. 9r-12; later printed by Ludolf, *Epistolae Samaritanae* (1688), 26–32. It is Bernard's, not Ludolf's translation (contra Sacy, *Correspondance*, 175 n. 1). It seems that Ludolf mistrusted that translation because he insisted for a long time on acquiring the original from Huntington, which was finally transmitted by the chaplain to the ambassador in the Netherlands, Edward Smith (The Haag 21/31 May, 1695, pres. Frankfurt July 3, 1695, UB Frankfurt Nachlass Ludolf, I Nr. 643).

sent that letter to Thomas Marshall, the English Oriental and Gothic scholar who had, in 1672, returned to Oxford as rector of Lincoln College, after more than twenty years from his service as chaplain to the Merchant Adventurers in the Netherlands.[111] It must have been quite an experience of "otherness" for Marshall and Edward Bernard when they worked in Oxford on Latin translations of the Samaritan letter, as well as of the Samaritan Pentateuch that Huntington had sent, together with over 150 Arabic and Hebrew manuscripts from the Levant. Bernard sent a copy of those translations to John Lightfoot.[112] Lightfoot was delighted to have what he understood as the "Samaritane Confession of their faith & Religion."[113] From the result of that work of translation, the English learned men understood that the Samaritans described themselves to their English addressees as "sons of Israel" who followed the profession of the prophet Moses. They started describing how they honored the Sabbath, how they celebrated the seven days' Paschal feast, "at what time our ancestors had left Egypt" (Ex 12:1–13,16), which culminated in a night service around Mount Gerizim. The ritual is described in detail, and small differences from those of the Jews are noted, for example the day of circumcision. They confessed to believe in the Lord, in Moses, the Law and in Mount Gerizim to which they always turned for prayer.[114] Their priests, they maintained, stemmed from the sons of Levi and from the sons of Aaron and Phineas, first mentioned in the Bible in Ex 6:25 as son of Eleazar, while the Jews did not recognize any priest from the "stirp[s] Pinchasica." They lived in the "holy city Sichem" and in Gaza. The Samaritans had a holy scripture, a *Codex Legis*, written in the holy Hebrew language which was inscribed "I Abisha, son of Pinchasus, son of Eleazar, son of Aaron the priest have written this holy scripture at

---

[111] Huntington to Ludolf, March 31, 1690, Smith (ed.), *Roberti Huntingtoni epistolae* (1704), n. 33, 46–53; ibid., n. 35, 55–57; Hamilton, *Copts*, 216.

[112] Bernard to Lightfoot, March 5, 1673/4, Lightfood, *Whole works*, vol. 13, 453.

[113] Lightfoot to Bernard, Much-Munden, April 29, 1674, Lightfood, *Whole works*, vol. 13, n. 75, p. 454f. Later Bernard even sent the original Samaritan letter and Lightfoot corrected some errors and omissions of Bernard's Latin translations of it, Lightfoot to Bernard, Much-Munden, s.d., Bodleian, Ms. Smith 45, f. 89r.

[114] This is one of the passages where Lightfoot in 1674 corrected Bernard's translation, printed by Ludolf, but only in 1808 did de Sacy correct that again, independently (Lightfoot to Bernard, Much-Munden, s.d., Lightfood, *Whole works*, vol. 13, 456–458. Sacy, *Correspondance*, 179 n. 3).

the door of the temple in the thirteenth year of the dwelling of the sons of Israel in the land of Canaan close to its borders."[115] This was the so-called Samaritan Pentateuch. The Bible does not mention a son of Phineas called Abisha, but because of this the seventeenth-century Samaritans claimed to possess a copy or a direct textual transmission of the Pentateuch from the hand of a great-grand-nephew of Moses himself.

The Samaritans themselves formulated several questions for their English correspondence partners which reveal much about their world-view, apparently characteristic for a quite isolated religious community that cultivated some specific convictions and traditions, such as hostility to the Jews. Their memory was organized in terms of a genealogical succession of ancestors and prophets, buried at Mount Gerizim. Their written tradition on this world seems to have been restricted to a small number of books. They asked, for instance, if the English could send them a copy of the "liber Joshuae Ben Nun." They still followed a calendar organized by lunar years.

In response, from Oxford Thomas Marshall wrote a letter to the Samaritans in Hebrew, starting with the creation of the world according to the books of Genesis, Deuteronomy, the book of Numbers, original sin and the expulsion of man from Paradise. By interpreting Levi 22:10 f. allegorically, he wanted to suggest that in celebrating the Sabbath on Sunday rather than Saturday, the Anglicans were close to the Samaritans and both were distinct from the Jews, the religious group and ethnicity to which the Samaritans were most related but from which they separated themselves strictly, recognizing them as their "enemies" as if the millennia since the split between the tribes had not changed anything.[116] In emulating a language supposedly understandable to the Samaritans, Marshall directly answered the questions they had posed: "We possess the book of Joshua, son of Nun, written in the holy language, we possess also a book of hymns and prayers and other books which have been written by the prophets of Jehova . . . We are the children of Japhet . . . we have also a great king powerful on land and sea. We possess an island of nations which is

---

115 "Ego Abishaa filius Pinchasi filius Eleazari filius Aaronis sacerdotis, scripsi hoc scriptum sacrum in ostio tabernaculi conventus anno 13 tio Commorationis filiorum Israel in terra Canaan circa limites ejus" (*Epistolae Samaritanae* (1687), 32 = Bodleian, Ms. Smith 28, p. 12a).
116 Sacy, *Correspondance*, 196.

called England." Marshall did not state that "they" are Christians, he just tried to find their shared Old Testament elements of faith and asks again for copies of their book of law in Hebrew or in Arabic, and he ended with an inquiry into their understanding of the Old Testament's prophecies about the Messiah.[117] The Samaritans' answer about the critical point on the Messiah was that they were still waiting for him to come. In addition, the Samaritans asked for some alms for the maintenance of the graves of the prophets at Mount Gerizim.[118] Others followed Huntington's interest in the Samaritans, such as Abraham Hough[119] and William Hallifax, Huntington's successor as chaplain in Aleppo.[120]

The English found themselves in an astonishing exchange with a people who claimed to live in an uninterrupted tradition from the time of Moses, different from the "normal" Jews dispersed after the destruction of the Temple of Jerusalem. Nevertheless, they applied the methods of historical proof to the matter. Huntington "wanted to see that most important monument of such a most remote antiquity," the supposed book of law written by Abisha. He returned five years later, in 1678, to Naplôsa and again looked up "this highly venerable codex in their synagoge." He had first judged the scripture to be about 500 years old.[121] But he returned and asked to be assured that he was really looking at the alleged original. They assured him it was so, adding without any sign of shame that the indication of the name and place of the scripture cited in their first letter to the English ("Ego Abisha ... ") had once been written on it. However, as Huntington inspected the codex, he could find no such inscription. He estimated its origin to be the year 756 of the Hegira, which would be 1355 CE.[122] This was quite disillusioning; the material testimonies of a people that claimed an uninterrupted history of 3,300 years,[123] were manuscripts that the seventeenth-century scholars dated to the fourteenth or even – for the

---

117 Sacy, *Correspondance*, 192–197, 197.
118 Sacy, *Correspondance*, 203–211.
119 Hough to Huntington: "Of ye Samaritanes now upon mount Gerazim" (1677), copy by John Covel, BL Ms. Add. 22911, f. 380r.
120 Hallifax to Smith, Aleppo, December 7, 1694, Bodleian, Ms. Smith 45, f. 79r.
121 Huntington to Ludolf, March 31, 1690, Smith (ed.), *Roberti Huntingtoni epistolae*, 48.
122 Huntington to Ludolf die Ascensionis Domini 1695, Smith (ed.), *Roberti Huntingtoni epistolae*, 55–57, 56.
123 BL Ms. Add. 22911, f. 380r.

Chronicon Abulphetachi Danefaei – to the end of the fifteenth century, and its apograph in Huntington's library was dated to the year 1596. The European scholars were also disappointed by the poor quality of the Samaritan computation.[124]

The cognitive effect of those decades of encountering the Samaritans was thus, to put it in emotional terms, a disappointment. More generally, however, it shows how the potential of empirical processes to cause the erosion of guiding assumptions was realized. It remained unclear what precise status these written testimonies, the Samaritan Pentateuchs, their letters in Hebrew, written in Samaritan characters, or in proper Samaritan dialect, might have, or how old this textual tradition really was. The very status of that group of "Samaritans" living around 1700 at Sichem was also unclear. Their historical link to the Biblical past faded out beyond the fifteenth century and what they declared in letters and orally could hardly fit within a European concept of history already shaped by the structure of uninterrupted teleological sequences of events, reigns and facts. Where were the meagre elements noted in the Samaritan chronology to be integrated here, and how to interpret the huge portions of void between one entry and the next if compared to the contemporary developments of Jewish, Christian, Arabic and Ottoman history? This led to doubts as to whether the Samaritans had always continued their existence as historical actors since the times of Moses. As a result, the English left much of this as an unsolved case; the specification of ignorance about the Samaritans had not led to an answer in the end, but to the erosion of the specifying assumption itself that there was "better knowledge" about the origins of Biblical times to be found in Palestine. Void was gained rather than filled. And experiences like this further corroded evidence of the framework of Biblical time and of the coherence of religious groups and

---

[124] Cf. the "Chronologiae Samaritane Synopsis a Cl. Edvardo Bernardo ex Manuscriptis eruta & Oxonio transmissa, Pinacion primum," in: *Acta eruditorum* (Leipzig), Mensis Aprilis 1691, 167–173, 171f. (a copy hereof with some corrections by Dodwell in BL Ms. Add. 21081, f. 161f.). Independently from the Leipzig print of that chronology, Huntington had sent fragments of it to Louis Piques who also corrected the Samaritan chronicle author to confound solar years with lunar ones for a part of the calculus (Louis Piques to Hiob Ludolf, July 22, 1691, UB Frankfurt Ms. Ludolf I Nr. 573): "Voilà d'habilles gens que ces samaritains tant vantés par nos sçavants idolatres des Nouveautés et des descouvertes bonnes ou mauvvaises si ils avoient veu ce que j'ay vû sur cela ... le compte des samaritains est faux et absurde."

theologoumena across centuries. While searching for proofs of the original roots and traces of *the* church of God within the regions currently under Ottoman rule, Europeans not only found, first of all, a plurality of religions and denominations, but experienced the limits of such an empirical search itself, of the knowability of *the* church.

Projecting European Nations into the Void of Ur-Religions

Traveling in the land of Canaan, or sitting in a rainy small town in Hertfordshire and longing to do so, receiving with great excitement handwritten Samaritan correspondence as if letters from the time of the Old Testament[125] – like Petrarch writing longing letters to Cicero and Virgil: this was a feeling that framed research into the unknowns of religious and cultural past. A third step beyond research into the present state of the Greek Church or into the Samaritan culture as eventual remains of the earliest biblical times was the search for civilizational roots even "beyond" Biblical times. The central motif was here the myth of a Phoenician-Druidic civilization which provided a link between the Mediterranean and the British. This shows how that ongoing hunt for religious roots and connections built into the unknowns of the present and past even went beyond the normal limits of Judeo-Christian horizons. While France had its own strong tradition of Druidism during the Renaissance, relatively speaking, one must stress that around 1700, "proto-national" French discourses referred to Druidic traditions much less frequently than the British. It is not by accident that the most famous French author to be encountered in that context is the *Huguenot* pastor of Caen, Samuel Bochart. Royal Catholic France could call upon a different array of historical traditions and underpinnings.[126] There was no such urgent need to find a pre-historical link between the Mediterranean and French history, as

---

125   "I cannot tell what to do to Mr. Huntington, whether more to honour him or to envy him … But I could half find in my heart to envy him for this, for that he hath the ocular view of those places in the land of Canaan, … as I have sitten here." (John Lightfoot to Edward Bernard, Much-Munden, April 29, 1674, letter LXXV, in: Lightfood, *Whole works*, vol. 13, 454f.).

126   The French Renaissance tradition of Druidism seems to subside after about 1620 (for authors like Pierre de la Ramée, Sébastian Rouillard, Jean Picard, Gosselin, Simon Dupleix cf. Asher, *National myths*; Pezron, *L'Antiquité* (1687) for example remains all in all concentrated on Judaeo-Christian sources).

France could claim an uninterrupted tradition of exchange with and proximity to the Orient.

This was different for the British. That a man like Marshall studied Gothic and Icelandic as well as Arabic, Hebrew and, as we have seen, Samaritan – languages which we would consider today as having no particularly common qualities – was not untypical.[127] At the end of the seventeenth century, the propinquity of languages was a major topic of educated discussion. Proper names were often considered to be of particular interest as they might point to civilizational roots beyond written historical narratives. One of the most remarkable themes was the identification of the British Druids with the Phoenicians and Eastern religious culture in general. Modern scholarship typically associates Druidism with Northern traditions.[128] In discussions around 1700, however, it was in fact an "Oriental," Mediterranean reference.

Until the seventeenth century, publications about Druidism usually consisted of just short notes and passages – Ludovico Vives' commentary on Augustine's *De civitate Dei* for example. Only with the large work *Geographia sacra* (1646) of Samuel Bochart did interest shift to a new level. The first publications dedicated exclusively to the history of Druids were written by Edmund Dickinson and Thomas Smith, both Orientalists and the latter also a chaplain of the Levant Company in Constantinople.[129] From the middle of the seventeenth century, the amount of paper filled by and printed with descriptions and histories of both, Druids and Phoenicians, exploded. The discrepancy between the amount of ignorance and the speculations about how to specify or to transform that into knowledge was much higher here than for the case of the Greek Orthodox Church and even of the Samaritans. Regarding the Greeks, the recognition of ignorance could have led to more manuscript gathering and a greater integration of texts and discourses, hitherto unknown or neglected, into the patristic canon and into theological argumentation. One could also at least travel to Sichem and copy a Samaritan Pentateuch. But even though neither strategy was possible in the case of the Phoenicians and Druids, the recognition of

---

[127] Cf. only Borst, *Turmbau*, vol. III/1; Coudert, *Language*; Demonet, *Les voix*, 63, 465 on Phoenician origins, for an example ranging from Phoenician to Irish to Japanese cf. O'Reilly, "Vallancey," 131f.

[128] Owen, *Druids*, 59–100; Piggott, *Druids*, 123–146; Asher, *National myths*.

[129] Dickinson, *Delphi phoenicizantes* (1655); Smith, *Syntagma* (1664).

· ignorance still likewise did not lead to muteness but to manifold speculative discussions that even resulted in a series of specialized treatises.
A historico-religious void deep in the past was discovered and filled
with narratives.

Dickinson's work was mainly about the Delphian Oracles and followed Bochart in focusing his chapter on the origin of the Druids on the
statue of Hercules Ogmius, and then, as usual, on etymology, linking
the terms Druid and Druidism to the Greek dryas/oak. For him, most of
the Druids' history remained unknown: "who was the founder of that
sect of philosophers, what the origin of the name and the religion [*quis
author istius sectae Philosophorum . . . quae nominis ac religionis origo
fuerit*]" was very unclear. Dickinson reasoned about language, names
and their probable relationships, rather than writing something
approaching the "history of Druids." Thomas Smith devoted a whole
small book to "the customs and the institutions" of the Druids. Still, the
amount of real information on the subject had not grown, even if he
cited the "archives of the Phoenicians" as being used by Porphyrius.[130]
Consequently, long pages were again devoted to the name of the
Druids.[131] He then speculated about the spread of elements of the
Druids' religion by taking their veneration for oak and similar trees
as signs for that diffusion, for example with the Romans.[132] As Smith
could not cite a contemporary source which would say much about
those rites, he defined the religion of the Druids simply as the full and
exact imitation of the life and moral character of Pythagoras.[133] When
he wrote a whole chapter about their learning, magic and understanding of the boundary between divine and human affairs, the only sources
he could count upon were the usual authors Lucan, Strabo, Caesar,
Cicero, Pliny and the rare church fathers who mentioned the Druids.[134]
Still, the passages which mention the Druids in those texts are normally
limited to just one sentence or adjective – the Druids as "Sacerdotes
Magi Gallorum," as wise men and so on. This provided a starting point
for ruminations about the content of the term "Magoi/Magic" in the
ancient languages and civilizations, and Smith could shine with citations from Arabic, Hebrew, Persian and Greek, arriving elegantly at

[130] Smith, *Syntagma* (1664), 22.   [131] Ibid., 6.
[132] "Quidni igitur in hac idololatrarum faece hanc impiam sub quercubus
sacrificandi consuetudinem, mala imitatione ab eodem fonte Druidas
deduxisse, juste censendum sit?" (ibid., 25).
[133] Ibid., 26.   [134] Ibid., 31f.: e.g. Celsus in Origines.

Zoroaster which he connected to Pythagoras. This means that his whole chapter on the wisdom of the Druids is actually a twenty page discussion of the notion of "Magi" and more general remarks about their power of "incantation" in the next chapter. Smith posits a role for the Druids as administrators of civil law based on common sense arguments without any real source.[135] Sometimes Smith evinced an awareness of his own way of conjectural processing, of "believing" rather than knowing.[136] He stated *that* the Druids knew Jurisprudence, Astronomy, and Geography, but he was unable to give an account of their content. Similarly he discussed the question of whether the Druids had used Greek, as the old Britons allegedly had.[137] One tradition holding that the Druids might have also conducted human sacrifices in their religion leads to a general discussion about that practice in religions – Abraham and Isaac and beyond. The idea that there was something like "excommunication" among the Druids brought Smith to discussing that practice among Jews, Maronites and others. As Caesar mentioned a "pontifex" among the Druids, Smith reflected on the order of the churches of the Jews and Christians. Some short passages in Ammianus Marcellinus and Caesar stating that the Druids believed in the immortality of the soul opens the way to a general deliberation about that belief in different religions. All these aspects give this treatise the character of a kind of early inter-religious comparison and ethnography. As he did not always clearly distinguish between when he was speaking about the Druids and when the topic was other religions and groups, the treatise tends to conflate the semantics of all groups, mostly because the title is *De Druidum Moribus ac Institutis*. Despite this appellation, if one put all the sentences and passages which really concern "the Druids" together, the text would not count 165 octavo pages, but perhaps five.

Theodore Gale, in his *Court of Gentiles*, solved the problem of lack of sources differently. He relied strongly on both Bochart and Dickinson. He provided an overview of the different historical schools of Philosophy and reserved one of the first chapters for "Phoenician Philosophy," discussing how it had supposedly been taken from the Jews. For this he drew heavily from Sanchoniathon's alleged Chronicle of the Phoenicians. In the chapter *Of the magi, gymnosophistae, druides and other barbaric philosophers*, he connected the Druids to the Phoenicians:

[135] Ibid., 61.   [136] "Conjectura," ibid., 66.   [137] This goes until ibid., 68.

These Druides, who in ancient times philosophised amongst the old Britains and Gauls, were indeed a peculiar, and distinct Sect of Philosophers, differing from al the world besides, both in their mode of philosophizing; as also in their Religious Rites, and Mysteries: yet we may not doubt, but that they received much of their Philosophie, as wel as Theologie from the Phenicians ... This Sect of the Druides began first in our Countrey of Britannie; and hence it was translated into Gallia.[138]

A major theme for Gale was what might have been the Eastern content of Druidism, and the first pages of the book mention "Sanchoniathon and Mochos, those great Phenician Sophists."[139] Gale revealed his method of reasoning in a quite conscious manner, writing about a "great Concurrence and Combination of Evidences, both Artificial and Inartificial ... Or suppose we arrive only to some moral Certaintie or strong Probabilitie."[140] In a poem that serves as prefix to the book, all the authors that are considered today as late medieval and Renaissance forgeries are named as sources for the Eastern Philosophy's History, especially for Phoenician History – Berosus, Sanchoniathon, Manetho and Hermes Trismegistos. Gale ranked them far below Moses and the Decalogue, but still regarded them as sources – and the "Great Bochart" is celebrated as the modern author who led the growth of "the tree of knowledge."[141]

Those themes and speculations were soon elaborated upon by many British scholars, and not just in printed books. William Baxter (1650–1723), a learned schoolmaster at various places in and around London,[142] was corresponding about many subjects with Edward Lhuyd (1659/60–1709), the famous keeper of the Ashmolean Museum in Oxford and important antiquarian and regional historian of Oxfordshire. In letters from 1703, he also speculated about the histories of ancient times not mentioned in the Bible or other written sources. He understood himself to follow a method that would not mix "mythologie with History," as he objected to authors like the abbé Pezron.[143] He considered some elements of ancient mythology to be

---

138   Gale, *Court of gentiles*, Part II (1676), 82f.
139   Gale, *Court of gentiles*, Part I (1672), *2ᵛ.   140   Ibid., *3ᵛ.
141   Gale, *Court of gentiles*, Part I (1672), **3ᴿ. Gale himself honors the sources, ibid., 28. On the famous Nanni / Berosus forgery cf. Joachimsen, *Geschichtsauffassung*, 160–63; Goetz, "Anfänge"; Baffioni and Mattiangeli, *Annio da Viterbo*; Crahay, "Réflexions"; Grafton, "Invention of Traditions"; De Caprio, *La tradizione*, 189–258; Baroja, *Las falsificaciones*, 49–78.
142   On him Mark McDayter in ONDB.   143   Pezron, *L'Antiquité* (1687).

purely allegorical, such as the "titans" whom he regarded as "Hieroglyphicall figures for the Meteors, especially of storms, clouds, & Hurrican winds. The Theology of ye antients being nothing but Physiology." Nevertheless, he constructed, like other Renaissance and Baroque authors, speculative material to fill the void beyond written sources.[144] He was aware of doing this, as is indicated by accompanying one of the usual lists of etymological comparisons between Greek, Hebrew, Latin and Gaelic words with the admission that, "And now I shall fill up this vacancy with my conjectures how ye 30 Rivers in ye Ravenna MS. should be restored & interpreted."[145] His points on "Druidism" belong in fact to the most speculative of his times:

And you having made mention of Druidism I cannot forbear transmitting my remarks. That not only Musaeus & Orpheus were Druids but also that these were ye most antient Theologi or Philosophers, that instructed ye Magi, as they ye Gymnosophistae, & Aegyptian Priests I nothing doubt, since ye doctrine of ye east is all ye same, & ye Phrygian or Scythian Empire lasted 1500 years all over Asia, even before ye date of our Empires, as appears by Trogus, his Epitomator Justin.[146]

While Gale had conceived of the Phoenicians as recipients of Jewish traditions, some thirty years later, Baxter now thought the Druids to be the oldest group of philosophers or theologians of all, to be identified with the mythical Orphic theo-philosophy.[147] He also stressed that the doctrine of the immortality of the soul and its wandering (μετεμψύχωσις) was not an Egyptian, but a Druidic invention passed over to the Egyptians.[148] Orphic theology, he argued, was Druidic

---

[144] "I have already given my opinion that ye northern Τρωγλοδύται were ye Γηγενεῖς, or Γίγαντες, who had indeed a very antient Empire over Asia, *Aegyptum usque per mille et quingentos annos*" (Bodleian, Ms Ashmole 1814, f. 184r). This first of the Baxter letters is from 1703, but for the last ones cited in the following, the date is mutilated, it must have been after 1705. Because of that, and as their character is more like a serial treatise than individual letters, I just cite the folio without distinguishing the letters.

[145] Bodleian, Ms Ashmole 1814, f. 185r.   [146] Ibid., f. 189r.

[147] The fragments of "Orpheus theologos" were introduced into Western discussion most prominently through the collection Estienne, *Poiesis philosophos* (1573); cf. on the texts West, *Orphic Poems*; Primavesi, "Henri II Estienne."

[148] Bodleian, Ms. Ashmole 1814, f. 189r: this against John Marsham, who asserted the "Immortalitas Animae" and that the transition of the soul from men to animals was an Egyptian invention: Marsham, *Canon* (1696), 225, 269. For the *prisca theologia* cf. Walker, *The ancient theology*; Muccillo, *Platonismo*.

theology and it was the oldest of the East and therefore of the World. For this he drew on Athanasius Kircher's ideas about the shape of the earliest "winged" deity which was discussed in Newtonian circles and extended the farthest by Stukeley some decades later. However, Baxter not only received Kircher's products of armchair Orientalism, but also current travel accounts such as "Monsieur Corneilles Travelles to ye Levant" in addition to experts on Northern languages such as Hermann Conring, Olaus Worm and Olaf Rudbeck.[149] As Baxter had it, the Druids taught everything: Musaeus, Linus, Orpheus were all "thracian Druids" and all the poetry and "sublime style" of the Greeks was derived from them.[150] Finally, Baxter even arrived at something like a reversal of perspectives, of center and periphery in his view of the prehistorical exchange:

I am glad you'l give an account of Druidism, the antientest Priesthood perhaps in the known world. For I take the MAGI, & the Aegyptian Priests to have learn'd all from the northern Colonies which for 1500 years planted themselves *Aegyptum usque*, before the date of the Grecian Chronologies. You know the verse, *Tradidit Aegypto Babylon, Aegyptus Athenis*.[151]

All that was not developed into a system and it bears signs of self-contradictions: "colonies" still suggested the Orient was the "mother-land/center," but the colonies now taught the Orient. Nevertheless, his correspondence with Lhuyd was one of many possible examples from that period that show how since the Civil Wars the pluralization of the religious arena in England had set discursive labors in motion. This also makes clear how the Renaissance tradition of *prisca theologia* was revived through the Anglican reception of Kircher, not only in the circles of the Royal Society for natural history, but also in the field of religious studies, and how that was still an attempt to root the present English Church in a very deep past which, all in all, was a void space. It

---

[149] Baxter took the name and shape of the *Ur*-Deity from the pseudo-Orphic Homerical Hymns in which the "first God" was called "Πρωτεύς & Πρωτογόνος & Ἔρως, whom Orpheus & ye Aegyptians make to be winged" as the epitheton for him is "Αἰολόμορφε [= of windy / airy shape]" (Bodleian, Ms. Ashmole 1814, f. 189r). He cited Kircher's *Prodromus Copticus sive Aegyptiacus* (1636) for the name κνήφ or Ράω or Ρά of the "winged Aegyptian god." Kircher, however, stressed the element of the serpent in the hieroglyphical picture of that sign of Deity, the "Hemepht, siue supremus intellectus" and "Anima mundi vita rerum" (cf. p. 254f., 262), while there is no comment on "kneph" or "Rao" in the *Prodromus*. Cf. below chapter "Science."

[150] Ibid., f. 208r.   [151] Ibid., f. 254r.

was apparently felt to be even "larger" in extension, as can be seen in Baxter's addition of 1,500 years before and beyond "our" Empires.

This discursive movement, while following some common rules of contemporary philological learning, seems not to have had any sharper limits or reservations in the case of Baxter. But others, such as Henry Dodwell, were more skeptical. As early as 1667, Dodwell wrote a long treatise to Thomas Smith questioning "the authority of Sanchoniathon" the Phoenician author thought to be "equall to Abibalus" and so approximately contemporary "to David and Solomon." No one, neither Baxter nor even Smith, questioned his credibility, which depended on "the pure testimony of Porphyry." Specifically, Dodwell argued using the silence of authors who should have known and should have had good reasons to use Sanchoniathon if his book had really existed in their times. Justin Martyr did not use him ("though he was a Samaritan, and had thereby opportunity to have known the famous writers of his neighboring Phoenicians"), nor did Theophilus Antiochonus, nor Tatianus ("so not long after him [sc. after the supposed lifetime of Sanchoniathon], though he [= Tatianus] was an Assyrian and takes especiall notice of the ancientest Graecian and Phoenician authors"). Dodwell evinced a remarkable knowledge of authors and sources specific to the Orient, far beyond the usual Christian patristic or classical learning. Only thanks to this extensive acquaintance with what could be known, the chronology of the authors' lifetimes, the places they lived, and the context and purposes of writing ("their design"), could his suspicion arise that this Sanchoniathon did not fit. What a disappointment; there was already so much darkness, so much ignorance about the ancient times in the Orient, and now Dodwell again subtracted an author, just as Lorenzo Valla had proven the *donatio Constantini* was a forgery.[152]

Dodwell's reasoning shows that one should not consider the process of "coping with ignorance" here as a simple consecution of three steps: (a) the emergence of an awareness of a space of ignorance; (b) the filling of that space through unchecked speculative constructions; (c) the disproval and deletion of those constructions. Rather, step (b) and step (c) are co-present at the same time within the discursive movement.

---

[152] Henry Dodwell to Thomas Smith, Trinity College, Dublin, May 23, 1667, Bodleian, Ms. Smith 49, here used copy Bodleian, Ms.Eng.misc.c.23, pp. 173–187. The final version of that treatise was only published much later Dodwell, *Discourse* (1691).

As Dodwell's skepticism and Smith's speculations were co-present, voices denying the ubiquity of the Phoenician etiology of British origins were now always to be heard. Nevertheless, belief in that connection remained strong through large parts of the eighteenth century. This was frequently accompanied by the conviction that the Druids had arrived earlier in Britain, and/or were more prevalent there, than in Gaul.[153] Writers pondered where the Greek/Phoenicians landed first, the Isle of Man being the most probable point.[154] Most important was to get as close as possible to the very origins of everything: "the most ancient of those people who came first into this island, were, as may be well presumed from the calculation of the encrease of mankind after the flood, within four or five descents at farthest from Noah or one of his sons."[155] The seventeenth-century authors felt themselves to be pioneers in this regard: "That the Beginnings of Nations, and the times next succeeding those Beginnings, as yielding least pleasure both to Writer and Reader, were generally neglected, and Men naturally hastened to those Ages, which being not so far removed, yielded a pleasanter prospect ... how few are there who have taken the pains faithfully to collect ... the scattered Records of Ancient Britain."[156] The depth of History was growing at that time, as has been stated long ago by Koselleck, or for the realm of Natural History, by Lepenies. In those theoretical accounts, however, the story of the *Verzeitlichung* or the growth of the timespan attributed to World History from the roughly 3,950 Biblical years before Christ to Buffon's 168,000 years – to cite just the most famous example –, is told only from the point of view of that upper-case concept of History itself, a *praeter-propter* materially teleological historiographical perspective upon the object "History/Time," instead of upon the experience of the observers themselves.[157]

If one tries to historicize their perspective and their work as unconscious agents and, so to speak, victims of *Verzeitlichung*, it is the feeling and experience of becoming aware of ignorance. The growth and opening of the time horizons produced, first of all, a look at empty

---

153 Sacheverell, *Account* (1702), 160, cf. Asher, *National myths*, 102.
154 Thomas Brown in Sacheverell, *Account* (1702).
155 Rowlands, *Mona antiqua restaurata* (1766), 45.
156 Sammes, *Britannia* (1676), f. A2R. Cf. for a fitting description of John Aubry's proto-archaeological work Parry, *Trophies*, 292.
157 Koselleck, "Geschichte"; Koselleck, *Futures past*; Lepenies, *Ende*; Hoquet, *Buffon*; De Baere, *Buffon*.

fields devoid of those who live in them. Here, this concerns the roots of religious history. Gaps of knowledge were felt – and filled.

English Chaplains versus French State Catholicism: Conditions and Functions of Non-Knowledge Communication

This decidedly incomplete set of examples indicates how, within different discursive branches of theology and investigations into religious rites, practices and history, there were similar developments between roughly 1650 and 1730. These included an acknowledgment of ignorance about many important aspects of Eastern religious traditions – the Greek Orthodox Church, the Samaritans, the religion of the Phoenicians –, the filling of those virtual blank spaces on the map of knowledge by the collection of texts and their integration into Western canonical writing – for Greek theology –, by pure invention or speculation – with the Phoenician-Druid case –, and then, around 1730, a kind of retreat through the determination of the functional unimportance of discourses or groups: the Greek Church after 1730, and the Samaritans as a group after the turn of the eighteenth century; or, in a somewhat dialectical manner, by reenforcing the border between West and East after integrating elements of orientalization into Western discourses and westernizing Eastern ones. The case of the Ludovician French conflation between Louis le Grand and St. Louis showed that religious memory and sacrality could also be a stabilizing factor, an element of cognitive closure.

It is important to correlate this epistemic level with its institutional and personal infrastructure and to seek the functions of those epistemic developments within the framework of the French and British Empires in the Mediterranean, and for the connections between the "centers" and the "Oriental periphery."

English Imperial Network: The Chaplains of the Levant Company

On the British side, the empire's specialists in religious questions were the chaplains to the factories of the Levant Company.[158] The best prosopographical account of that group of men is still Pearson's work dating from 1883, which extracted from the minute-books of

---

[158] Russell, "Introduction," 8.

the Company a total of forty-nine elected chaplains and more rejected candidates: nineteen for Constantinople between 1611[159] and 1707, fourteen for Aleppo between 1625 and 1706, sixteen for Smyrna between 1636 and 1706.[160] Pearson stopped with that date as the minute-books of the Levant Company are lost for the period after 1707, and he noted only chaplains for those three factories. Certainly, however, there were also chaplains serving in the eighteenth century and apparently not only in the major trading places. In Algiers for example, we know of the chaplains George Holme, the famous Thomas Shaw, later a Mr. Tonyn, and Thomas Bolton for the years ca. 1699 to 1750.[161] Unfortunately, the state papers do not reveal much about the presence of a chaplain at Tunis, Tripoli or Fez.[162] The most significant of those chaplains – important as founders of Oriental and Arabic studies[163] and/or because they published a great deal and were highly active in British religious discussions – are well known and we have already encountered them in those functions above:

Edward Pococke (elected March 25/31, 1630 as chaplain to Aleppo)
Thomas Smith (elected June 2, 1668 as chaplain to Constantinople; June 20, 1684 to Smyrna)
John Covel (elected March 17, 1669–70 as chaplain to Constantinople)
Robert Huntington (elected August 1, 1670 to Aleppo)
Edward Smyth (elected May 3, 1689 to Smyrna)
John Luke (elected November 6, 1674 to Smyrna)
Thomas Shaw (appointed in 1720 to Algiers)

Other divines, such as Lancelot Addison (1632–1703), chaplain in the service of the governor of Tangier, Andrew Lord Rutherford, Earl of Teviot (1663–1670),[164] the chaplains of British ambassadors like the abovementioned John Crawford, serving in Venice, travelers and also the English armchair Orientalists such as Dodwell, Lightfoot, Baxter,

---

[159] The first chaplain dates back to the final decade of the sixteenth century but continuity only starts with 1611.

[160] Pearson, *Chaplains*. Wood, *Levant Company*, 222–224 mentions the chaplains briefly.

[161] PRO SP 71/4, f. 149r; 71/6, f. 101r, 343r; 71/8, f. 290–296 and Zizi, *Shaw*, 27–29.

[162] I have seen PRO SP 71 / 1–26; likewise, in Pennell, *Piracy*, no chaplain is mentioned.

[163] Toomer, *Eastern Wisedome*; Russel (ed.), *"Arabick interest"*; Hamilton et al. (ed.), *Republic of letters*.

[164] Hamilton in ONDB; cf. Stein, "Tangier," 996.

Marshall interested in the Levant Christians also belong on this list. Finally, the focus on the British has to be relativized, because the network that entangled British ecclesiastical affairs with the Levant also included people from different nations, around 1700, especially some Halle Pietists such as Heinrich and Hiob Ludolf.[165]

One may ask if this group of chaplains tended to belong to a certain strand within the English Church.[166] Several problems make this difficult to answer in a precise quantitative form. Firstly, not much is known about some of the chaplains.[167] Contrarily, others wrote so much that there may be scholarly debate about how to classify them. From the about eighteen people of the ninety-two listed by Pearson whose religious preferences have left some evidence, an uncommonly large part apparently tended to be traditionally episcopalist and mostly pro-Stuart oriented or, after 1688, overtly non-jurors, or at least sympathetic with those views. This was the case for Benjamin Denham, John Covel (1638–1722), Henry Denton (1640–1681), the non-juring Bishop of Gloucester, Robert Frampton (1622–1708), the non-juring Bishop of Llandaff, St. Asaph and Norwich, William Lloyd (1636–1710), the non-jurors John Hughes, William Hind, Thomas Smith (1638–1710) and William Wase.[168] Of William Hallifax (1655–1721/22) is known that his religious convictions were marked by a strong inwardly directed religiosity, which put him in a close relationship with Halle Pietism.[169] The connection between the protégé of Ann Stuart's husband George of Denmark,

---

[165] Moennig, "Die griechischen Studenten" and Wilson, "Heinrich Wilhelm Ludolf."

[166] Pearson, *Chaplains*, 9 seems to deny this.

[167] Cf. Runciman, *Great Church*, 292.

[168] Cf. for biographical information in order of mentioning Neudecker, "From Istanbul," 179, 194 (Denham); Leedham-Green in ONDB, Covel writes of himself at the age of 80: "I found not the least encouragement from any part in England, some cal'd me Jacobite some Willebite . . . I haue all along vindicated one Church of England, yet as to any Absolution, and the Exercise of the Power of the Keys, I haue set down honestly any owne free and frivole thoughts" (Covel to George Wheeler, May 28, 1717, draft, BL Ms. Add. 22911, f. 218r); Hamilton in ONDB (Denton); Cornwall in ONDB (Frampton): He was deprived from his office with the other non-juring bishops in 1691, cf. Spurr, *Restoration Church*, 84, 94, 229; Overton, *Nonjurors*, 69–74; Handley in ONDB (Lloyd): Spurr, *Restoration Church*, 94, 177, 191–193,199, 217, 365f., Overton, *Nonjurors*, 25, 48, 73, 231, 282, 422, Sirota, "Trinitarian Crisis," 49, Quantin, *Church*, 318f.; Overton, *Nonjurors*, 480f. (Hughes, Hind); Harmsen in ONDB (Smith), Pippidi, *Knowledge*."

[169] Cf. Gray in ONDB; Gray, *Chaplain*.

Heinrich Wilhelm Ludolf, the English chaplains and early Orientalists has already been mentioned.[170] A former Levant chaplain like Edward Fowler (1631/32–1714), Bishop of Gloucester, is often classified as Latitudinarian but he also held some positions similar to non-jurors like George Hickes (1642–1715).[171] Robert Huntington (1637–1701) was a juror after 1688, appointed Bishop of Raphoe in 1701, but he had been a good Royalist and retrospectively very anti-Cromwellian during the Restoration period and remained unusually open to conversation with Catholics in Ireland, a rather non-Whiggish marker at that point.[172] If one compares the proportion of the small number of 400 identified non-jurors in the contemporary lists compiled by Overton[173] to the whole body of British clergy, with the proportion of those later becoming non-jurors within the group of Levant chaplains, one can be assured that there was indeed such an over-proportion right from the middle of the century. An analysis of book ownership among the Levant merchants themselves, first of all the "public" libraries of the factories in Aleppo, Smyrna and Constantinople, which were supervised by the chaplains – for each of which a catalogue has survived from around 1700 – confirms that picture. The earliest library of the factory in Aleppo evinces a quite clearly conservative Anglican, episcopalist and for some items even monarchical, stamp: several works remembering Charles I's beheading as a martyr's death were present, for instance. Catholic neo-scholastic, Thomist and Jesuit works are far more present next to Anglican divines than any theological brand that one might deem close to Puritanism or still more "left-wing." This was so even though the library had been first established not under the Stuarts, but by Frampton in 1655 under the Cromwell Commonwealth.[174] This is not to say that there is a complete equivalence between Levant chaplains and non-jurors, but there was at least something like an elective affinity between both groups. The investigation thus confirms for the second half of the seventeenth century and for the first decades of the eighteenth century what Brenner has stated concerning the religious convictions of the Levant Company until

---

[170] Cf. above n. 165.
[171] Spellman, *Latitudinarians*, 119; Spurr in ONDB. On Hickes Harmsen in ONDB.
[172] Hamilton in ONDB; Littleton, "Ancient Languages," 155, 159f., 165; Feingold, "Oriental Studies," 491f.; Toomer, *Eastern Wisdom*, 281–287; Bodleian, Ms. Smith 50, f. 183r (February 15, 1700).
[173] Overton, *Nonjurors*, 467–496.
[174] Cf. Zwierlein, "Coexistence and Ignorance."

1650: their "anti-Puritanism" and "conservative political orientation" continued.[175] This correlates with what had been said in the first chapter about their ideas about economy. The Levant Company was not only conservative in their sticking to an old form of bullionism and regulated trade, they were also conservative in religious matters.

What did those chaplains think, believe and preach as representatives of the English Church during their stay in the Levant?[176] At least four sermons by Thomas Smith, preached in 1669 during his stay in Constantinople have survived: on Mt 6:33, on Mt 15:22f. on Christmas, on Rm 6:17 and on Deut 12:29.[177] The first remark is that those sermons could have been preached anywhere in Europe, there are nearly no specific references to Constantinople, to Christian/ Muslim encounters, to daily events or similar things that one would perhaps expect. Yet on second thought, that very "normality" is revealing.

The first sermon is on the relationship between the kingdom of God and the kingdom of the world. Smith emphasized that "the Lord has prepared his throne in heaven, his kingdome ruleth over all. All the monarchs of the world are but his subjects."[178] Clearly against proto-Deistic arguments, Smith underscored that God's order of things through Providence did not mean that He was disconnected from all "affaires of the world" and that men were subject to "fatal necessity" as if "everything [was] left to itself." Instead, "God does often interpose and appear by the stupendous arts of his Prouidence, manifest his power from heaven, & by flight & unexpected meanes brings mightly things to pass."[179] He also described another aspect of the link between this world and the kingdom of God from the human perspective, stressing the connection between the kingdom of grace and the rewarding kingdom of God, where the fruits of men's "righteousness" were to be harvested. Smith's basic theological conceptions followed the mid-seventeenth-century formulation that was prepared within the Reformed Church through federal theology – the double covenant of grace and of works – without using the complex Calvinist doctrine which was of less importance to

---

[175] Brenner, *Merchants*, 378.
[176] For some examples by Henry Maundrell and by Henry Teonge cf. Hamilton, "English Interest," 42.
[177] Bodleian, Ms. Smith 131, pp. 68–169.
[178] Ps. 103, 19; Bodleian, Ms. Smith 131, p. 70   [179] Ibid.

Anglicanism.[180] This implies an anthropology that credits positive capacities to post-lapsarian human existence as it reentered Protestantism with the Philippist movement, and as it was most strongly developed later by Latitudinarians.[181] In this accordance, his sermon on the doctrine of the law and of obedience to government, following St. Paul's letter to the Romans, shows a distinct narrowing of religion and morality and a belief in the ability of man to perfect himself through obedience to the law.[182] This was a way of thinking which was only possible within an anthropological framework that smoothened the impact of Original Sin. It was also a standpoint that led, in the long run, to a conflation between theology and philosophy that created problems in Enlightened thinking.[183] Smith was still putting conscious limits to that.[184] Moreover, while prudently pondering morality and religion, he also found his way through the Scylla and Charybdis of necessary obedience to the secular powers (*exousíai*) according to the classical Paulinian locus (Rm 13:1) and the problem of how far that might reach. He addressed the injustice of the Roman emperors who imprisoned Christians because of their religion, but he lets his audience understand that every secular power had to be obeyed, especially if it did not concern religious matters. Although he only referred to early Christian times, the identification or conflation between the two time periods, as was typical within Anglican thought,[185] could allow inferences from one to the other. Hence, with some caution, one may conclude that Smith

---

[180] Weir, *Origins*; Smith might have reacted to tendencies present for instance with John Toland.

[181] Spellman, *Latitudinarians*. The Pera library, of which only a catalogue from 1711 survives (PRO SP 105/178, pp. 300–303), in contrast to the earlier 1655-/1688 catalogue for Aleppo, shows for that later time a good deal of Latitudinarian influence (Hooker, 1710 additions of Barrow, Stillingfleet, Tillotson) next to an older Laudian-Arminian basis. However the content of the library at the time Smith was preaching is unknown.

[182] "Relligion in ye practick part of it being nothing but morality improued to ye height: so far from being an enemy of it. Yet it is in effect ye same but advanced to higher degrees of perfection, & enforced upon us by a more pervasif law" (Bodleian, Ms. Smith 131, p. 142).

[183] Cf. for that Zwierlein, "Glück."

[184] "The true christian is much better as being the most complete moralist in the world" (Bodleian, Ms. Smith 131, p. 148).

[185] Cf. literature above in n. 70.

wanted his flock to understand that obedience was due even to the Ottoman Sultan, at very least in secular affairs.[186]

A tiny allusive element of expansive missionary language emerges in the comparison Smith makes between Christ's coming to the human world and the New World's discovery by Columbus, which blends colonial conquest and early Christianization together. However his conclusion, from this comparison, that humans had to respond to the hope that Christ offered, remained more on a spiritual level. The central concept of the first sermon, righteousness, shows how he man-euvered in a very experienced manner between a theology of grace and a theology of works. Smith did not make use of his great familiarity with Oriental languages and literatures in his sermons, yet he chal-lenged the capacities of his merchant audience to a significant degree. On Christmas Day, for instance, he reflected on how to correctly conceive of Mary's virginity. While some metaphors he used were quite mechanical and even slightly prone to misunderstanding,[187] his literary skill was manifest in his citations of the twelfth-century Byzantine canonist Theodore Balsamon, Patriarch of Antioch,[188] Homer in the original language, Tacitus, Minutius Felix and the early French bishop Eucherius. This was not easy off-the-cuff preaching, but a good deal of Oxford in Constantinople.

If one can take those sermons as representative, the Anglicans in Constantinople around 1670 were enjoined, in their inward and spiri-tual morality, to perfect their individual abilities and to work on their place in the kingdom of Christ, but also to obey peacefully the external rulers in the world. That had nothing in common with the confession-alist, apocalyptical tones used some forty years earlier by John Harrisson to urge the British government to wage a final holy war

---

[186] "St. Paul gives in the Romans these rules of obedience, that they should be subject to the secular Powers. ... Tis certaine, that their prisons were ful of christians, but their onely fault was their religion." (Bodleian, Ms. Smith 131, p. 143).

[187] "Making her but halfe an one, & scarce that neither, as if he [sc. Christ] onely passed through her like water through a conduit Pipe. But the concurrence of the B.V. was necessary to him being made man" (Bodleian, Ms. Smith 131, p. 97).

[188] Ibid., p. 93 "Οὐρανομήκης ... ἐχρημάτιζε," cited from Theodore Balsamon: Ἐπιστολη ... πρὸς τούς Ἀντιόχενους = "Epistola de jejuniis quae peragi debent per singulos annos, missa ad Antiochenos," cf. Migne PG 138, 1335–1360, 1352B.

against the Spaniard as servant of the Antichrist on the Barbarian coast.[189] The somewhat Quietist spirituality of the sermons is congruent with the very low-scale missionary engagement of the Anglicans. For the Greek translations of the thirty-nine Articles, of the *Book of Common Prayer* (1569, 1638, 1644) as well as of the Greek-Latin catechisms, our knowledge about reactions on the Greek side is very meagre. Mostly it is thought that they were produced more for purposes of Greek language training of the English than for missionary purposes.[190] In Aleppo, Huntington and Hallifax seem to have been more active, using Pococke's translation of Grotius' *De veritate religionis Christianae* (1640/60) and of the Anglican Liturgy (1674) into Arabic,[191] and there is some evidence that they distributed such books among the people, which the Jesuits bought back to hinder heretical influence.[192] Some of the Greeks who had studied at the Greek College at Oxford did, indeed, use their knowledge and relationships acquired to introduce Greek pupils to Anglican theology, such as Georgios Homeros, who had been in Oxford for thirteen years and was then schoolmaster in Smyrna. In 1713/14, he asked John Covel for books such as Richard Hooker's *Of the lawes of ecclesiastical politie: eight books* (first ed. 1593) and Gilbert Burnet's *An exposition of the Thirty-nine articles of the Church of England* through Humfrey Wanley.[193]

---

[189] "As the state non standeth the daungerous or rather desperate estate of Gods Church at this daies the house of Austria being growne soe stronge, ... the Spaniard [tempting to make] himselff an absolute Monarch over all ... [he wishes instantly] would christian princes ... divert the wars from them & finallie to overthrowe the whole states seate of Antichrist, & restore a true Christian Monarchie as in tymes past, to endure to the end of the world, what is that wee daiely praie for thy Kingdome come, and as Daniel saieth is that which shall never be destroyed, but shall breake in peeces and consume all other kingdomes, and it shall stand for ever Amen." ("A Proposition to his Majestys and the State, Renewed by John Harrisson, late Agent of his Majesty into Barbary," PRO 71/1, f. 493r-v – cf. Matar in ONDB).

[190] Cf. Cuming, "Eastern Liturgies," 234; Hering, "Orthodoxie," 125f. For an attempt to communicate the thirty-nine Articles to the Ecumenical Patriarch in 1677 by the then acting chaplain in Constantinople Edward Browne in 1677, cf. Williams, *Orthodox Church*, xvf.

[191] Trevor-Roper, "The Church," 109; Littleton, "Ancient Languages," 155–161; Robert Huntington to Boyle, Aleppo, September 8, 1673, Bodleian Ms. Smith 28, f. 43.

[192] Jean Verzeau S.J. to cardinal Giuseppe Sacripanti, propraefectus S. Congregationis de Propaganda fide, Aleppo July 19, 1698, Libois (ed.), *Monumenta Proximi-Orientis*, vol. 5, Nr. 122, pp. 322–325, 325.

[193] BL Ms. Harley 3779, f. 184 – Arabadzoglu, "Σχέσεις."

Records in the archives of the Congregation of Propaganda Fide and his probatory record after death show the merchant Rowland Sherman to have been massively engaged in missionary work and promoting translations of Anglican and other works into Arabic in Aleppo, mostly during the 1720s and 1730s, importing the British imperial forms of missionary activity as a member of the Society for the Propagation of the Gospel in Foreign Parts, founded in 1701 – which is usually only associated with activities in the American colonies.[194] All that was never really competitive in manpower and in continuity when compared with the continuous Catholic activities. All in all, what can be seen from the chaplains and the little evidence of their preaching in the Levant, suggests that the British churchmen in the Levant were far more subservient to a flow of information backwards to Britain and to the transformation of ignorance where a demand was present in their homeland.

The times of Revolution and of Interregnum had exposed the English Church to a very serious threat to its identity, as there was no head, and Spurr has stressed how it developed, precisely through that experience, a strong sense for the survival and paradoxical subsistence preserved in the bishops, churchmen and in the Communion with Christ as such, even without a king.[195] It was conviction in the necessity of ecclesiastical continuity, even in times of internal struggle, that also guided the Orientalist chaplains. High Churchmen already tended to think of the English church as "transcend[ing] all political structures ... that the church's roots as an independent society reached back to the apostolic church."[196] Non-jurors necessarily even had to conceive of the legitimacy of the church beyond and without royal supremacy.

While most of the important figures like Smith, Covel, Huntington and Frampton had been in the Levant before 1688 and even before the Exclusion Crisis, the late attempt of the non-juror bishops to form an alliance with the Greek Orthodox Church, and with the Patriarch of Jerusalem as the common head (1716–1727), shows that the ecclesiological tendencies present in the group had the potential to be not in service and even in contrast to the Empire's political strength and unity. The newly-elected Archbishop of Canterbury had to write on

---

[194] Zwierlein, "Coexistence"; Heyberger, *Les Chrétiens*, 476.
[195] Spurr, *Church*, passim and xiiif., 1–12 for the problem of the "identity" of the church.
[196] Cf. Cornwall, *Visible*.

September 6, 1725 to the Patriarch of Jerusalem, Chrysanthos, clarifying that the non-jurors were schismatic, not rendering their necessary fidelity to the king, and violating ecclesiastical unity. Thereby, the Archbishop wrote, they were denying the king of the laws of the kingdom and the whole "Imperium Britannicum."[197] Obedience to the king, the Empire, the English Church, and its representatives in the Mediterranean cities – here the chaplain Thomas Paine – formed a unity which had to be respected. Contrary to that, the non-jurors had just tried to divide and separate the political empire and the church – not only with their concept of the "two societies" within Britain, but also through "confessional foreign relations." The Oriental interest in potential confessional alliances, early Christian, early Biblical and even the *ur*-religion was fed by the instability of the church's identity and its need for firm and deep roots, whereas the politico-religious situation in England produced chasms of uncertainty and insecurity. This could have, in the non-juror case, anti-imperial functions. There was certainly a well-pronounced conception that the British infrastructure in the Levant had to be Protestant in conformity with the king's confession, which becomes visible, for instance, in the separate royal instructions for a newly elected ambassador.[198] It becomes visible when Protestant voyagers in the Levant from countries without their own consular network asked Whitehall for documents of protection. But in comparison to the French in the Levant and also to the Protestant shaping of Britain's own settler colonies in the Americas,[199] the

---

[197] Mansi 37, 593f.

[198] Cf. Instruction by King William to William Trumbull, newly elected ambassador at the Porte, August 5, 1689, point 3: "In all the time of your residence there you must be carefull to maintaine a good Correspondence with all the Ambassadors and Agents of Christian Princes, especially with these that shall be in a nearer degree of Amity & allyance with us, embloying your selfe likewise towards the good of all Christians in generall, of what Nation, degree, quality or opinion soeaver they be, and more particularly those of the Reformed Religion that shall desire your protection, endeavoring to procure them Justice and fitting favour upon all occasions wherein they shall apply to you." – it is significant that this article, stressing the "Reformed religion," is newly entered by William after the Glorious Revolution in the text of the instructions otherwise copied without any change from one appointment of an ambassador to the next and indifferent to the religious question under the Stuarts (PRO SP 105/145, p. 153). The instructions issued at the same occasion by the Levant Company to the ambassador did not contain a clause concerning religion.

[199] Cf. only Porter, *Religion versus empire?*

chaplains' function and outreach remained directed toward their own merchants and toward the home country.

Therefore, this brief look at the institutional structures and network of people that supported the formation of discourses and cycles teaches us an instructive lesson: What seems to be, in the examples chosen for the British, on the discursive surface smooth transitions from ignorance(s) to different kinds of "filling the voids" or decisions upon the negativity of knowledge are revealed to have changing functions across time. The major development has, in fact, a somewhat dialectical character. Although the future non-jurors and Laudian Orientalists belonged to England's and the empire's mainstream in the mid-seventeenth century, and while their networks, knowledge acquisition and religious-confessional impulses structurally overlapped with the political outreach and the Empire's direction itself – from the times of the Exclusion Crisis and after 1688, the Oriental English Church, partially a non-juror network on the one hand, and the empire on the other, and thus also the function of the discourses produced by them, tended to opposition.

The French State in the Orient: Protection of the Christian Faith

All this was very different in the French case. The extension of Catholic missionary orders in the lands under Ottoman overlordship and farther East was pronounced during the seventeenth century, and it was perhaps not even first of all a French enterprise, even if the activities of père Joseph have become emblematic.[200] Rome and the *Congregatio de propaganda fide*, founded in 1622 – or rather, finally stabilized after the first attempts in 1599 –, as well as the other Catholic powers, were driving forces in those attempts at Catholic missionary work, mostly aimed at Christians already present in the Levant, the Copts, Armenians, Maronites, not – or at least not openly – the local Muslim inhabitants.[201] Recent research has focused primarily on the

---

[200] Cf. for all that for Syria the fundamental Heyberger, *Les chrétiens.*

[201] Cf. *Monumenta Proximi-Orientis*, vol. 5, 245 (1697). Starting in the late sixteenth century, there were also ideas to systematically convert Muslims, first the learned Arabs, and those projects also circulated among the papers of the leading cardinals responsible during the first attempts in 1599 to found the *Congregatio de propaganda fide* as Cinzio Aldobrandini. But once arriving in the Levant, missionaries normally remained quite prudent with that in order to not upset Ottoman authorities.

Roman perspective, on the *Congregatio* and the missionary orders.[202] Here I am comparing the epistemic treatment of religion within the trade empires. The point of view is restricted to the intersection between the state and commercial communication with the religious sphere – while admitting that the complex intervowenness of Catholicism's Roman center with the French claims in religious matters in the Mediterranean as a whole was a different story. It suffices to recall here the successive establishment of the two most important orders in the Levant missions, the Capuchins and the Jesuits. The Jesuits were installed in Constantinople (1583/1603), Aleppo (1617), Damascus (1643), Tripoli (1644), Seyde (Sidon, 1644), Aintoura/ Keserwan (1657), Trabzon (1688), Erzeroun (1688), Yerevan (1687), Chamaki (1682), Julfa (1682), Cairo (1697) and Rasht (1746).[203] The Capuchins established their missions in Constantinople (1587), Dalmatia (1606) and then mostly under the guidance of père Joseph. The Capuchins of the province de Paris were in Constantinople (1626/ 1629), Chios (1627), Smyrna (1628), Naxos (1628), Syros (1637), Andros (1637). Capuchins of the province of Brittany worked in Seyde (Sidon, 1625), Beirut (1626), Tripoli of Syria (1629), Mount Libanon (1628), Damascus (1637), Cyprus (1633), Antalya (1633). Capuchins of the province of the Touraine functioned in Aleppo (1626), Baghdad (1627), Niniveh/Moussoul (1636), Isfahan (1628), Cairo (1626) and Ethiopia (1638). The Capuchins in Greece were also multiplying their stations during the seventeenth century (Kuşadası 1641, Nafplio 1655, Athens 1657, Milos 1661, Crete 1670, Chania 1674, Naoussa/Paros 1675, Parikia/Paros 1683, Patmos 1684, Thessaloniki 1689). Barenton has estimated that just during the twelve years of the prefecture of père Joseph, about one hundred French Capuchin fathers went to the Orient.[204] The chaplains of the consular chapels, who often served also as pastors of the nation, have to be added to understand that at the end of the seventeenth century, there

---

[202] Cf. for the state of research Heyberger, "Pro nunc"; Pizzorusso, "Congrégation."

[203] Libois, *La compagnie*, 23–40; the most important series of documents edited is the *Monumenta Proximi-Orientis* (until now 6 vol.) within the *Monumenta Historica Societatis Jesu*.

[204] Hilaire De Barenton, *France catholique*, 91–99; da Terzorio, *Missioni*, vol. 3, 4 (Smyrna and Greece), 5, 6 (Turchia Asiatica), vol. 10 (Africa) with 1–218 (Egypt, Abissinia), 317–340 (Marocco), 562–629 (Tunis, Algier, Tripoli, Tabarca).

was a quite dense network of Catholic order clergy in the Orient, a huge part of it, if not the largest, being French.

While the reflections of Anglicans about missionary motivations and possibilities in the Levant remain very scarce around 1700, the manuscripts of Eusèbe Renaudot alone contain a half dozen specialized treatises on the question of how to organize the Catholic conversion of Oriental Christians.[205] Apart from the mission, the Catholic clergy in the Levant administered the sacraments for the (French) Catholics dwelling there. The number of those Catholics is hard to estimate; sometimes "the nation" in a small vice-consulate consisted of only two or three people. In Cairo, however, Lemaire estimated the number of weekly mass attendants in his chapel at sixty French people in 1714.[206] At least in terms of numbers the French certainly had a far more elaborate clerical infrastructure in the Levant than the three or four chaplains active at the same time in the factories of the Levant Company.

The right and claim of the French king as protector of the Christian faith gained some clear shape only with the fifth renewal of the French-Ottoman capitulations in 1673.[207] The earlier treaties (1535, 1569,

---

[205] Treatises and chapters on the possibility of a reunion of the Greek with the Roman church in BNF NAF 7459; the texts in NAF 7468 were designed to train missionaries in Oriental theology, cf. specially pp. 312–366 (322: "Quae doctrina Missionariis necessaria sit") and pp. 368–399 ("L'estude que j'ay faite durant la plus grande partie de ma vie de la Religion et de la discipline des Eglises Orientales, et les entretiens que j'ay eus avec plusieurs Missionaires, m'ayant fait connoistre diverses choses tres importantes par rapport aux Missions anciennes et modernes"), pp. 400–415 and 416–435 (46 points useful "ad Missionum propagandae conseruandaequei fidei"); pp. 526–542 ("De fide apud Orientales haereticos et Schismaticos propaganda"); pp. 544–594 ("De ratione agendi cum Orientalibus Haereticis, vel schismaticis, ut ad Catholicam fidem adducantur"); pp. 596–634 ("De Missionibus ad Orientis Ecclesias faciendis"); pp. 688–702 ("De Fide apud Orientales haereticos et Schismaticos propaganda"); pp. 704–739 ("De ratione agendi cum Orientalibus Haereticis uel schismaticis ut ad Catholicam fidem adducantur"); pp. 740–767 ("De Missionibus ad Orientis Ecclesias faciendis ubi agitur praecipue de necessaria Apostolico minimo doctrina").

[206] AN AE B I 317, f. 199r.

[207] Around 1900, the French protection of the Christian faith in the Levant was conceived as a solid institute of modern international law (Famin, Histoire; Féraud-Giraud, Juridiction; Turpaud, Juridiction; Verdy du Vernois, Frage; Lammeyer, Protektorat). But this was a usually ahistorical systematical point of view postulating a tradition of a legal institution existing since François I. For aspects of the "real" early modern circumstances of the French protection cf. Heyberger, Les chrétiens, 241–273 (Syrie); Windler, Diplomatie, 178–187 (Maghreb); Debbasch, La nation, 77–108 (Tunis).

1581, 1604) remained very imprecise and the formulations were restricted to granting the free exercise of the Catholic religion to merchants and to the custodianship of the sacred places in Jerusalem.[208] In order to ensure the free passage of French pilgrims to Jerusalem, the 1604 edict used the words "under his [the French king's] safeguard [*aveu*] and protection" (Art. 4). From the point of view of Sultan Ahmed I, it seems that he was simply delegating parts of his own power, as he declared himself in the preamble to be "protector and governor of the holy city of Jerusalem."[209] Taking the formulations very seriously, the subject of that protection is not clearly defined in the 1604 text. It could be interpreted as referring to either the Sultan, or the French king. These powers of guardianship applied just to the French king, there was no mention yet of any extension of that religious protection to the purview of French consuls in the Levant. Still, the earliest contemporary description of the consular office, the 1667 manuscript *Traicté des Consulz de la nation françoise aux pays estrangers* by Pierre Ariste, merely declared that every consul has the right to have a chapel with churchmen to administer the sacraments.[210] While the spread of Catholic missionaries occurred during the reign of Louis XIII and under Richelieu, only with the Colbert administration did the protection of the Christian religion in the Mediterranean become consciously "constructed" as a means of politics and assigned to the consuls. The *Mémoire pour la réstitution des lieux saints de Hierusalem* (ca. 1680) inserted into the journal of the sieur de la Croix (ca. 1630–1704), once secretary of the French ambassador at the Sultan's court,[211] shows how the French now conceived of their claim in a far more generalized way. Therein, the right of the "Empereurs de France" to protect the sacred places of Jerusalem was based on the conquest of this kingdom by Godefroy de Bouillon, alleging that the French rights of protection were older than the Ottoman monarchy itself.[212] This medieval

---

[208] Testa, *Recueil*, vol. 1, 17. The wording in its contemporary French rendering ressembles that of the later edicts of the Wars of Religion cf. Carbonnier-Burkard, "Préambules."

[209] Testa, *Recueil*, vol. 1, 142f.

[210] Pierre Ariste, "Traicté des Consulz de la nation françoise aux pays estrangers" (1667), BNF Ms. fr. 18595, pp. 213–215.

[211] On him cf. Sebag, "Sur deux Orientalistes."

[212] The places covered by the protection were the *Grotto* of Bethlehem, the two hills attached to it, the Stone of the Anointing in Jerusalem, the *Via Dolorosa*, the Church of the Holy Sepulchre and its two domes.

prehistory of that concept of guardianship was a French construction.[213] Only in the fifth renewal of the capitulations, in the 1673 treaty negotiated by the Marquis de Nointel[214] was a more specific understanding of the religious functions and rights formulated, as can be seen in the articles headed "Articles nouveaux."[215] These start with the lines

[we decree] that the bishops and other churchmen [] of the Latin Sect who are subjects of France, of whatever quality they might be, shall be in all places of our Empire as they have been before, and that they *shall exercise their functions* wherein no other person should trouble and hinder them.[216]

The other articles which regarded religion again concerned free passage to Jerusalem (Art. II) and some specific points responding to issues in Galata, Smyrna, Seyde and Alexandria. Nointel was not satisfied by that provision and had aimed for a much wider phrasing, aspiring for French royal protection to cover all Catholics or even all Christians in the Ottoman Empire, be they French subjects or not.[217] There would have been no special status for the Holy Land anymore. Even if those attempts were stopped by the prudent Ottoman politics of temporization and negotiation, the result was still something akin to the translation of a Gallican concept of a national French church into the Ottomans' own political language by referring to the "Religieux de Secte Latine, qui sont *Sujets à la France*." This indeed established something like a French hegemony over the role of protector in religious affairs since, with this legislation, there was a clear advantage for all religious orders, and the church as a whole, to send *French* friars instead of subjects of other countries.[218]

French consuls now always served as mediators in conflicts within the Christian community and between the Christian orders and the Muslim heads of government.[219] The 1673 text became a constantly

---

213 BNF NAF 10839, pp. 131–141. De la Croix provided examples of the importance of the French protection in that place by narrating the well-known disputes between the French Franciscans, the Armenians (in the 1620s) and the Greek Orthodox (after 1669) about the keys to the sacred places.

214 Vandal, *L'Odyssée*, 43–45.   215 Famin, *Histoire*, 224.

216 Dumont VII 231–234, 234.

217 Vandal, *L'Odyssée*, 89, 97–99, 109–112; Dupront, *Mythe*, vol. 1, 508.

218 Louis XIV was quite satisfied with the result, cf. Rey, *Protection*, 321 n. 1.

219 Heyberger, *Chrétiens*, 241–271. Many cases are narrated in Homsy, *Capitulations*, 233–412; cf. for a good depiction of the close collaboration between consuls, Capuchins, Jesuits, also for conflicts between both

recurring point of reference, which was interpreted in an increasingly broader manner by the consuls and other agents of the crown. In 1695, the Sultan, when reconquering the island of Chios which the Venetians had taken in 1694, prosecuted the Christians on the island, destroying their churches or transforming them into mosques.[220] Taking that case as an opportunity, he planned to promulgate an edict (Catacherif) more generally

that will forbid the French missionaries to frequent the Christians and the other Levantine sects of the Grand Signor's subjects. It seems that the edict contradicts the capitulations, *at least their application in practice [au moins à l'usage]*, since the French have been established in the Levant in form of a nation, and what the Turks have tolerated without contradiction since the earliest times until now. This concerns not only the exercise [sc. of the members of the French nation] of the religion under the protection of the king, but also the work of the missionaries for the promotion of the religion and the conversion of the schismatics, in what functions the latins have always been even assisted by the servants of the Sultan.[221]

The ambassador Chasteauneuf warned the court about this possibility. The French were now highly conscious of the difference between the restricted written words of the capitulations and the extensive interpretation of the first of Nointel's new articles by the French in their daily practice ("... au moins à l'usage"). While the Ottomans had probably intended by "functions" only the giving of sacraments to the members of the French communities, the French subsumed the expansive counter-reformation impulse of mission under the Word.[222]

The protection of religion had become a highly reflexive part of French politics. The king was "the only protector of the religion," and he was "recognized in the whole [sc. Ottoman] Empire as such

---

Monumenta Proximi-Orientis, vol. 5, 137, 213, 231, 245, 246, 357, 364, 368, 407–409, 454, 456.

[220] Terzorio, *Missioni*, vol. 4, 38–47.

[221] "Memoire – Commerce de Leuant – Pour donner une idée de l'Estat present des François en Leuant" (1695), AN AE B III 235, Nr. 53; cf. for similar problems Rey, *Protection*, 341f. on the establishment of the Jesuits on Chios since 1590 cf. Laurent, "L'âge d'or," 220f.

[222] Nointel's negotiations with the Ottoman Porte and the other articles of the 1673 edict which precisely stipulate specific concessions and details for churches in Galata, Smyrna, Seyde and Alexandria suggest *e contrario* that the term "function" in Art. 1 was not intended to have such a wide and general meaning.

by the missionaries of all the nations, even by those of the Propaganda." The French had gained hegemony over the politics of protection by "excluding the emperor and the Spanish king," and the protection of religion would also serve "other political aims, the commerce, the credit of the nation and the glory of his Majesty."[223] Around 1700 the actors in Versailles and Paris were very consciously calculating the extra-religious functions of religious politics from a *raison d'état* point of view. This meant that the protection of the Christian religion had really become a tool of empire.

As this increasingly formalized power of religious protection at the end of the seventeenth century was derived from the king's sovereignty over his subjects, it led to conflicts not only with the Ottoman authorities, but also in a dual way within the Christian context, between the king and the Roman church/pope,[224] and between the king and the Gallican clergy independently from Rome. Consular jurisdiction normally only dealt with problems of civil and criminal law with regard to trade and the commercial affairs of the merchants of the nation.[225] Each consul had the formal position of a judge ("la souveraine jurisdiction sur la nation françoise"),[226] but joined the deputies of the nation as body of the court.[227] A small conflict within the vice-consulate of Rosetta in Egypt during the year 1727/28 may illustrate how the institutionalized protection of the Christian faith had now acquired the character of para-governmental French rule and self-understanding under the umbrella of Ottoman overlordship. Rosetta was close to the sacred places and so the vice-consul, at that time François Rosset (1663–1731),[228] despite his low rank, was quite important in everyday affairs.[229] A sequence of local conflicts between Rosset and Julien Rubatel, the president of the hospice of the Holy Land, was ended by the replacement of Rubatel, first by two intermediary fathers sent by

---

[223] AN AE B III 235, Nr. 53.
[224] Some remarks on that in Heyberger, *Chrétiens*, 267–271, but not concerning problems of jurisdiction.
[225] The best systematic description of the consul's jurisdiction remains Debbasch, *La nation*, 195–226
[226] Ariste, "Traicté," BNF Ms. fr. 18595, p. 178.   [227] Ibid., p. 180.
[228] Mézin, *Consuls*, 526.
[229] AN AE B I 968: "Exposé du differend entre le R.P. Rubatel et le Sieur Martinenq, et Réflexions à la Requête de ce religieux du 21. Mars 1727. Exposé de ladite affaire."

the consul in Cairo, then by the new chaplain who arrived on April 22, 1727. Rubatel was not willing to accept his deposal and replacement.

Rosset, in his legal declaration concerning the problem, started with the question of the *iudex ad quem*, stating that normally the order of "court" compositions went from one composed of three merchant members of the nation, to the consul together with the deputies of the merchants' nation and, if they were not available, to the two principle French merchants of the municipality as a declaration of the king from May 25, 1722 prescribed.[230] As in the Rubatel case, insufficient merchants were available because of the many losses from a recent outbreak of pestilence, and the vice-consul Rosset decided to pass the case directly to the court at Cairo, composed of the consul and the deputies of the merchants living there. Even if the chaplain was paid by the nation, the dispute was about the behavior, capacities and deeds of a churchman in his religious functions. Thus one might have expected a church court to be more competent and canon law to be more appropriate. Nevertheless, the case ended up before a consul-merchant court.

Rubatel defended himself from the perspective of a non-papist church and from a moderated concept of monarchical rule. He referred to the fundamental laws of the kingdom and the Gallican Liberties, stating that even the pope could not excommunicate the officers (!) of the king. This indicated a Gallican sense of sovereignty of the state extending into religious affairs, alongside hints of a quasi-conciliarist conception of the church's supremacy over the pope. This he based on free citations of the proverbial passage about the Church's supreme Petrian binding power surpassing the state's authority according to Mt 16:19 ("Quaecumque ligaveritis super terram erunt ligata et in coelo, et quaecumque solveritis super terram erunt soluta et in coelo") – but only to deny its normative value. For that rejection he invoked classical examples deriving from the doctrine of the *correctio fraterna* and its elaboration within later neo-scholastic teaching about the authority of the ecclesiastical power over the secular. These included St. Ambrose, who set limits to the powers of Theodosius, St. Babylas before the Emperor Philip, and St. John the Baptist in front of Herod the Tetrarch. The church, he maintained, deliberately and piously obeyed the state, but while obeying, the church did not lose its rights. Rubatel

---

[230] AN AE B I 968; Debbasch, *La nation*, 216 n. 3: that composition remained unchanged until 1778.

also listed among the *lois fondamentales* the "royal protection of the church."[231] The principle of the monarch's Catholicity first became a fundamental law with the Leagueist Edict of Union in 1588, but the institution of the protection of Christianity in the Levant was not linked directly with that. Rubatel understood his reference to the royal protection of the church as a reminder to the consul to refrain from violating its rights. Even though his primary goal was to minimize the consul's jurisdiction over him as a churchman, he nevertheless had to admit to, rather unwillingly in that moment, the imperial potential of the institution of protection, as he placed himself and the church's case in the Levant under the umbrella of the French *lois fondamentales* – not a Roman court, not the authority of canon law, and certainly not the Ottoman ruler.

The institution of the protection of Christianity was anchored to the royalty itself. This meant that the French politico-economic structure fed off the network and personnel of the religious orders – certainly not as "official officers of the crown," but nevertheless as a para-state instrument. The institutionalized French consular network with its administrative and jurisdictional functions placed elements of the growing French state onto the Mediterranean borders, implementing a precise Gallican vision of state and religion, and they partially incorporated the network of missionary churchmen into it.

Comparison and Conclusion

Ignorance of given religious groups and their doctrines, alongside the corresponding obtained or created knowledge tends to belong more to the realm of epistemic than to operative non-knowledge, with an identificatory function – or at least its potential – within religious and national discourses. This is very much the case for the British and less for the French. For the British, the flow of information was more or less unidirectional, from the Levant to Europe. It started with recognizing and specifying ignorance in England within purely British discussions about the relationship of church and state, and of the episcopal system. The leading group behind that, predominantly Oxford and Cambridge

---

[231] "La loy salique, c'est à dire, la succession en ligne masculine, *la protection roiale à l'eglise*, et la déffense du Roiaume. Les libertés de l'église gallicane sont l'ancien droit gardé en France en ce qu'on peu sur le droit nouveau." (AN AE B I 968 [unfol.]).

churchmen and Orientalists, might seem to be a very small and specific set of intellectuals, but as the position of the chaplain in the Levant was always a transitory position to higher positions within the English Church, they were quite important. The relationship between that group and "the empire" changed remarkably due to the inner-British (r)evolution before and after 1688, while the cohesion of the conservative group remained quite stable. The same assemblage, and thereby also the discursive entanglements that they constructed between Levantine Christians, the Phoenician origins of civilization and the Anglican Church and English nation, shifted from generally supporting to relatively opposing the leading discourses of the empire. The identificatory function of those epistemes was lost for "the nation" and "the empire" and remained active only for that group. While Byzantine theology, the contemporary seventeenth-century Samaritans and the Phoenician-Druidic connection had all been virtually unknown, during the decades from 1650 to 1730 there was a dramatic increase in narratives and texts – but then, for most of those cases, what had been in a state of nescience before was transformed into deliberately recognized negative knowledge. The British now knew that, if there was a great deal unknown about the Greek Church, it was not important for their own work because the amalgamation of the Greek Orthodox Church with theirs made no sense and was not practically feasible. The Samaritans had been visited and what had been first nescience, then specified ignorance about much of their history, then likewise transformed into negative knowledge about the unimportance of the contemporary Samaritans and a gradual acceptance of the insignificance of their Pentateuch and their rites for urgent Anglican affairs. So, in all the cases above, epistemic non-knowledge changed its function from an identificatory one to one purely cognitive. In the British case, the other direction of discursive flow, the missionary impulse, was present, but only in an embryonic form.

The fact that it was only the Port-Royal Jansenists – they themselves a harassed minority in France – who took part in the "Orientalization" of the Eucharist discussion in a similar way to the Anglicans, fits into the picture well. Their use of and interest in the Greek Church had the French function of profiling the Jansenists as good defenders of the Catholic vision of the church's history following Bellarmino against the Huguenots. And the collaboration of the Huguenots with the British chaplains shows how the confessional element remained a point of imperial competition around 1700. Beyond the aforementioned inner-French function for the Jansenists, this Eastern orientation was part of their work on religious self-definition and a search for possible alliances, as the activities around the ambassador Nointel at Constantinople show. The position of the Jansenists and their relation to the empire is quite analogous to what has just been seen with the British and the non-jurors. The core of French Catholicism had no use or need in the same sense for Byzantine or Samaritan theology, not to mention Phoenician Druidism, as the post-Civil War English Church had. Interest in all that did exist and had been cultivated beforehand in France (Postel, du Choul ...), but the Catholic Gallican Church had a great many more potential of identificatory traditions than the English Church, whose survival through the Civil War period, the Restoration, and after the dynastic change of 1688, was exposed to a very different threat that the French episcopacy did not experience even during the sixteenth-century Wars of Religion. The French could allow themselves a cognitive closure to some point in religious matters. Clearly, the discursive flow was more oriented *from* France and the religious orders defining themselves partially as French *to* the Levant, at least in the primary conscious guidance of the missionaries' action and of the consuls that protected their work.

Admittedly, a great deal could seem similar for all the Europeans in the Mediterranean. They all gathered manuscripts. They all were interested in the local Christian communities and also in the Jewish and the Muslim forms of faith and religious worship in the region. It *was*, certainly, a very densely connected transnational *respublica litteraria*. Nevertheless, from the *functional* point of view of the two empires chosen here, all that what has so similar forms still reveals a significant difference between the French and the British.

Ignorance, its specification and the consequent gathering of knowledge belonged to, in the French case, a realm of (non-)knowledge with largely operative functions. There were some hundred French missionaries active in the region, and the French government and French-Ottoman diplomatic negotiations tended to integrate those primarily church agents into "the empire" itself and thereby also their rather expansive impulse within the field of religion. This was not or far less the case for the English. In brief, the entanglement between France and the Levant in the field of religion is more an entanglement on the level of power, while the entanglement on the epistemic level is less significant.

Contrarily, the British entanglement was predominantly on the epistemic level until 1730/50, and less distinctive on the power level.

For the general questions about a history of ignorance, the case of religion stands for the fluid forms of passing from operative to epistemic non-knowledge. It also provides an example of how everyday non-knowledge communication, that is, the more short-lived and transitory forms of it, were linked to the institutional level of imperial structures and frames of thought. Shifts in religious non-knowledge communication on the functional level, from supporting to later opposing the empire, indicate, for the English, the possible divergence between the more short-lived and the more slowly changing institutional patterns of communication. For all the examples analyzed here, despite their different outcomes and endpoints, a remarkable congruence of the cyclical form and the chronology of the epistemic process itself between roughly 1650 and 1750 is evident.

3 | *History: How to Cope with Unconscious Ignorance*

What general concept of the history, society, constitution and "culture" of the Barbary states, and even more general of "the Levant," had imperial actors in the Mediterranean? British and French merchants, consuls, chancellors, chaplains, travelers and scientists integrated into the network of European merchant colonies were living in that world, or at least between that world of "the Levant" or of "Barbary" and their homelands. They were often recognized as experts of those regions during their time there as well as after their eventual return. Likewise, administrators and decision-makers in London and Paris/ Versailles, from the kings down to the simple clerk, continually endeavored to be familiar with the specificities and particularities of the Mediterranean, as they did for the other outposts and markets of the world. This general knowledge and the respective unknowns corresponded more or less with the wider notion of the early modern *historia*, that still encompassed geohistorical knowledge, natural history and knowledge about the different peoples, state forms and constitutions of the world. General historical knowledge, or a lack of such knowledge, is situated even closer than religion to the "pure" epistemic side on a scale between operative and epistemic (non)knowledge. But for early modern men of politics, even knowledge about the medieval past of a country normally *was* considered to be important for everyday decision-making as is evident not only from all kinds of propaedeutic texts designed for the education of nobles, princes and *bourgeois* men of politics, but also from early modern governmental practice itself, in which knowledge about dynastic histories, noble lineages, feudal successions, territorial property, privileges and past rights granted, all were considered absolutely indispensable. Knowledge about the character of a people and a region that was to be ruled over, or with which one might be in contact, was deemed necessary long before and beyond the arrival of a more theoretical concept of the political use of history that propagated the analysis of past events, alliances, behaviors and causalities for deducing rules that might be applicable again to rule the future. This means that a lack of historical knowledge about a region where Europeans were living in, where they were interacting with its inhabitants through commerce, war and treaties, must have been a problem – in fact, a far greater problem than for people of politics today according to their respective standards. It is, therefore, perhaps surprising how much they did *not* know, how much they lived in ignorance, even if they were in perfectly functioning everyday commercial and political contact with their "non-European" counterparts. More specifically, historical knowledge about those regions was largely obfuscated, both qualitatively and quantitatively, or even in a state of nescience. It was only by epistemic stimuli, which will be defined in the second part of this chapter, just around 1700, that nescience was transformed into conscious non-knowledge and new dynamics set off, trying to fill the voids. In view of the standards of the time of how politics functioned, these forms of ignorance are even more striking than those within the field of religion.

The Forgotten Arabic Middle Ages

Within the French administration of foreign affairs, many directive *mémoires* about a region concerned trade and politics. There was, however, also another broader, more general type of these documents: *Mémoires historiques* or simply general memorandums. Such *mémoires* might be produced at the order of a new secretary of state, who would sometimes ask all consuls to give a broader account of the history and state of affairs for their districts when he took the office. On other occasions, experienced consuls wrote them, often when returning home from assignment. One may see them standing in a very long tradition stretching back to the final summary accounts written by Italian ambassadors during the Renaissance of which the Venetian *relazioni* are the most famous. British imperial agents wrote such manuscript reports quite rarely within internal administrative contexts, but they sometimes prepared comparable printed texts as descriptions and histories of a respective country or its religions.

Rarely do we possess exact information about the production of these texts. The French manuscript *mémoires* normally do not contain any reference to authors or works that were used. But in the first volume of the Morocco *Mémoires et documents* of the *Archives des affaires étrangères*, some helpful traces have survived. The first extensive *mémoire* archived here is by Jean-François-Raymond Guy de Villeneuve (1728–1780).[1] As direct diplomatic contact ceased between Morocco and France during the years 1727 to 1767, Villeneuve's work probably represents an attempt to establish a general orientation as a new starting point. When he prepared this *mémoire* about Morocco, he used documents and letters sent by "M. de St. Didier," obviously stemming from the archives, the *dépôt de la Marine*. The archivist of the *dépôt*, Laffilard, had prepared at some point an inventory of all pieces regarding the history of Morocco from 1533 to 1742.[2] At the occasion of Guy de Villeneuve's request for information or close to the redaction of his *mémoire*, a somehow canonical list of authors on Morocco was compiled. These printed works were thought to serve *mémoire* writing in the service of the French foreign affairs:

"Ortelius Sanson
Jean de Leon
Marmol
Daper
Histoire de Taphilet
Diego Torres histoire des Cherifs
Moüette Histoire de Maroc
Sanuto
Mercator
du Thou
du Val
Histoire de Tafilet ecrite par un agent du roy d'angleterre en Afrique
Herbelot Bibliotheque Orientalle
De la Croix Histoire d'Afrique tome 2
Thomas Corneille dictionnaire Geographique
Baudran
Dauity de l'Afrique

---

[1] The final "Observations sur le Maroc. Details historiques" (AE MD Maroc 1, pièce 6, f. 166r-242r) are dated by a different hand on f. 166r "Vers 1778," but it had already been received favorably in the Versailles administration before March 1777 and Calonne ordered that he should compose a similar one about Tunis, cf. April 5, 1777, AN Mar C7/137, f. 2. Guy de Villeneuve had been an advocate at the Parlement de Toulouse, and had worked in the *bureau des Consulats de la Marine* after 1762. He had experience in the Levant as vice-consul of the Dardanelles (1776), on a mission to Smyrna (1779), and as general consul of Morea, the Peloponnese peninsula.

[2] AE MD Maroc 1, pièce 1, f. 4r–158r.

Maty Dictionnaire
D'Herbelot Bibliotheque orientalle [sic, cited twice]
Vansleb
Villaut
Relation des Costes d'Afriques"[3]

Such a list serving the work of a French consul can be compared with a similar English source. In 1733 Thomas Shaw started to finalize his *Travels*, a geohistorical description of the Barbary coast that would become a standard work of reference for decades. The first proof prints date from that time,[4] although the first complete edition only appeared in 1738. In a letter to the then president of the Royal Society, Hans Sloane, Shaw sent him "the list of what I have consulted, that if there are any others that [I] am not acquainted with, you will have the kindness to inform me":

1. Ptolomy. 2. The Antonine Itinerary. 3. Strabo. 4. Pliny 5. Solinus. 6. Mela. 7. the minor Geographers published by Dr. Hudson. 8. Johannes Leo. 9. Marmol / The French Translation / 10. Gramaye. 11. Peutingers Tables. 12. Besides what is occasionaly be met with in Amianus Marcellinus, Appian, Procopius, Livy, Salust, Hirtius and some other of the Classic Authors. These I have collected; and if you remember any other, that may give me any light, or Information, I should receive it with all imaginable gratitude.[5]

---

[3] AE MD Maroc 1, f. 161r–v: "Autheurs qui ont écrit sur la Mauritanie et Maroc." This means: Ortelius, *Theatrum* (several always augmented versions since the Antwerp 1570 first edition, cf. below n. 9); Léon l'Africain, *Description* (Villeneuve might have used an Italian edition following the Ramusio edition, a French translation or the latin 1632 Elzevir translation); Mármol, *Descripcion* (1573-1599); Sanson, *L'Affrique* and Torres, *Relacion* (1586) – but Villeneuve probably used those three sources in the French version *L'Afrique de Marmol* (1667); Dapper, *Beschreibung* (1670), presumably used in the French translation Dapper, *Description* (1686); *Tafiletta* (1669); *Tafilette* (1670); Moüette, *Relation* (1683); Sanuto, *Geografia* (1588); de Thou, *Historiarum libri* (1620); du Val, *Diverses cartes et tables* (1677); d'Herbelot, *Bibliothèque orientale* (1697); de la Croix, *Relation universelle* (1688); Corneille, *Dictionnaire* (1708); Baudrand, *Geographia* (1682), french translation as Baudrand, *Dictionnaire* (1701ff.); d'Avity de Montmartin, *Description* (1637); Villault, *Relation* (1669); Wansleben, *Nouvelle relation* (1677, several editions).

[4] Cf. BL Ms. Sloane 4053, f. 59r, 64r: First prints of the Chapter I "Of the Kingdome of Algiers" within the correspondence with Sloane from 1733.

[5] Thomas Shaw to Hans Sloane, Queens College Oxford, October 4, 1733, rec. October 25, 1733, BL London Ms. Sloane 4053, f. 62v (spelling as in orig.). In the printed book of 1738, Shaw used Abulfeda in Gagnier's edition and translation and also al-Idrīsī's *Geographia nubica* (mostly the abridged version).

Shaw, Professor of Theology at Oxford, a highly renowned Orientalist of his time and member of the Royal Society, has to be placed on a very different level of specialized learning than Villeneuve, and if one reads Shaw's final product closely, one discerns that he used several other sources than the ones cited here. But his list of authorities of 1733 is nevertheless quite representative for what formed and could form the backbone of any geohistorical description of the Barbary Coast at his time.

Both lists, prepared by practical actors in the Levant for a long time as well as later active in the "centers" of imperial communication, did not contain any "original" Arabic or Ottoman authors neither in 1733 nor in 1777, if one excludes the famous Leo Africanus who had dictated his *Descrizione* in Italian. Shaw was, moreover, a renowned Orientalist. This points already, right from the beginning, to a perceptional bias that can be hardly overestimated for this period.

Views on the Void

I will take those two lists of works and authors used by the French government, by consuls like Villeneuve and by Shaw as a starting point to consider what those books did *not* contain, as they apparently formed something approximating a standard canon, as Maghreb and African historians confirm.[6] If one looks at the content offered by that canon of historical works, the case which illustrates perhaps the best the problem to be treated here is the complete, and also completely unrecognized, absence of the Arabic Middle Ages.

Gaps in Early Modern Histories, Mémoires, Descriptions of the Levant
Since the early nineteenth century, the quantity of professional research on Christian-Muslim contacts, the transmission of manuscript chronicles, and the study of the historiography and geography written in the Middle East and North Africa from the end of the Roman Empire until

---

[6] When Masonen, *Negroland* concentrates on the early modern historiography on Western Africa, he uses exactly the same authors. Cf. similarly what Jones, "Decompiling Dapper" has reconstructed as sources used by the Amsterdam armchair compiler Dapper. The amount of original elements in other manuscript narratives of the Barbary coast as in BL Add. 8312, f. 1–52v is very small. On several similar Spanish manuscript relaciones cf. Bunes Ibarra, *La imagen*. On the printed works cf. Thomson, *Barbary*; Tourbet-Delof, *L'Afrique barbaresque*; dos Santos Lopes, *Afrika*; Katzer, *Araber*.

the rise of the Ottomans has grown incredibly dense. So much so that today there seems to be a completely comparable "upper case" continuum of *History* for those regions as is the case for the Western and Northern parts of Europe. Time has always run at the same speed regardless of geography. People lived in those regions, and they must have necessarily produced events, institutions, remains and, therefore, the content of History. But with what mindset did early modern actors come to the Barbary shores and the Levant? What was *not* in the virtual rucksack of knowledge proverbially strapped to the back of a typical French or British consul, chancellor, merchant or learned traveler? As every history of ignorance always, implicitly or explicitly, measures the form and shape of ignorance in the past against a measure of past or contemporary knowables, this will provide here a contrast with the huge amount of differentiated and specialized knowledge about the "medieval" Middle East as is represented today and easily reachable through encyclopedias like the *Christian-Muslim relations: a bibliographical history* or the traditional *Encyclopaedia Islamica*. Reconstructing the contours of that void is only a first step to better understand the varying forms of tentative attempts to recognize and to cope with those ignorances within the different epistemes around 1700. Ideally, it would be worthwhile to characterize the contours for all possible ignorance dimensions of time, space, events, ethnic, and different thematic and functional aspects of knowledge on past societies, their constitutions and cultures. But for reasons of practicability, I will concentrate on a *period* that was nearly completely absent in the mental map of the early modern actors of the seventeenth and eighteenth centuries in the Levant: the post-Roman Middle Ages until roughly 1500.

Early modern authors who wrote about any part of Africa invariably referred to Leo Africanus, the link between the fading Ibero-Arabic world of the *Reconquista* and the later sixteenth century.[7] However, Leo's text has no great historical depth. As the title *Descrittione* betrays, it belongs to the genre of "present-state-of . . . " descriptions; it is ordered largely according to a spatial movement of author and reader through Africa. Only sometimes are passages included that narrate the prehistory of this or that city, region or people. One could

---

[7] Léon l'Africain, *Description*. The Italian text available in early modern times was Leo Africanus, *Della descrittione* (1563).

not easily reconstruct a long-term multi-century history for each region from the scanty material Leo Africanus supplied. Despite being the standard point of reference for African history in general as well as for North African history in particular, Leo Africanus did not, ironically, provide any "History" in its narrower sense at all. Leo still named Ibn Khaldun once, citing him from memory as not having read "for ten years any book on Arabic history." Yet this author, whose *History of the Berbers* would become in the nineteenth century the conventional starting point for the reconstruction of basic dynastic and narrative history of the high and late medieval Maghreb, disappeared from the "menu" of Western authors until around 1750. By chance, the quite precise indication of Ibn Khaldun's name and work was omitted in the early modern text tradition of Leo's *Descrittione*, so the early modern Leo, as edited by Ramusio, did not even transmit the non-knowledge about the work ignored.[8]

For their descriptive narratives, the geographical works on Villeneuve's canon (Ortelius, Sanuto, Sanson, Mercator, Duval, Corneille, Baudrand, Maty) relied either on Leo and the later authors mentioned below or contained no substantial description at all and made no contribution to historical coverage either.[9]

Louis Mármol and Diego Torres, two central late sixteenth-century Spanish authors, depended largely on Leo Africanus as recent research has started to show, but they also display great differences from the Arabo-Italian author. For example, Mármol included a very detailed account of the history of the late medieval dynasties in Morocco and al-Andalus Spain in his second book. But Mediano has suggested that all his sources were probably European, written in the vernacular or Latin. If he sometimes cited the "escriptores arabes" or authors such as

---

[8] Cf. Léon l'Africain, *Description*, vol. 1, 34f.; Épaulard's translation/edition is a collation of the Italian 1526 manuscript Rome Biblioteca Nazionale, Fondo Vittorio Emmanuele 953 and the Ramusio text (on the former cf. Rauchenberger, *Johannes Leo*). But in Ramusio's text, the only one diffused in early modern times, Khaldun's name had become unrecognizable ("nell'opera di Hibnu da me sopradetto," Leo Africanus, *Della descrittione* [1563], vol. 1, 5D; Leo Africanus, *Descriptio* [1632], 42).

[9] Neumann, "Imagining European community," 433 n. 50; Bodenstein, "Ortelius' Maps"; Sanuto, *Geografia* (1588) adds only very small sections of his own to Leo's description. Sanson, *L'Affrique* relied on Leo, and Gramaye. The descriptive parts of the Mercator-Hondius Atlas were very short anyway. Cf. Norwich, *Maps*, maps 10, 11, 12, 15, 21, 27, 29, 34, 38, 41, 44.

"Abdul Malic," he meant Leo Africanus or referred to those Arabic authors second-hand through Leo. The second book was thus mainly compiled from Paolo Diacono, Celio Agostino Curione and various Byzantine authors that had been translated into Latin or the vernacular by the sixteenth century (Zonaras, Comnenus, Choniates, Cedrenus). For the Muslim times of al-Andalus, he relied on the *Historia arabum* and the *Historia de rebus Hispaniae* of Rodrigo Ximénez de Rada and the *Crónica de los Reyes Católicos* of Fernando del Pulgar.[10] Mármol can thus be considered as a "channel" between the medieval tradition of European-Islamic confrontation and exchange and the early modern period. Curiously, seventeenth-century British and French Imperial writers almost never went beyond Leo or Mármol. If there was a link to the medieval Latin traditions of knowledge about the Arabic world, it was in that palimpsest method of transmission through sixteenth-century European writers. Arabic authors and historians themselves had no textual influence on any of this. Moreover, those portions of knowledge about medieval times always concerned areas directly across from the Iberian Peninsula, Morocco and Fez. The parts of North Africa that, under Ottoman rule, were transformed into the regencies of Algiers, Tunis and Tripoli in the sixteenth century are either entirely absent or, at best, hardly represented.

Diego Torres' work, which borrowed large parts directly from Mármol, covers the Spanish-Muslim confrontation in North Africa between 1502 and 1574. This is prior to the region's definitive Ottomanization, and there is also no reference to any historical event before 1500.[11] Likewise, Haëdo only starts with the year 1504 and the deeds of the famous Barbarossa brothers.[12]

The most important historian for the Barbary Coast itself beyond Morocco was Jean-Baptiste Gramaye with his *Africae illustratae libri decem*, published in 1622 in Tournai (second edition 1623). Gramaye had been already active as a historian in his home of Brabant. Starting in 1618, he became prominent as a Maghreb traveler, diplomat and protonotary apostolic.[13] The *Africa illustrata* has 353 pages containing

---

[10] Rodríguez Mediano, "Mármol"; Martínez-Góngora, "El discurso."

[11] The editor García-Arenal of Torres, *Relacion* (1586) provides a list of the passages literally borrowed from Mármol and Góes on 304f.

[12] Haëdo, *Topographia* (1612); Haëdo, *Epitome* (1612), translated as Haëdo, *Histoire*.

[13] Ben Mansour, *Alger*.

the history of all kingdoms and places of the Maghreb, including *Numidia* in the sense of ancient geographical toponymy.[14] He used Leo Africanus, Mármol, Löwenklau, "Ricius," Thevet, and he even cited certain Arabic historians by name, although it is not absolutely clear if he knew them directly or indirectly through Leo.[15] Some passages seem more like general state descriptions, such as his political analysis of the different offices within the Algerian political system or geographical descriptions of the different Mediterranean provinces. Taken altogether, the work contains on the one hand a large compilation from classical sources. On the other hand, however, it also provides a great deal of information about his contemporary period from the sixteenth century, much of it taken from Leo Africanus, even if Gramaye added large amounts of his own firsthand experience and analysis. Nevertheless, what is missing in his work, with the partial exception of Morocco, are the Arabic Middle Ages. Gramaye's discussion of the *Historia numidica* stops with the year 710 and the invasion of the "Saracens."[16] His description of the region of Tunis deals with ancient Carthage in depth, but then within the space of one and a half pages, it jumps from the Roman era to the conquest of Tunis by Charles V in 1541 and its reconquest in 1574.[17] When he does discuss "medieval" times (roughly between 700 and 1500), it is church history with lists of bishops, monasteries, or information about the Migration Period.[18] In other words, it does not concern the North African region proper. The Ibero-Arabic section closes with Vitiza and the *Compendium Annalium Afro-Turcicorum* with quite original content gathered by Gramaye himself, and starts again with the year 1502 on one and the same page.[19] This "jumping" from antiquity to the sixteenth century, characteristic of the whole work, was not due to a

---

[14] Gramaye, *Africa* (1622).

[15] For instance Gramaye, *Africa* (1622), 21: "quae ex Arabum historijs, & signanter libro generationum Africanarum per Ibnu-Rasu annotata sunt." No comprehensive study of Gramaye exists that covers his whole life, the *Africae illustratae libri* and labors as Counter-Reformation diplomat, which has left important traces in several European archives. For a short sketch of his biography cf. Ben Mansour, *Alger*, 35–45.

[16] Ibid., 135.   [17] Ibid., 150f.

[18] The short section included in his early work on *Asia*, written before he had visited Africa himself, likewise concerned only what in early modern times was thought of the "Arabs" of Antiquity, with an addition of some putative customs gathered from Joannes Boemus and Scaliger. Gramaye, *Asia* (1604), 519–530.

[19] Ibid., 81.

particular conviction or humanist prejudgment about the allegedly benighted Middle Ages. Previously, as professor at the University of Louvain, Gramaye diligently dug into medieval sources for Brabant history, citing unpublished charters and manuscripts regarding the dukes and cities of his homeland.[20] He clearly had an idea of a *medium aevum* for Western European history. The absence of an Arabic Middle Ages in his later work therefore seems to stem from pure ignorance, and even in some way a preconscious nescience: one does not find any expression by Gramaye of a sense that some eight hundred years were missing. And Gramaye was not only a historian. He was also a leading proponent of a combined European military force against the Barbary corsairs, he was a politico-religious diplomat in the service of the emperor and the courts of Spain and France, and he even negotiated with the Baltic maritime powers regarding the Barbary corsairs. He is therefore a representative European-Mediterranean actor for the earlier period of European interaction with the Levant before the Navigation Act and the Marseille edict. He possessed a highly detailed understanding of the recent history of North Africa, of "his own times," or rather the present, and also of the Africa of antiquity – the Roman and Ptolemy's Africa – blending and merging those two Africas with each other. While doing so, however, the pre-Ottoman Arabic times were completely lost.

There were not many other authors on Villeneuve's and Shaw's lists, and they could either claim not to be specifically well informed about the North African region at all (de Thou), or they focused on the present or the most recent past.[21] Herbelot's famous *Bibliothèque* (1697), marks a new era in Oriental studies, but as it belonged to the genre of early Enlightenment encyclopedism, it provides entries on authors in alphabetical order with short discussions about their works; there was no historical narrative.[22] The printed *Relation*

---

[20] Cf. Gramaye, *Thenae et Brabantia* (1606); Gramaye, *Arscotum Ducatus* (1634).
[21] Even Wansleben, who used Maqrīzī on the first four pages for a short historical introduction, then follows the genre of a travel narrative.
[22] This was expressed by the contemporaries themselves: "Dr Herbelot do's not in this particular do what I would haue him. And besides his Alphabetical method is so confounded . . ., that one is quickly tired, I went there often. I had read your 2d vol. the History of the Ommiad & Abbassid Calephs, & that of the Ginghizkamain Kings in him & I do not know that I was ever more weary in [deleted: of] my life. Yet his is a most excellent work" (Wotton to Ockley

universelle de l'Affrique (1688) by Phérotée de La Croix (ca. 1640–1715)[23] likewise concentrated on the present state of the regions.

Adding some early modern authors that were not on those two arbitrarily chosen lists does not change the picture concerning the point here made. Travelers, consuls, chaplains, Orientalists, translators like Paul Rycaut,[24] Thomas Smith,[25] Thomas Hyde,[26] and Theocharis Dadichi,[27] either published nothing, or concentrated on language, manuscript editions and the benefits that studying Oriental traditions purportedly had for Biblical studies. Alternatively, they published travel narratives, "the present state of" type descriptions or narratives of the recent history like Pierre Dan,[28] Lancelot Addison,[29] Johann

19.7.1718, Cambridge UL Ms. Add. 7113, n. 18). Cf. Dew, *Orientalism*, 169, 191–196.

[23] Signing on the title only as "sieur de la Croix," he is not to be confused with the sieur de la Croix, sécretaire of the ambassador Nointel, nor with François Pétis de la Croix father and son.

[24] Anderson, *An English Consul*, 210–247; the most historical of Rycaut's works, the *History of the Turkish Empire*, covered the period 1623 to 1677.

[25] Both works of Thomas Smith on the Greek Church and on "the Turks" were descriptions of their "present state."

[26] Thomas Hyde was, as Orientalist and Oxford librarian, a renowned expert of the Levant, Barbary and Ottoman affairs and served the government for several translation services (cf. SP 71/3, f. 158, Oxford, May 25, 1690 and f. 323r; Toomer, *Eastern Wisedome*, 298). He did not publish a real historical narrative work.

[27] Dadichi was a learned Greek whose father Rali was in contact with Salomon Negri and the Halle mission. He regularly translated documents from the Barbary and Ottoman powers for the English government into French in the 1720s/1730s (cf. SP 71/7, 71/17, 71/18, 71/22, 71/23, 71/28, 71/29 and passim). He apparently possessed an important collection of manuscripts but did not use his skills for publishing any Arabic work or a historical treatise, cf. Thomas Hunt's letter to Rawlison, February 9, 1743/44, Bodlein Ms. Rawlinson letters 96, f. 130r: Dadichi had worked on an edition of the New Testament in Arabic.

[28] Dan, *Histoire* (1637).

[29] Addison, *West Barbary* (1671) starts only with the recent Morocco history since the year 1508. The first part (up to p. 55) is identical with "The narrative of the revolutions of Barbary" in BL Ms. Sloane 3495. This part is strongly dependent on Mármol. It does not seem that Addison – whether he is the author of Sloane 3495 or whether he just used that manuscript for his book – had any independent source. For the second half of the book, Addison relied on his own experiences as chaplain at Tangiers, as he did for his *Discourse of Tangier*. This textual comparison shows that even someone like him who had lived in North Africa for seven years still possessed no "deeper" insight into pre-sixteenth-century North African history. The medieval part of Mármol was even cut off deliberately as "This Narrative of the revolutions of Barbary shall not bee derived beyond the Annals of our owne Memory" (BL Ms. Sloane 3495, f. 1r).

Frisch,[30] Laugier de Tassy,[31] Jean-Baptiste La Faye, Jean-Baptiste Tollot, Franz Ferdinand von Troilo, d'Arvieux[32] and Gottschling.[33]

The nescient leap over the non-existing Middle Ages is a typical pattern found in printed descriptions, histories, as well as in the French general *mémoires*. It is present in 1731/33 with Christian Gottlieb Ludwig: In his diary on the history of Tunis and its revolutions (*mutationes*) he made that "jump" from antiquity to the sixteenth century and Charles V, and he even exposed the prevailing ignorance:

None of the African kingdoms was troubled more by revolutions than the kingdom of Tunis. In Antiquity, i.e., in the times of the Carthaginians and the Romans, the historians testify that it was moved by several struggles. I do not want to speak here of the time after the birth of Christ and of the Middle Ages because of ignorance and the obscurity [sc. of that history], but I want to enumerate some revolutions that happened during the times of Charles V until the present day.[34]

A 1729 *mémoire* on the history of Egypt by the vice-consul Rosset of Rosetta/Egypt is similar. He developed a historical narrative starting with the beginning of populating Egypt by the children of Cham, the

His existing correspondence in Bodleian Ms. Tanner 35 only shows him as a fervent reformer of his district as Dean of Lichfield in 1683/84, but does not reveal more about his Oriental interests.

[30] Frisch, *Schauplatz* (1666) largely makes use of Pierre Dan and adds several slavery stories and accounts, but no comprehensive history.

[31] Tassy, *Histoire* (1724) never reached back before the sixteenth century and the Barbarossa/Charles V episodes (only a passage taken from Gramaye on p. 55).

[32] La Faye, *Etat* (1703) and Tollot, *Nouveau voyage* (1742) are pure travel accounts; Troilo, *Orientalische Reise-Beschreibung* (1676), 616–655, an Algiers captivity narrative. D'Arvieux, *Mémoires* (1735) contain small historical introductions in some chapters, but they never go back beyond the sixteenth century barrier either.

[33] The works of Gottschling, *Thunis* (1710); Gottschling, *Fez und Marocco* (1711); Gottschling, *Algier* (1720) are purely compilatory and belong to the genre of state description; for Algiers it starts with Barbarossa.

[34] "Nullum inter Regna Africae tantis fuit obnoxium mutationibus quam Regnum Tunetense. In Antiquis temporibus tempore scil. Cartaginensium et Romanorum, variis agitatum fuisse vicissitudinibus Historici testantur. De Tempore post Christum natum et in mediis seculis nunc ob imperitiam et obscuritatem loqui nolo sed tantum e tempore Caroli quinti usque ad hodiernum diem varias enumerare volo mutationes." (Ludwig, "Observationes miscellaneae durante itinere Africano scriptae," UB Leipzig Ms. 0662, entry December 21, 1732, p. 124).

son of Noah, but after reaching the period of the Arabic conquest in 644, he jumps to the Ottoman conquest of Egypt in 1517 within one and a half sentences.[35] Even long after the golden age of the Oxford Orientalists, after Galland, Renaudot, Pétis de la Croix and Cardonne, the jump over the Middle Ages remained the usual. In an anonymous *mémoire* for the French ministry entitled *Détails historiques* of 1777, for example, that leap from the destruction of Carthage to the sixteenth century is performed within one and the same sentence.[36] Some thousand years are left without mention. The same is true for an undated but post-1776 historical *mémoire* about Algiers: the chronological distance from the seventh to the sixteenth century is covered, again, in one sentence.[37]

An analysis of the titles listed by Villeneuve and Shaw reveals the absence of the Arabic Middle Ages, not visible at first hand, and this is completely consistent with the findings in the *mémoires*. The "jump" from Ancient to Modern times was customary and, due to the lack of knowledge, a necessary procedure. This happened usually without any mention and explicitness, so, in state of nescience. Was this lack of some hundred to even a thousand years for the regions between Fez and Cairo a general feature of a mental map shared by all individuals concerned or were there some specialists that were better informed, mainly the Orientalists?

---

[35] "Mémoires d'Egipte, pour Monsieur de Maillet. En 1728," AN AE B I 968 [not fol.], filed after letter of Rosset to Paris, Rosetta July 12, 1729.

[36] "Après la destruction de Carthage, dont il existe encore quelques ruines aux environs de Tunis, l'Afrique fut divisée en six provinces qui passèrent sous la domination des Romains, et furent ensuite conquises par les Califes mahométains, les Gouverneurs que ces Princes y avaient établi, s'emparèrent peu à peu de l'autorité souveraine, et Maroc devint le siège principal de leur domination vers le commencement du seizième siècle. Mahomet Gouverneur de Tunis pour l'Empereur de Maroc qui était occupé à se déffendre contre l'Espagne, se rendit indépendant, et prit le titre de Roy, il laissa trente quatre enfans que Muley Hassan, qui était le dernier de tous, fit poignardés, Araschid fut le seul qui eu le bonheur d'échaper à ses cruautés, il se mis entre les mains d'Airedin Barberousse fameux Pirate, alors Maistre d'Alger, lequel fit une descente à Tunis à la tête des troupes du Sultan Soliman." ("Détails historiques" 1777, AE La Courneuve MD Afrique 7, fol. 3r–v).

[37] "Les Grecs furent Maîtres de la Mauritanie jusques en 643. Les Arabes Mahometans, arrivés par l'esprit de fanatisme, désolèrent l'afrique: plusieurs s'y établirent. Le Gouvernement d'Alger [appartenait] successivement à des familles et à des Peuples différents. Il paroit qu'en 1505 les Algérines se gouvernoient en République" (AN K 1355, n. 22).

Arabic History in the Blind Spot of Early Modern Oriental Studies

What Shaw and the Marine archive list (probably for Villeneuve) do not mention and what was not used by any of the authors listed above are the contributions to Arabic historiography by early modern erudite Arabists such as Erpenius, Pococke, the Maronite Abraham Ecchellensis and Hottinger. This is probably due to several reasons. First, there was some divide between the practical culture of the consular network as represented by Villeneuve and the academic culture of Orientalists, even if the chaplain and professor Shaw is just the example of how those spheres were merging, albeit more on the British side than the French. Shaw certainly knew the work of his Orientalist precursors, but a further reason for the absence of these Arabists' work is that they did not produce works about the North African region. And finally, even if we extend the regional focus to medieval al-Andalus on the one hand and to Egypt and the Middle East on the other, representations of the Arabic Middle Ages remained extremely fragmented until the eighteenth century. Erpenius' 1625 translation of the Coptic historian Girgis al-Makīn was a revolution. For the first time the Arabic perspective (even if written by a Coptic Christian) on the Arabic expansion and elements of the crusades became accessible to the West. However, it was concentrated exclusively on the regions of Syria and Egypt. Furthermore, Erpenius stopped his translation at the year 1118 (the later parts were not edited and translated until 1955).[38] Into this same edition Erpenius added the twelfth-century *Historia arabum* by the Archbishop of Toledo, Rodrigo Ximénez de Rada, a rare compilation that contained excerpts from several Arabic histories and chronicles. De Rada's work covered the period from Mohammed to the year 1145, but for the later centuries it focused exclusively on the Spanish Arabic kingdoms in al-Andalus.[39] Pococke's work on Arabic history started in the *Specimen historiae* of 1651 with a very small extract and Latin translation of the Arabic version of the *Mukhtaṣtar* of the Eastern Jacobite Patriarch Bar Hebraeus (d. 1286).[40] In 1663 he edited the

---

[38] Erpenius, *Historia* (1625). It stops with Lib. III, cap. 9 on p. 300. Cf. den Heijer, "Coptic Historiography," 88–95. Al-Makīn relied heavily on the earlier thirteenth-century Coptic historian Ibn al-Rāhib (edited only in 1903) and on Eutychius, the latter edited by Abraham Ecchellensis in 1661.

[39] Maser, *Historia Arabum*, who underlines that Rodrigo's work is the only European work on the Arabs of that kind that has been preserved (p. 70).

[40] Pococke, *Specimen* (1650). Cf. Toomer, *Eastern Wisedome*, 160–162. Pococke's text itself ended with a short mention of the diversification of Islamic

full Arabic text, provided a complete Latin translation, and added a brief supplement with an enumeration of the "kings" and caliphs of the Abbasids, the Ottomans, the successors of Tamerlan and of Persia.[41] For these lists, he used the many Arabic manuscripts he possessed from Abulfeda, Al-Maqrīzī, Ibn Duqmāq, Al-Jannābi, Ibn Yūsuf, Muhammad ben Abī Surūr al-Ṣiddīqī and an unspecified *historia chalifarum*.[42] While Holt and Wakefield have shown that the manuscripts of several other Arabic historians were deposited in the Bodleian during the seventeenth century,[43] there was almost no text written by a European user of the library apart from this short list of kings and caliphs compiled by Pococke that referred to them. Nor was the Maghreb covered either as Pococke did not consult any historical manuscript (such as Ibn Khaldun) about that region. The amount of historical content added by Abraham Ecchellensis[44] and Hottinger's *Historia orientalis*[45] was also not of importance in terms of either the region or period in question. Abulfeda's historical work remained

religion into different branches up to the end of the ninth century (a.h. 270). For the complex relationship between Bar Hebraeus' Arabic text, his *chronicon syriacum* written in Syriac and for its character as a word-by-word transposition of the *Kāmīl* of Ibn al-Atīr for the high Middle Ages (period 447–599/1055–1202) cf. Conrad, "Bar Hebraeus"; Micheau, "Le Kāmil d'Ibn al-Athīr."

[41] Pococke, *Historia* (1663): after p. 367 and the "Index [nominum]" followed on Hhh1^R–Hhh4^V an "Index annorum Hejrae" pointing to each year with the page number which helped to transform a text, whose very pre-modern organizing principal was simply the dynastic succession of reigning caliphs, into an early modern history organized along the abstract sequence of the flow of time.

[42] Pococke, *Historia* (1663), B2^{R-V}. Cf. Wakefield, "Arabic Manuscripts." For the identification I am referring in the following to Uri, *Codices*, pars I and to Wüstenfeld, *Geschichtsschreiber*. Abulfeda manuscripts had already been deposited in the Bodleian with Kenelm Digby's library and Pococke possessed himself one. Pococke also brought Maqrīzī to the library; the fourth text in his list is the *Gemma pretiosa de gestis Regum et Sultanorum*, written by Ibrāhīm ben Muhammad Ibn Duqmāq (d. 809) in a.h. 801–805/1398–1402; Pococke used the Bodleian Ms. Arab. 680 which contains a continuation to the year a.h. 906/1500; Al-Jannābi (d. 999/1588), also deposited by Pococke (Pococke 176, Bodleian Ms. Arab. 785); Ahmad ben Yūsuf ben Ahmad's work, a continuation of Al-Jannābi of the year 1008/1599: Ms. Bodleian Arab. 771 = Pococke 246; al-Ṣiddīqī: Bodleian Ms. Arab. 832 (1042/1632) = Pococke 80.

[43] Holt, "Arabic Historians"; Wakefield, "Arabic Manuscripts."

[44] The supplementum on the sequence of the Arabic "reges" in Ecchellensis, *Chronicon orientale* (1651/1729), 205–216 is less than to be found with Pococke. Cf. Heyberger, "L'Islam."

[45] Hottinger, *Historia* (1651), cf. Loop, *Hottinger*, 190f.

unpublished in early modern times.[46] Even Abulfeda's more famous work, the *Geography*, owing to a series of coincidences, was never published entirely before the very end of the eighteenth century. The translations available during the period of this study ended with his description of Egypt, and his next chapter on the Maghreb remained in manuscript form.[47] Other very small portions of Abulfeda's work had been integrated into Ramusio's sixteenth-century Italian translation[48] and, probably translated by Abraham Ecchellensis into Latin, into Melchisédech Thévenot's late seventeenth-century collection of travel narratives, but these concerned the Middle East and India.[49]

For England, Oriental studies went into a decline for decades following Pococke's death, men like Ockley, Gagnier and Shaw being exceptions. Concerning the reconstruction of Arabic history, matters were not entirely different in France. The work of specialists like Ockley, Antoine Galland and Pétis de la Croix, indicate that there were attempts to translate or at least consult various works of Arabic historians, between 1670 and 1720/30. However, virtually nothing of that was incorporated into a more or less continuous historical narrative in a Western language that would have been easily accessible to consuls,

---

[46]   Cf. only Reza, "Abū al-Fidā'." Jean Gagnier edited and translated in 1723 only the small part concerning the life of Mohammed.

[47]   John Greaves had published a Latin translation of the chapter on Transoxania (1650) and Hudson – mentioned by Shaw – had published in his *Geographi minores* in 1711 the parts on Arabia and Egypt. The chevalier d'Arvieux had rendered those sections into French as an appendix to d'Arvieux, *Voyage* (1717). Jean Gagnier edited those parts again: Abulfeda, *Descriptio* (1740). In a letter draft in Bodleian Ms. Map. Res 73, Gagnier writes to "his Lordship" – probably to Lancelot Blackburne, Archbishop of York as Lord Almoner that financed Gagnier's professorship – that he had already spent all subscription money for the seventy-two pages printed and that he would need at least £200 more to print the rest of the "Geography of Abulfedah." Therefore, this edition also remained a torso: the existing copies stop in the middle of the text.

[48]   A small longitude/latitude table for locations from Anatolia to Uzbekistan (Samarkand) is to be found in Ramusio, *Delle navigationi* (1559), vol. 2, f. 18r.

[49]   Thévenot had obtained part of that Latin translation from an "Arabe de nation" professor of Arabic at the Roman university *La Sapienza* using a Vatican manuscript (Thévenot, *Relations* (1663), vol. 1 [pagination is not consecutive], 18–22) which must have been Abraham Ecchellensis, who held the chair of the Arabic language at the Sapienza from 1636 to 1644 and again in 1652, and that for Syriac from 1653 to 1663. He knew Thévenot personally from his stay in Paris, cf. Heyberger, "Abraham Ecchellensis," 39. On Thévenot's project, proposed to Colbert in 1669 and never realized, of a complete edition of Abulfeda's *Geography* cf. McClaughlin, "Une lettre."

merchants and men of politics in the network of the Mediterranean trade empires.

The many manuscript descriptions and histories left by François Pétis de la Croix and by the sieur de la Croix have still to be investigated thoroughly.[50] An initial perusal suggests that neither produced anything that differed significantly from what was published by travelers and consuls like La Faye, Laugier de Tassy or Tollot in their own narrative works. For some parts of his historical study of the Ottoman Empire, the sieur de la Croix used the Ottoman historian Hezārfenn ("Hussein Effendi"), a disciple of Kātib Çelebi, and so his work then represents to a small extent a palimpsest transmission of Ottoman manuscript history writing that was otherwise not accessible in Western Europe.[51] However his passages on Tunis, Algiers, and Tripoli and the Barbary coast are original only on what concerns recent history. They contain nothing on the medieval period[52] or, when he contrasts the "old" pre-Ottoman constitution of Tunis with the "modern" of his own time, he merely copied from Leo Africanus: the "old" constitution was the one described by Leo for the time around 1500.[53] François Pétis de la Croix the younger, a renowned Arabist, royal translator, and professor at the *Collège royal* for oriental languages made the usual "jump" from Tripolitane antiquity to the recent history of Tripoli around 1500 or, more precisely, from a brief mention of the Arabic conquest of Africa in 788 to the Algiers of 1500 in his descriptions of both of those cities within a few

---

[50] Cf. the Bio-bibliography Sebag, "Sur deux orientalistes français."

[51] "Journal du sieur de la Croix secrétaire de l'ambassade de France à la Porte Otomane" (partie I), BNF NAF 10839, pp. 219–223: a conversation with "Ussein Efendi Historien turc" about metempsychosis; partie II : BNF NAF 10840, f. 78: a very short history and description of the Ottoman Empire which seems to be a summary of Hezārfenn's *Tenkīhü't-tevārīh* (written 1671–1673) of which he took a copy on September 3, 1679 (cf. Wurm, *Ḥüseyn b. Ġaʿfer*, 126). The work belonged to the "bestsellers" of Turkish chronicle writing at that time, cf. Tezcan, "Ottoman historiography," 198.

[52] Journal du sieur de la Croix, partie III: BNF NAF 10841, pp. 184–365 "Du Roiaume de Tunis": all historical passages start not earlier than around 1500.

[53] Sieur de la Croix, "L'Égipte Ancienne, et Moderne – Seconde Partie. De l'Égipte Moderne" (1704), BNF NAF 4989, pp. 243–368 contains chapters on Tripoli, Tunis and the Barbary Coast and is very similar, but not identical with the Journal. The only chapter that claims to refer to "medieval" times, chapter 18 "De la grandeur des anciens Rois de Thunis" is a nearly word-by-word copy from Leo Africanus (Jean-Léon l'Africain, *Description*, vol. 2, 385–388).

lines.[54] It is surprising that he was not trying to add more medieval history, as it seems that he worked on French translations of unidentified Arabic and Ottoman histories of al-Andalus, Morocco, Algiers, Tunis and Tripoli, of which nothing was published and not much has survived in manuscript either.[55]

Antoine Galland, usually known as the translator of the *1001 Nights*, was sent with the French ambassador Nointel to Constantinople primarily for exchanging knowledge about Greek Orthodox Eucharist theology, as has been seen in the last chapter. He also used this time there for the collection of manuscripts and he continued to work on many translations of Arabic, Persian and Ottoman texts, among them historical works. With just two exceptions, all of those translations remained in manuscript and had apparently no wide circulation. Moreover, the only printed one about Ottoman history was a discussion of the recent revolt of the Janissaries in 1622.[56]

When the Cambridge Orientalist Simon Ockley worked on his history of the Arabs after Mohammed, he seriously endeavored to produce something new and original from unpublished sources. In order to do so, he traveled to the Bodleian at Oxford in 1706 and first asked the son of Edward Pococke, Edward the younger, for advice about which

[54] François Pétis de la Croix, "Relation abregée de la ville et Royaume de Tripoly de Barbarie," presented to Pontchartrain, Paris, April 26, 1697, BSB Munich Cod.gall. 729, f. 3ᵛ–4ʳ: jumps from Antiquity to Charles V within one sentence, the rest of the description contains no longer passages about the past. François Pétis de la Croix, "Alger" (presented to Pontchartrain 1695): jump from 788 to 1504 within two sentences.

[55] Sebag, "Sur deux orientalistes" lists a "Histoire des Arabes d'Espagne depuis le VIIe siècle jusqu'au XIVe siècle" by an unknown Arabic author, the "Rawd al-Quirt'ās – Histoire de Fès et du Maroc depuis 145/762 jusques en 726/1326" of Ibn Abī Zar (Wüstenfeld 392, BNF Ms. fr. 25288), an Ottoman "Histoire de Tripoly de Barbarie, a Histoire d'Alger" and an Arabic "Histoire de Tunis depuis le XIe siècle jusqu'au XVe siècle" by an unknown Arabic author which de la Croix would have all translated. But only a translation of the "Futūh' al-Shām – Histoire de la conquête de la Syrie par les Arabes" of Al-Wāqidī survives, cf. Sebag op. cit., 116 n. 4. Ockley had already made use of that work. A partial French interfoliated translation of Al-Mas'ūdī's *Kitāb akhbār al-zamān* of Pétis de la Croix in BNF Ms. Arabe 1473.

[56] Cf. Abdel-Halim, *Galland*, 214–243 and the list on 476f. His translation of the 1648 work by Hādjdjī Khalīfa as "Chronologie mahométane" and of the fifteenth- and sixteenth-century annalists Khodja Sa'd al-Dīn and of Mustafa Na'īmā as well as his "Les Ayoubites ou histoire de Saladin" remained in manuscript.

manuscripts to use. Pococke the younger had left Oxford some thirty years previously, but he pointed Ockley to the list of authors provided by his father in the supplementum to the *Specimen historiae*.[57] He had also taken some notes and catalogue excerpts with him to Mildenhall. That list – as well as d'Herbelot's *Bibliothèque* – shows that the Orientalists knew well what they did not know, or, better: that they knew well where one should search if one would like to produce a historical narrative of the kind planned by Ockley. The non-knowledge became specified. Ockley should look for:

> Abulfedae Historia
> Epitome Yafti ab initio Heg. ad annum 750
> Potuho lDiarbecri / Photuho lMesr both, I think, tārīkh
> Al Jannabij historia
> Al Jauhari historia Regum Chalifarum &c.
> Jellalodin Ossyoti historia Aegypti
> Ebn Hamedi annales
> Tārīkh al-Shuṭī, Tārīkh a l-Ṭabarī
> Tārīkh al-Nuwayrī, Tārīkh al-Qud'aī
> Ebn Mesudi, Marā a l-Zamān ab an: Hej: 480 ad 530.
> Zebdol Pecrati.
> There [sc. are] also some other Historiae Chalifarum,
>   whose authors names I have not.[58]

Obviously, Pococke's son remembered the authors used by his father well, and he also named other historians not used by the elder Pococke. Some of the names he provided are hardly recognizable, others not identifiable at all, but others also show that the knowledge of several works were linked to one or another precise copy in one library, usually the Bodleian.[59] One may doubt if the list with its very unstable

---

[57] Cf. above n. 42.

[58] Edward Pococke Jr. to Simon Ockley, Mildenhall, April 20, 1706, UL Cambridge Add. 7113, Thanks go to Abigail Krasner Balbale and Bernard Heyberger for help with the Arabic titles.

[59] Abulfeda is a clear case. The "Epitome Yafti" obviously refers to the specific Ms. Bodleian 672 = Pococke 357 (Wüstenfeld 429, 1) which is a version of the *speculum principis* of Abd Allāh b. As'ad al-Yāfi epitomized by al-Dhahabī; "lDiarbecri" should refer to Husayn al-Diyārbakrī (Wüstenfeld 526, d. 966/1558), there are three Mss. in the BL; which history of Egypt was meant with "lMesr" seems unclear, perhaps the *Tarīkh-I Misr el-kadim* of Souheïli Efendi (1030/1621). "Al Jannabi" is clear (cf. above n. 42), "Al Jauhari" seems to indicate the title of the work of Ibn Duqmāq as indicated by Pocock instead of the name of the author. "Jellalodin Ossyoti" refers to Gelaleddino Alsoiuthi

transliteration was helpful to Ockley. Even the specification of what was known by name but unknown by content was not precise. Ockley probably found his own way more quickly through the Bodleian manuscript collection. His work really relied on the Bodleian manuscript sources, but it did not go beyond the year 87/706. Therefore, it did not change anything concerning the ignorance about the medieval History of the Arabs and of North Africa.[60]

This is a very short list meant to help in one of the rare moments when an early modern Western author, Ockley, tried to write a History in a Western language on Arabic history based on hitherto unused sources. However one could understand even the whole *Bibliothèque* of Herbelot as a very long list and vast specifier of unknowns, and not as an encyclopedic codification of knowns. Many or even most of the 8,204 entries were names of authors and works copied from the encyclopedic work or "card catalogue" *Kash al-zunūn* by Kātib Çelebī (that contained 15,000 entries) which was present in two manuscript copies that had been sent to Paris by Nointel and by Guilleragues in the 1670s and in 1682. At both times Galland had served as collector of manuscripts in Constantinople.[61] The entries were only to some degree fruits of Herbelot's original research or reading. Even today it is still

---

(Ms. Bodleian Arab. 777 and 780 = Marsh 107 and 293 which is in fact Al-Suyūṭi, cf. below). "Ebn Hamedi annales" probably refers to the same author and work that Pococke named "Ibn Yusuf" (cf. above n. 42). Al-Mas'ud's work had entered the Bodleian with Huntington's manuscript collection (cf. Wakefield, "Arabic Manuscripts"). "Tārīkh al-Shutī" refers probably to the *Tārīkh al-khulafā* of Jalāl al-Din al-Suyūṭi (d. 1505, Holt, "Arabic Historians," 451 nr. 19). "Tārīkh a l-Ṭabarī" to the *Tārīkh al-rusul wa'l-mulūk* of Abū Ja'far al-Ṭabarī (d. 923, Holt, "Arabic Historians," 450 nr. 1 points to the presence of fragments only in seventeenth-century Bodleian library Mss.); the Tārīkh of al-Nuwayrī (d. 1332) refers to the historical work of which Reiske made a first Latin translation in 1748 (Wüstenfeld 399), but there is no trace of a manuscript in the Bodleian; For "Ebn Mesudi, Marā [sic, instead of Akhbār] a l-Zamān ab an: Hej: 480 ad 530" cf. Holt, "Arabic Historians," 450 nr. 3 who points to a Selden Ms., but cf. Bodleian Ms. Arab. 815 = Huntington 168). I am not sure about the identification of "Zebdol Pecrati": Herbelot uses for several historical works the name "Zobdat"; nor of "Tārīkh al-Quḍ'aī": one could think of historical works of al-Qāḍī al-Fāḍil or al-Qāḍī 'Iyād if the author is meant, or one could think of al-Kindi's *Kitāb al-Wulāh wa-Kitāb al-Qudāh*.

60 Cf. Kararah, *Ockley*, 84f. The two works are Ockley, *Conquest* (1708) and Ockley, *Saracens* (1718/57). Ockley, *Account* (1713) contains as the only "historical" part a biographical account of the then currently reigning Muley Ismael (pp. 82–97).

61 Dew, *Orientalism*, 179 n. 33 on Galland's involvement.

sometimes difficult to trace what his translations of names into French were meant, either by him or originally by Çelebī, to refer to. To a large extent, it was a huge list of probably existing texts and authors that were unknown.[62] Herbelot's *Bibliothèque* was something like an enormous query form or list of what still had to be researched, it was not a manual compilation of what was already mastered and easily available in Western Europe.[63]

Changing Knowledge Infrastructure? – Enfants de Langue and Libraries

One may ask how the academic system and the republic of letters reacted to that growing awareness of ignorance and sense of the lack of knowledge. There were some institutional reactions that one could label today as forms of orchestred "cultural politics." An intensive but brief moment of institutionalization through the serial translation of major works in Oriental languages was put into practice during a reorganization of the *enfants de langue*. The royal librarian and president of the French *Académie des sciences*, the abbé Bignon, started to use the consular network for cultural politics in close collaboration with the ambassador in Constantinople, Louis-Sauveur, Marquis de Villeneuve (1728–1741). In 1669, Colbert established the *enfants de langues* as one of the measures to professionalize commercial politics and the capacities of the consuls during negotiations with the Ottoman powers. Every three years, six children were sent to the Capuchins in Constantinople to teach them "Turk" and Arabic to serve afterward as dragomen or to become even consul, vice-consul, or chancellor. This number was later doubled, and for some time it was Armenian youths, not French, who pursued that carrier path. In 1721, the number was fixed at ten and it was decided that only French boys could be eligible. They were first trained at the lycée Saint Louis in Paris, then in Constantinople, where Father Romain, the school's headmaster, became an important figure. Starting in the late 1720s, but most

---

[62] Leiser, "A figurative meeting."

[63] Cf. Dew, *Orientalism*, 168–204; Laurens, *Aux sources*, 37 – 8158 / 8204 / 8600 according to what one counts as lemma. While Laurens was eager to demonstrate the "exactitude et le sérieux du travail d'Herbelot" (which was not disputed; that was not the point), Dew shows well that the *Bibliothèque orientale* "was by no means a trouble-free instrument of Western knowledge or a rational panorama" (ibid., p. 175).

particularly from 1731 to ca. 1737, a system had come into being, in which the more experienced youths translated long manuscripts, most of them concerning Ottoman history.[64] These efforts produced over 120 translations in easily readable French, each sent from Constantinople to the secretary of the Marine, Maurepas.[65] This, therefore, might have been a window of opportunity to close the gap of the unknowns. But only five of the manuscripts today preserved in the Bibliothèque nationale were translated from Arabic, nearly all the others from Ottoman Turkish. The two Arabic historical texts translated between 1733 and 1737 were a compilation from and continuation of Abulfeda and Ibn Yūsuf, that is, two authors Pococke had at least already used even if not translated completely.[66] Most of the Ottoman works translated did not concern the early dynastic history of the Ottomans, instead they were chronicles of the times *after* the conquest of Constantinople. In sum, the French translated not medieval history but rather "contemporary history," with the exception of a few works on Egypt.[67] All the rest of North Africa, a region of distinct importance for the politico-economic system of the Levant trade, was almost entirely absent in these translations.[68] Still not much information is known about the use or impact of those translations. None of them was published in print.[69] One who probably profited from those works during the 1730s, was Denis Cardonne, who followed the dynasty of the de la Croix in the office of the royal translator-in-chief and who, therefore, had direct control over the manuscripts in the library. In 1765, he published a *Histoire de l'Afrique et de l'Espagne* which now, finally, could claim to have been written by consulting many unpublished manuscripts about the Arabic "Middle Ages," among them, likewise finally, Ibn Khaldun. Even Cardonne continued

---

[64] Hitzel (ed.), *Istanbul*. For the Venetian precursor cf. Rothman, *Brokering Empire*, 165–186. Cf. also the papers of Venture de Paradis, BNF NAF 9137 (unprinted); CCM J46–J48.

[65] Cf. Berthier, "Turquérie." The number of translations done was surely higher than the list of manuscripts now in the BNF. There are traces of similar, but decentralized attempts beyond Constantinople among dragomen and consuls.

[66] BNF Ms. arabe 1539–1541 and Ms. arabe 1559.

[67] More than the half of all titles and nearly 90% of all historical works in Berthier's list concern the "recent" history from around 1500.

[68] Only one report of the conquest of La Goulette and Tunis in 1573/75 in BNF sup. turc. 926.

[69] Berthier, "Turquérie," 292.

to refer to Mármol as one of four Western authors consulted aside from the manuscripts. This meant that, even the work of the *enfants de langue* did not cause a great change in how the Arabic Middle Ages were forgotten. Their translations went largely unused in both the consular network itself and in Paris.

Another way to control those findings is to look at the content of libraries and the individual possession of books or manuscripts by merchants and consuls in the Levant. One might suspect that individuals in the Levant possessed more manuscripts and better knowledge about Oriental culture and history than in the far-away imperial centers of London and Paris. However, a short survey of what remains of library catalogues rather proves the opposite: academic Orientalists were probably in closer touch with pre-Ottoman Arabic history behind medieval English college walls and in baroque Paris libraries than the Levant merchants trading silk with Armenians in Aleppo. Among the 113 books that the Levant Company's library in Constantinople (Pera) contained in 1711, only Hottinger's *Historia orientalis* falls into the category of learned Arabist production. Ludolph's *History of Ethiopia* and his *Lexicon Aethiopicum*, Kircher's *Lingua Aegyptiaca* and four Ottoman items[70] suggest a certain expansion of the library's coverage into realms of academic Orientalist studies, but otherwise, its holdings focus completely on Anglican and Catholic theological works, ancient authors, as well as English works on state affairs and general philosophy. The Smyrna library (109 titles in 1702) held Seaman's *Grammatica Turcica* (1670) and Marsham's *Canon Chronicus aegyptiacus, ebraicus, graecus* (1672), but these were the only "Oriental" texts among otherwise exclusively Western theological, historical, geographical and scientific works.[71] The only works with bearing upon the ongoing progress in academic Arabist learning at the Aleppo library (228 books in 1688) were the Arabic translation of Bellarmino's catechism (Rome 1627), the Arabic translation of Grotius' *De veritate Christianae Religionis* promoted by Boyle and Huntington, Seaman's *Grammatica linguae Turcicae* (1670) and Matthias Wasmuth's *Grammatica Arabica* (1654). Otherwise, the library's holdings consisted, again, solely of Western theological, historical and political

---

[70] "Catalogus librorum Bibliothecae Peranae" (Anno 1711), SP 105/78, pp. 300–303: A "Turkish Grammar," "Turkish Testaments," "Sepher Jari Mosheh" and "Sepher Kaphota" are listed.

[71] SP 105/145, pp. 301–304.

works.[72] Edward Pococke was present in the Aleppo library only as Hebraist with his commentaries on Micah and on Malachi (both Oxford 1677). Still, none of the three "public" libraries of the large English factory cities contained, for example, a copy of his *Specimen historiae Arabum*, and none possessed works of any seventeenth-century English Arabist studied by Holt, Toomer, Hamilton or Russell. Even Herbelot's *Dictionnaire* was not present in any of the three libraries. Likewise, the larger private libraries of English merchants – holdings of which we have some traces thanks to the inventories registered in the factories' chanceries after the death or bankruptcy of a merchant in Constantinople, Aleppo or Smyrna – almost never contained any Oriental books or manuscripts beyond a "Turkish" grammar, dictionary or a mostly English copy of the Koran that could have helped to inform its owner about pre–sixteenth-century Arabic history. This was true already for Lewes Roberts in the 1620s[73] and it was still the case in the eighteenth century.[74] An exception is the curious figure of the Aleppo merchant Rowland Sherman (d. 1747) who owned a considerable number of Oriental manuscripts – mostly Greek but also some Arabic and Ottoman ones.[75] While most

---

[72] SP 105/145, pp. 157–164. Cf. on the Aleppo library Zwierlein, "Coexistence and Ignorance."

[73] Cf. Zwierlein, "Coexistence and Ignorance."

[74] Library inventories of: William Lateward (SP 105/178, pp. 46–48, Constantinople, 1702/03, 57 titles: some items of higher philosophical character as Gassendi, Descartes, but no Oriental item, only an "old Greek bible"); Thomas Savage (Galata, February 9, 1708/09, SP 105/178, pp. 100–102: "A Turkish word Book" is the only item in a collection of 72 nearly completely vernacular titles in English, Italian, French); George Norbury (Pera/ Constantinople, March 2, 1710/11, SP 105/178, p. 305f.: 42 almost only vernacular titles, no oriental item); Woolley & Cope (SP 105/178, pp. 407–410: "seaman's Turkish & Latin Grammar" and "History of the Turks 2 vol. by Knoll & Rycaut" are the only relevant items of 71 books saved from fire by being thrown into a cistern); Lister Bigge (Aleppo, March 18, 1747/48, SP 110/ 73/2 (2), p. 52f., transcribed by Ambrose: Levant Company, pp. xc–xcvii: with the exception of a copy of the Koran, no oriental work among a nearly completely vernacular collection of 103 titles); Leiler Bigge (Aleppo, October 23/24, 1747, SP 110/73/2 (7), f. 51v–53v contains an unspecified "Arab book" as the only oriental item among 124 titles); George Wakeman (Aleppo, May 17, 24, 29, 1758, SP 110/73/2 (3), contains among 539 volumes in French, English, Italian, some Latin books, two volumes of "Turkish History" and an English version of the Koran as the only "oriental" works).

[75] SP 110/73/2 (7), f. 663–73v. Cf. White, "Brothers" and Zwierlein, "Coexistence and Ignorance."

merchants only stayed a few years to a decade in the Levant, he lived there for sixty years and apparently developed "[u]nlike most factors ... an interest in local culture and languages."[76] For the French side, what books French Levant merchants and consuls possessed is harder to survey. All in all, it seems that there are fewer inventories of deceased French than English merchants. This is probably because the French merchants usually spent even less time in the *échelles* than their English counterparts, due to the rigorous regulations instituted around 1700, and thus fewer French merchants died abroad.[77] The renowned Tunis merchant Nicolas Béranger apparently owned only a small amount of books.[78] In almost none of the library catalogues seen in the chancery records does a book in an Oriental language, or a book by a European Orientalist, appear.[79] For one of the most learned consuls around 1700, Claude Lemaire, we only know from such an inventory that he left "200 volumes of used books by diverse authors" which his son Louis took into possession in Cairo in 1722.[80] All this indicates for the actors of the French trade empire a situation similar to the English. If they possessed books in the Levant, they were overwhelmingly Western ones. The work of the academic Orientalists, fed by manuscript collections in the Levant, did not have a reciprocal effect on general knowledge available to and demanded by merchants and consuls in the

---

[76] White, "Brothers," 547.

[77] Cf. the inventories mentioned in Grandchamp, *Tunisie*, vol. 8 (1681–1700), 784, 911, 918; vol. 10 (1705), 244, 315

[78] Only French books, among them 31 volumes of the *Journal des Sçavans* and some religious works, cf. Grandchamp, *Tunisie*, vol. 9, 386, 395: inventory and auction of his effects in 1707. Indeed, in the "Mémoire pour servir à l'histoire de Tunis depuis l'année 1684," Béranger did not cite any printed work, cf. the edition Sebag, *Régence*.

[79] Cf. the list of books Pierre Armény de Bénezet (d. 1775), consul at Algiers and Tripoli/Barbary, AE Nantes, Tripolie de Barbarie 706/PO/1/carton 39: much of French Enlightenment publications from Voltaire's *Candide* to Rousseau, Pascal and Montesquieu, but no oriental book. Inventory of Nicolas van Masseik, Aleppo, 1784: 58 books, only in vernacular European languages, as far as the very general notation allows one to judge: AE Nantes, 18 PO/BO n. 40, f. 40v. Only books on geometry and arithmetics in the inventory of the scribe Etienne de Mourguz, Tripoli/Barbary, 1711, AE Nantes 706 PO/1/carton 45, f. 85v. The comparatively rich library of the former ambassador Andrezel (1727) contained many guides to architecture, travel narratives as the voyages of Thévenot, a "description de la Morée" and the *Affrique* of Mármol, Busbecq, "L'Estat de l'Empire Turc," but no item in any Oriental language, AE Nantes 167 PO/A/ 36bis.

[80] Cairo, February 24, 1722, CCM J 35.

Levant, at least not before 1750. Merchants and consuls were very rarely themselves readers of Arabic or Ottoman manuscripts, least of all, it seems, of pre–sixteenth-century historical works.

A cursory glance at the "knowledge infrastructure" confirms that medieval Arabic History was almost completely unknown, in Europe as well as by the actors in the Levant themselves, at least until around 1750. This is to be taken as the most apparent case of a complete lack of a knowledge unit or item. For other elements – knowledge about social realities, customs, learning, political and constitutional structures – it would not be so clearcut in the way that the simple non-presence of centuries of history indicates, but knowledge was likewise absent or porous. Consequently, the unknown Arabic Middle Ages should serve as a most striking example, a "totum pro parte."

Filling the Void: Structure Instead of Content[81]

People can act in a state of nescience despite that nescience. What stands in the place of the unknown then, if we do not suggest that individuals can orient themselves through a total *tabula rasa*, that they can be guided by pure void? The answer lies in the use of preformed structures instead of content. General theoretical and methodological reasoning within postcolonial studies, on Orientalisms and similar concepts have dealt with the idea that Europeans have relied upon cultural prejudgments when observing foreign civilizations. In a certain way, this could be considered part of a history of ignorance concerning the inhuman negative functions of consciously or unconsciously ignoring. I will not enter into a methodological debate at this point.[82] I rather concentrate on the cognitive part of the problem and not on the function of (d)evaluation and judgment, even if the one is linked with the

---

[81] The argument of this paragraph is developed in greater detail in Zwierlein, "*Conversiones.*"

[82] It is not the aim of this study to enter into a long discussion of the paradigmatic Said, *Orientalism*; the very engaged Ibn Warraq, *Defending* nevertheless makes convincing points about the shortcomings of the East/West dichotomy in Said's work. For a more affirmative use of the "orientalist" paradigm for the premodern texts cf. Curtis, *European Thinkers*. The lessons drawn from Said are now the mainstream approach, as testified by Goffman, *Ottoman Empire* and collective volumes such as Maclean (ed.), *Re-Orienting.*

other, as different functions of language and cognition are likewise always processed simultaneously.[83]

As the subjects of this study normally just "jumped" over the Arabic Middle Ages, there was no structure that would have replaced the missing centuries in their historical concept. For the histories and "present-state-of"descriptions for the early modern period, a great deal still remained in very basic forms and was written on "porous ground" as already stated. An important structural element used, instead of unknowns or in combination with bits of knowns, was the idea of North Africans' alleged inclination to "revolutions," a very attractive notion for men of politics, consuls and merchants. This idea had its roots in the well-known cyclical concept of history which had developed since the Renaissance and is often treated as a part of an early form of a "philosophy of history."[84] It emerged within political theory to understand the "life" of states for planning their present and future. It was important for each political observer to know the general structure of state activities, their ups and downs and to potentially be able to predict that movement. This general idea was reprojected into history writing. Either Polybius' *Historiai* (VI, 4–9) and the Aristotelian version (1279a/b, 1286b), or its first original reformulation by Machiavelli, was the major starting point,[85] and it was soon generalized by Bodin and his followers beyond the narrower framework of the cyclical succession of constitutional states as in book VI of his *Methodus* and Chapters IV,1 and IV,2 of his *Respublica* into a firm topos of "changes or revolutions of states – *De conversione* or *de mutationibus rerumpublicarum*" within the emerging academic and para-academic political theory.[86] The cyclical pattern itself, serving as a structural code derived by political theorists *from* historical examples, could also be reintroduced *into* newly produced historical narratives as an ordering principle. This led to the result of when empirical knowledge about a period, events or the framework and development of a non-European polity was scarce, or even non-existing, at least

---

[83] Bühler, *Sprachtheorie*. No purely representational speech act is possible without elements of expression and appellation, likewise no "pure" cognition without elements of emotional judgment is possible.

[84] Cf. Schlobach, *Zyklentheorie* for the French sixteenth- and seventeenth-century authors; Dubois, *La conception*, 465–500 and Seifert, *Rückzug*.

[85] Machiavelli, *Discorsi* I, 2; Sasso, *Machiavelli e gli antichi*, vol. 1, 3–65.

[86] Quaglioni, "Scienza politica"; Dockès-Lallement, "Les républiques"; Bianchin, "Conversiones rerumpublicarum."

something could remain: the organizing structural pattern itself. It could be used in this way rather unconsciously. Bodin was, in fact, a completely Eurocentric author. Asia, the Far East and the Americas played almost no role in his political œuvre which was so fundamental for early modern political theory.[87] He saw the inhabitants of America, the "Scythians," Australians and Asians beyond the Euphrates as subject to more adverse climatic conditions, which allegedly consigned them to a lesser state of development. He only referred to these peoples as examples for those natural laws that applied everywhere in the old pre-Grotian Roman law sense of *ius naturale*.[88] For the question of the *conversio rerumpublicarum*, he could not use examples from Brazil or Asia, because, according to Bodin and his fixed application of Galeno-Hippocratean climate theory,[89] those people were missing a fundamental predisposition to that applicability. There was, allegedly, no complex social stratification into *populares* and *optimati* that would allow for the constitutional distinction between aristocracy and democracy. Consequently, the transition from monarchy to other forms of governance, or, conversely, the transformation into monarchy, was impossible.[90] The only extra-European territories that played a constant role within Bodin's political œuvre were the Ottoman Empire, the North African regions and Ethiopia.[91] For those, no distinction of theoretical importance between European/extra-European, aside from general derogatory labels like "Barbari," was drawn. The southern Mediterranean shores were structurally, even if not qualitatively, completely comparable to Western or Northern European examples. Bodin thus juxtaposed the North African "kingdoms" with the Florentine Republic, characterizing the latter as specifically prone to "tragedies" (*tragoediae*) thanks to the accounts of Machiavelli (*Istorie fiorentine*), Antonino da Firenze and Poggio Bracciolini (*Historia*

---

[87] Solé, "Le comparatisme," 411.
[88] Bodin, *Methodus*, 464 (chap. VI, nr. 127) – "propterea fratres sororibus, parentes liberis, matrimonio iungi natura vetuit, vel potius, lex omnium gentium communis; quae vel apud barbarissimos Americos inviolabiliter servari dicitur." That is the reversal of Inst. I.2pr. and Dig. I.1.1.3 where matrimony is classified as a universal norm.
[89] Most strongly in the book V of the *Methodus*. Cf. Tooley, "Bodin" and Crahay, "Pays, peuples" on Bodin's idea of the superiority of those who lived between the 42nd and 51st parallels of northern latitude.
[90] Bodin, *Methodus*, 470 (chap. VI, nr. 133).
[91] Solé, "Le comparatisme," and Le Thiec, "L'Empire ottoman."

Florentina).[92] The history of Florence, he argued, was a succession of sedition and revolt, one after the other, during the aristocratic and democratic forms of its governance. Only with the ascension of Cosimo I (1537), and the transformation into a monarchy, had the state found greater stability according to Bodin.[93] The same "law" of monarchies possessing more cohesion was then to be demonstrated by comparing the only other people that were said to surpass the Florentines in terms of unruliness, the peoples of North Africa, for which he returned to Leo Africanus. According to Leo, the inhabitants of Segelmessa first chose a popular constitution after the fall of the royal government. A great deal of sedition and many revolts followed because no citizen could support the reign of his fellow citizen. This led in the end to the complete dissolution of any constitutional form of the society at all.[94] For Bodin, this served to prove again the superiority of monarchical regimes.[95] Both peoples, Florentines and (North) Africans, were comparable because of their strong inclination to revolts, to "tragoediae." In so doing, Bodin used the vision of a cyclical structure of a succession of events ending catastrophically transferred not only from the Polybian theory of *anakyklosis* but also from the realm of tragedy theory. The necessity of failing and what was more commonly called from the times of Donatus the "catastrophe" was a structural pattern of movement on the *theatrum mundi* connected to the specifically revolt-prone spirits of such peoples if they chose non-monarchical forms of government.[96] Political theorists who followed Bodin in developing the topos of *De conversione rerumpublicarum* – Italian authors such as Albergati, Sammarco, the French Le Roy, Grégoire, Duret, the Germans Avenarius, König, Pregitzer, Müller, Hassel and Conring – became even more "Eurocentric" and Classicist, largely referring only to examples from Greco-Roman antiquity. Its application to non-European circumstances actually took

---

[92] Bodin, *Respublica* (1586), 383. [93] Bodin, *Respublica* (1586), 381–383.
[94] Bodin, *Respublica* (1586), 383.
[95] "Quamobrem Affricae populi, cum populares & optimatum status ferre non possent, monarchiam vbique fere constituerunt." (Bodin, *Respublica* [1586], 383).
[96] For the metaphor of the *theatrum mundi* cf. Cavaillé, *Theatrum mundi*; Christian, *Theatrum mundi*; Blair, *Restaging*, 153–179; for the reception of ancient tragedy theory, the catastrophe and its transfer to other discursive realms in early modern times cf. Briese and Günther, "Katastrophe," 161f., 175–179.

place outside political theory, within the writing of history, cosmographical and travel narratives. Regarding the Levant, this emerges in its early forms with the description of the Barbary Coast by the Trinitarian Pierre Dan.[97] Chapter I, 1, 2 of his well-known *Histoire de Barbarie et de ses Corsaires* (1637) bears the title "The Barbary [sc. region] is a bloody theatre where a great quantity of tragedies is played." Beyond the very general baroque topos of the world as theatre, the Bodinian theory of the Barbary people especially inclined to "tragedies" becomes visible later. The inhabitants of the Barbary regencies, he posited,

originate from the traitor Cham, the second son of Noah, who execrated that unfortunate person for the high crimes by which he found him to be burdened; and by that they [i.e. the Barbary people] inherited his damnation as well as his place of residence – if their country is the same land of Cham of which the Holy Scriptures are speaking. Whatever has been the case, what kind of denaturated monsters Africa might ever generate, it does not produce more cruel ones than the peoples of this region that one calls Barbary: To such a degree that one may say of them that they are effectively the most barbaric of all men.

It is also possible that for this reason that damned region has always been the fervent theatre where libertinage, heresy, impiety have represented to the whole world an infinity of bloody and tragic acts. Several people that have inhabited the region from time to time, all infidel and inhuman, such as the Goths, the Vandals, and the Saracens, have played the principal personages of that tragedy.[98]

Dan explicitly excluded the Phoenicians, Romans and Greeks from that category of people prone to tragedies and he wanted to characterize only the Barbary inhabitants as the once settled Saracens, that is, Arabs or Berbers as such. This is in fact a reference to the alleged inclination toward "everyday" revolutions fixed in the Arabic character. On a second level, Dan also received the concept of the *conversiones rerumpublicarum* in the sense of the rises and falls of the seat and possession of the sovereign authority. Several passages clearly show a familiarity with contemporary political theory's technical terms.[99] This leads to a three-level concept of politico-social movements in history: the general natural inclination of a people to incessant small

[97] Dan, *Histoire* (1637).   [98] Dan, *Histoire* (1637), 2f.
[99] Dan asked about the "Diverses revolutions *en* la souveraineté de Tunis" which is a quite prudent and precise wording, aware of the abstract notion of sovereignty as developed in theoretical discussions since 1576 (Dan, *Histoire* [1637], 106f.).

revolutions (level 1); then, from time to time the transformations of the state's political framework (level 2). Finally, enormous revolutions like the Arabic invasion of North Africa at the juncture between antiquity and the Middle Ages, resulting in civilizational oblivion (level 3). That third level is close to the big climacterian changes that Bodin thought to happen every 496 years. But Dan, not following the Pythagorean-Platonic speculations on numbers instead referred, like Louis Le Roy, to the idea that the round form of the earth was not just a sign of its perfection (as had Pseudo-Aristotle in *De mundo*), but also an indicator of the constant "flux and reflux" of all secular things, of an "inconstante revolution des choses de la terre."[100]

From one of his notebooks, it is also evident that Edward Pococke was himself cultivating the concept of socio-political revolutions in an even narrower scientific form via an anti-astrological tradition that united Calvinists like Erastus with Bodin against the Philippists around 1600. It is also visible as an interpretative pattern in one of the rare "Orientalist" works that really *was* widely diffused among the Levant merchants, William Seaman's *Turkish dictionary*, a work whose preface Pococke apparently helped edit. Therein, the rise and fall of Middle East kingdoms and empires, most of all the Ottoman, is linked to the rise and fall of Christ's kingdom in a way echoing the *conversio* topos.[101]

So, around 1650, the cyclical model was present in the French as well as in the English Orientalist and Levant trade circles in a distinctly technical form in political theory and in the writing of history. Moreover, within that guiding perceptional framework, the identification of the Barbary people as particularly prone to revolutionary behavior following Bodin/Dan became an implicit and sometimes explicit leitmotif commonly employed by the trade empire actors. Lancelot Addison entitled his short account of Moroccan history – both the 1666 manuscript and the 1670 printed version – as *a Narrative of the revolutions of Barbary*. It certainly featured many examples of revolts, sedition and power struggles, but the same was also the case for the sixteenth/seventeenth century history of any European state. Nevertheless, neither Guicciardini, de Thou, nor Camden conceived

---

[100] Dan, *Histoire* (1637), 82.

[101] *An Regum et regnorum mutationes Cometae certo praenuntient?* in: Bodleian, Ms. Pococke 428, f. 35r-v and the preface on fol. 5r–9v to be compared with the different printed version Seaman, *Grammatica*, f. A2v–A4v.

their works of proto-national European historiography predominantly as a succession of "revolutions." Even a specialized history of the French Wars of Religion would run under the antiquizing title of *Civil Wars* (Davila). For the Barbary state, it became the emblematic term for the quintessence of Barbary history itself; its peoples were putatively habituated to "incessant troubles" and the "prognosticks of [] revolution[s]" were an integral part of political life.[102] Discrete examples of unrest, such as that of consul Thomas Baker who wrote a *Narratiue of the Reuolutions in Tripoli in November and December 1672* were then contributing to the collection and accumulation of "revolutions" in the European historical memory of the Levant.[103] From Cairo, consul Truilhard submitted his reports for several years under the title "suitte des Reuolutions d'Egypte."[104] When a new civil servant took office in Paris or in the Levant, and the central administration ordered a new general survey of that official's remit, its expectation of what that history might be is telling: Maurepas requested for instance a *"mémoire … about the revolutionary events and other matters which might have a relationship to the history, the religion or the customs and manners of the country."*[105] The abovementioned vice-consul Rosset of Rosetta in Egypt organized the succession of events depicted in his historical accounts of the early eighteenth century as a series of revolutions. "Mehmet Pacha de Gedda … must have amassed a great deal of goods during seven years of government and considerable revolutions."[106] In these texts, the notion of "revolution"

---

[102] BL Ms. Sloane 3495, p. 9f = Addison, *West Barbary* (1671), 23f.

[103] SP 71/22, box 2, f. 62–67: "A Narratiue of the Reuolutions in Tripoli in Nouember & December 1672" (another copy of the same text on f. 68–72).

[104] AN AE B I 320, f. 214r–215v: the narrative starts with "Au Caire le 10e aoust 1729," f. 216r–217v "Suitte des troubles du Caire," f. 265r–267r "suitte des Révolutions d'Egipte," f. 276r–278v "suitte des Révolutions d'Égipte." The reporting vice-consul, writing a day-by-day account employed that structural pattern as the only form that could give an overarching sense to the "chaos" experienced.

[105] AN AE B I 1116, f. 63r: Alexandre Lemaire to Maurepas, Tripolie of Syrie, March 25, 1726.

[106] "Des Affaires concernant l'Égipte, et la nation françoise," 1710, non pag.; "De la Révolution arrivée au Caire en 1726," sent with letter of February 23, 1726; "De la Révolution arrivée au Caire en 1726," sent with letter March 4, 1726; cf. also the "Mémoires d'Égipte pour Monsieur de Maillet en 1728" with the paragraphs on "Révolutions en 1720" and also frequently the theatre metaphors: "En Février 1726 la scène commança par le massacre de Ibrahim Effendi," AN Paris AE B I 968.

becomes something like a stable state of political conditions in motion. It is true that there *were* in fact important "troubles" in North Africa, particularly in those years when effective sovereignty passed from Ottoman overlordship to a succession of dignitaries of the rank of Bey or Dey from local dynasties, who needed only formal approval from the Sultan. In Tunis this began in 1705 with the reign of the Husayinides, in Tripoli in 1711 with the reign of the Karamanli; similarly, from 1711 the Janissaries directly elected the Dey of Algiers instead of an outside appointment by the Ottoman Porte and the Ottoman Pasha lost everywhere his effective powers. It is true that, in the 1720s and early 1730s, important conflicts and wars still continued between these city states. However, warfare was also nearly endemic between European states until at least 1715, and contemporary historians of the European Wars until that of the Spanish Succession still did not use the concept of "revolutions" to characterize the specific inclination of a European ethnic group or "nation" to the quickly changing state of constitutional settings.[107] This form of perceiving and describing the present state of affairs was really a structural pattern that was a fundamental element of the professional administrative *horizon d'attente* within the imperial network. When the former consul of Cairo, Maillet, who was now directing affairs from the Ministry in Paris, read a draft version of Rosset's 1728 *Mémoire d'Egipte*. He wrote back in 1729 about some points that Rosset should change in a revised version:

Art. 2 The revolutions that have been happening for several years have their place in your report. But they should be preceeded by similar events that have been happening since the conquest [sc. of Egypt] by sultan Selim, with a dissertation that has to show that since time immemorial there have been two factions in that country opposed one to the other which the Turks have supported there for always finding their firm support with one of the parties to hold down and diminish the credit of the other when it would rise too

---

[107] The general terms of "troubles," "wars" or the Classicist "civil wars" is far more usual until the first half of the eighteenth century. Even the "Glorious Revolution" which marks the starting point for a more philosophical concept of revolution was still described by Clarendon, *History* (1702/03) as "Rebellion and Civil Wars," while only early Enlightenment French observers made stronger use of the idea of an epochal shift linked to "révolution" (Dorléans, *Histoire des révolutions* of 1693, Bayle's *révolutions d'état*, Voltaire juxtaposing English *revolutions* to *séditions* in other countries, cf. Koselleck, "Revolution," 719–721). The southern Mediterranean periphery in its post-Bodinian framing seems thus chronologically as a precursor.

much and for preventing one of the subjects from usurping the absolute authority and revolt against them which has not happened since the conquest of sultan Selim.[108]

Maillet thus asked that the "empirical" report about Egypt's history be written in a straightforward manner following the deductive conceptual pattern of the revolutionary habit of its inhabitants cultivated in Paris. Rosset answered that such a history of all the revolutions would be too long to be useful and not easy to produce.[109] He added that one could extend it even further back to the Mamluk era. The succession of revolutions and the changing balances of power would be the same for the pre-Ottoman period. There was the idea of a constant predisposition to revolutions, and the idea was self-explanatory. There was no need for an empirical supplying of the general constant pattern, the structural matrix was enough to understand the region and the people. Moreover, this structural pattern to fill the void remained stable for a long time – not in the form of elaborated academic reasoning like in the *conversio* literature, but in semantically linked simplified forms that directed the perceptions and the narration of professional observers.

During the 1730s/1750s, an important change within the values attributed to the Barbary states' governments took place. With notable surprise, several observers from the imperial networks noted a quite enduring stability of the regencies.[110] This engendered a reversed perception which culminated in the late example of Venture de Paradis, who could characterize the Barbary regencies in the late 1780s as

---

[108] AN Paris AE B I 968, unfol., March 28, rec. August 13, 1729.

[109] Rosset, "Réponses auxdites demandes [sc. of Maillet], en addition au Mémoire de 1728," ibid. (unfol.).

[110] Consul André-Alexandre Lemaire described the government of Algiers in 1731, modified again in 1751, as an aristocracy with the Dey as *primus inter pares* who acted as an enlightened monarch, as servant of the state, succeeding to organize now "une politique très rafinée et des mesures assez justes pour empêcher les fréquentes révolutions qui ont tant de fois agité cet État." The effect thereof lent the state great stability: "Depuis environ [blank in the Ms.] ans les choses ont entièrement changé de face et quoyqu'on voye toujours subsister en aparance le même fonds de Gouvernement, l'on peut dire cependant que le sistème de politique est tout différent et qu'il s'est rectifié sur des *règles beaucoup plus sages* et plus propres à assurer l'État des gens en charge et à reprimer l'insolence du peuple." ("Observations sur le Voyage des Eschelles de Barbarie et du fond du Levant en l'année 1731," enlarged in 1751, AE La Courneuve MD Alger 13, f. 127r–144r plus f. 199r which belongs to the same *mémoire*; shorter first version of 1731 in AN MAR B VII 311).

governments full of Roman virtue and stability opposed to the depraved European civilization which had disintegrated into factions and was now itself becoming revolutionary.[111] The categorical difference between the premodern "little revolutions" and what would become the epochal age of progressive Revolution was not yet evident, so the Barbary regencies could serve as a late-Enlightenment counterfactual mirror for Europe, as had been the case for China or Persia in earlier decades. These reevaluations were not necessarily a sign of higher empirical intimacy with the cultures of North Africa, even if Venture de Paradis was a former *jeune de langue* and highly renowned Orientalist. Rather, the reversability of the topos from the revolutionary disposition of a people to their being an example of high stability shows how structural "code" could prevail over content. The general idea of what one could expect and what the constitutions and the peoples of the Barbary coast represented forged a pattern that could structure even unknown parts of history. In any case, one could expect that it had always been like this and that the future would not deviate from this model.

Both could be done in a state of nescience, simply by ignoring centuries of history and both broad and finely-grained empirical data about the "Barbary people" on the one hand; and on the other by filling the void with the somewhat depraved, faded and altered topos of the revolutionary character of those people.

Growing Awareness; Structure Replaces Content; Standardization and Spatialization

The emergence and first flourishing of Arabic and other Oriental language studies, alongside the growth of the "knowledge infrastructure" of libraries and institutions such as the *enfants de langue* with their chronological increase of "serial translation," are indicators of an awareness of the unknowns touched upon here. However, the latter phenomenon was rather late (under Bignon-Villeneuve around 1730). No *mémoire* writer explicitly noted that he did not know exactly how the political circumstances had really been, nor did he question whether or to what extent the character of the Barbars was inclined to revolutions. *Mémoire* writers took it as a given. They assumed it to be true,

---

[111] Cf. BNF NAF 9134; Venture de Paradis, *Tunis et Alger*, 209f., 219f.

just as the cyclical pattern predicted. They merely used the pattern. However, around 1700, in the texts themselves and in the processes of their production – as far as the sources provide evidence thereof – a new awareness of ignorance arose in different forms concerning geography, history, society and culture, and it became increasingly explicit and recognized by authors and imperial agents. First, a theme already touched upon in the previous paragraph – Dan's portrayal of the end of the Christian era and the Arabic invasion as the extinction of a civilization – has to be addressed, and will be in the next two sections. How did early modern individuals conceive of the potential eradication of important aspects of a civilization's history in general since the Renaissance? How did European travelers and observers describe and imagine the alleged "post-oblivion" low-level state of culture they claimed to consistently encounter? Second, if, through that experience, the explicitness of ignorance, uncertainty and the constant effort to frame the borders between known and unknown increases in the narratives, the question then is what the driving forces were behind that process.

The Oblivion Recognized

In the paragraphs above, I discussed several well-known reasons why most European merchants and consuls might have been "ignorant", in the common sense meaning, of the different forms of more elevated Levant cultures and most of all of their past strata and remains, why and how even the most learned Orientalists had blind spots concerning large fields like Arabic and Ottoman historical writings, or at least transferring what they might have read in manuscripts in Leiden, Paris and Oxford into Latin or vernacular narratives that would have been easily accessible for agents of the trade empires, right until 1750. A more nuanced understanding demands noting that the long duration of European "blindness" was rooted not only in European incapacities or disinterest, but also in the North African situation after 1500. Ottomanization had created a rupture within the local Arabic cultures of learning themselves. While many schools attached to mosques survived, higher forms of philosophical training apparently found little support under Ottoman rule. Ghalem confirmed that during the sixteenth and seventeenth centuries, "chronicles, annals and other historiographical genres are rare ... there is no interesting work of historiography [sc. newly written in early modern Algiers]. One cannot deny that the Ziyyānīde and Hafsid historiographical tradition" had been lost. He furthermore pointed to the important fact that Algerian learned men such as Muhammed Abū Rās had noted themselves in the eighteenth century that "we are living in times when knowledge and its institutions are in decline. History, literature, genealogy, the chronicles have disappeared. The learned men are not interested anymore in these historical genres as was the case during the past."[112] Echoing Ghalem's portrayal of Algiers, Gringaud has written of a "historiographical desert" in pre–nineteenth-century Constantine.[113] Abdesselem has likewise shown for Tunis that historiography of significance and length was rare during the early modern period. Ibn Khaldun, himself a Tunisian and author of perhaps *the* most important work on medieval "Berber" history, became forgotten in his city of birth. A manuscript copy of his work had to be reimported from Fez "in the middle of the eighteenth century."[114] This seems quite typical, as Morocco, being independent from the Ottomans, was a stronghold for the maintenance of Arabic traditions after the Reconquista. Still, even there the number of dynastic histories and lists of biographies produced between the sixteenth and the eighteenth centuries is quite small, and hardly any of them were used before the middle of the eighteenth century by Europeans or in contact with Europeans.[115] The situation was surely different in Constantinople and Cairo,[116] but the perceptional gap and bias between the European Imperial actors and the Arabic and Ottoman cultural production continued on that epistemic level for a long time.

Nevertheless, there was a certain cultural "Renaissance," for instance in Tunis around 1700, which is today usually reconstructed

---

[112] Muhammed Abū Rās, "'Ajā'ib al asfār wa latā'if al-akhbār," ms. 1633, Bibliothèque nationale Algiers, transl. and ed. by L. Arnaud, Revue Africaine 1878–1884, cited by Ghalem, "Historiographie," 115.

[113] Grangaud, *La ville imprenable*, 307. [114] Abdesselem, *Les historiens*, 51.

[115] Lévi-Provençal, *Les historiens* comments on Ibn el-Qādi (960/1552–1025/1616), el-Ifrānī (1080/1669-ca. 1145/1732) and an anonymous and states "il n'y a au total, du début du XVIIIe siècle à la fin du XIXe, que deux savants marocains qui aient pris la peine d'écrire l'histoire de leur pays" (p. 142).

[116] But still the first Ottoman historian to use Khaldun again in early modern times was only Kātib Çelebi, cf. Fīndīkoğlu, "Ibn Haldunism" cit. after Wurm, *Ḥüseyn b. Ga'fer*, 155. Fleischer, "Royal Authority" is confirming that there is no earlier clear trace of literal reception of Khaldun.

mostly on the basis of the information found in the rare contemporary Arabic historians and biographers, mostly the *Dhayl* by Khūja, a part of his history consisting of a large sequence of biographies of local learned men.[117] It is curious to note how these Tunisian writers, religious learned men, marabouts, "intellectuals" shortly before and during the Husayinid Renaissance around 1700/20 must have lived exactly at the same time in the same place where the European travelers, consuls and chaplains worked and observed – even the most curious, learned and experienced of them as Thomas Shaw or Christian Gottlieb Ludwig. Nonetheless, a few scattered exceptions aside, there are no traces of reciprocal perception in their works. For other issues – politics, economy, corsairing – concerning Tunis and the other regencies, even modern historians rely extensively and sometimes even exclusively on European texts, but for the "cultural life," European sources reveal, from the point of view of current Maghribian historiography to be nearly without value. The "cultural life" of North Africa was not only forgotten for the period of the Middle Ages, but it remained obfuscated, not understood, or just not observed and discussed even during early modern times. This, even though European empires had already established a tradition of continuous observation, measuring and narrated accounts of the *economic and political* affairs of the countries of North Africa. In view of this astonishing juxtaposition of what one might consider as the coexistence of closed epistemic enclaves or epistemic parallel societies, it is important to ask to what degree and how these forms of ignoring and ignorance became recognized in a reflexive form.

Christian Gottlieb Ludwig's still unpublished account on the cultural situation in Tunis is probably the most detailed by a European that has survived for that period, this question, and the Barbary regencies.[118] I will therefore use it even if he did not belong to the British or French trade empires in a formal sense. As a Protestant member of a group

---

[117] The best account of Tunis' cultural life, education and science remains Abdesselem, *Les historiens*, 21–97 who never uses a European source; Sebag, *Tunis*, 235–251 mostly refers to Abdesselem, and, as he, to Khūja's *Dhayl* and to Ibn Abī-Dīnār. Cf. similarly Chérif, *Pouvoir et société*, vol. 1, 295–336.

[118] "Observationes Miscellaneae durante itinere Africano scriptae," UB Leipzig Ms. 0662. The manuscript is divided in a diary and a second appendix, separately paginated, both in chronological order of the days. I have profited from directing and reading a Master thesis by Marianne Timpe on the diary (March 2012).

headed by Johann Ernst Hebenstreit sent by the King of Poland and Elector of Saxony August in Dresden, they traveled to the North African shores under British protection.[119] Their mission belonged to the early "scientific" research missions as they were financed now increasingly more often by the courts and academies.[120] How did he gauge the existence of learning, and how did he assess the lack of knowledge, both on the part of the Berbers and of his own compatriots?

Ludwig found a very good and rare partner for conversation and source of information in Tunis, the physician of the Dey or one of them. This was a former Spanish Trinitarian monk who had converted in Tunis not to Islam, as Christians always feared, but to Judaism. We do not know anything more about the biography of that "Joseph Carillo" as he figures in Ludwig's writings, but he apparently not only read and spoke Arabic and Ottoman, but also had access to manuscripts and probably to the Dey's library itself. It is to some extent astonishing how seldom one finds mentioned in earlier travel narratives an intensive, ongoing contact between a European narrator and a learned person in the Barbary regencies. This was different for Constantinople, as the accounts of the group around Nointel – Galland, Pétis de la Croix – show.[121] Thanks to the "observationes Carilli," some of them being fruits of oral exchange, some probably narratives in Latin given to Ludwig, which summarized local Arabic material that Carillo had at his disposal,[122] Ludwig could give an account of the "studies in Tunis."[123] He commented in Western vocabulary on the theology and "legal" thought – the Hadith as "juris civilis canonici et moralis corpus" –, and vividly described the elementary schools. The simplest schools would be held in simple "boutiques," small rooms on the first

---

[119] Cf. consul Charles Black to the Duke of Newcastle, Algiers, February 7, 1731, SP 71/7, f. 234r; vice-consul Jon Berswicke, Tripoli, October 15, 1732 and letter of Hebenstreit to Duke of Newcastle, Algiers, October 10, 1732, SP 71/ 23, f. 74–78; consul Lawrence and Hebenstreit to the Duke of Newcastle, Tunis, April 16/17, 1733, SP 71/28, f. 331r, 333r. They had also letters of safe-conduct of the other powers, lodged with the French consul in Tunis and met the French consul in Tripoli, but were otherwise nearly exclusively in contact with the Protestant European representatives.

[120] Döring, "Afrikaexpedition."   [121] Cf. above n. 51.

[122] Cf. for the identification of one fifteenth-century source of Carillo/Ludwig Zwierlein, "Natur/Kultur-Grenzen," 28–30.

[123] "Observationes quaedam de Tuneto D. Carilli" (November 14/15, 1732) and "Continuatio Observationum D. Charilli" (November 24–26), UB Leipzig Ms. 0662, part I, pp. 97–103, 114–115.

floor of houses, eight to ten feet wide but not nearly as long, quite similar to the everyday small shops that sold clothes, silk, food, fruit and cooking utensils. School was held in the same rooms as the shops and right in their midst. One man would teach ten or more boys, writing something on a wooden table that the class would then read aloud together. Those who wanted explore literary culture more deeply, and Ludwig stressed that he had known several merchants who pursued that goal, would frequent schools attached to the mosques. Ludwig then gave an idea of the different disciplines and fields of knowledge that were present in Tunis around 1730. He stressed the public/private division of that culture: natural philosophy was allowed to be taught publicly in the mosque schools only in terms of "natural theology" and "metaphysics," known in Arabic as "ʿAllim." Philosophy as such, medicine, arithmetic, algebra, geometry, astronomy were not publicly taught but were discussed in "private academies" at home and among circles of curious people.[124] He referred to a culture of History writing and emphasized that there were even contemporary authors writing the "provincial History of their times," and that there was a culture of poetry writing and recitation, the reception of Egyptian poetry and that different poetic genres – love poems, heroic and satirical poems – were cultivated.[125] Concerning philosophy, he noted that locals followed Democritus' atomist principles while rejecting the Aristotelian teaching concerning the origin of matter, for medicine they followed the peripatetic doctrine. Plato and Euclid were still taught. He portrayed how the teaching of philosophy was allowed even if it came from before Muhammad's times, legitimated by the idea that

---

[124] "Hinc Philosophia nullibi publice docetur quae ad physica attinet. Enim vero Theologia naturalis sive illa Metaphysicae pars quae de Deo ejusque attributis agit, quam Allim vocant, publice in scholis acutissimeque explicatur. Jure an injuria scientias naturales Fidei zelotypi damnent in medio relinquo, hoc tamen illic et experientia innotuit, quod ubi plus calet minerva, ibi plus tepet fides. Medicinae artem nullum discere vel profiteri methodice vidi. Arithmeticam, Algebram, Geometriam, Astronomiamque non omnino negligunt, quae Scientiae licet publice non doceantur, privata domum Academia et curiosorum conturbernio tractantur" (ibid., p. 99f., November 14, 1732).

[125] Ludwig writes of a "Mahomet Veluzir" who had written the History of the currently reigning Husayn Ali. This must refer to Muhammad Sa'ada (1677–1757), cf. Abdesselem, *Les historiens*, 193–205, even if it remains unclear how Ludwig understood and wrote the name. The same Mahomet Veluzir (so Sa'ada) is named together with "Hamuda Carvi" as the most prominent poet. Cf. UB Leipzig Ms. 0662, p. 100.

there might have been a partial recognition of the truth embedded into those preprophetic writings. This was an argumentation similar to how one had legitimated, since the times of Christian Humanism, the moral and even religious potential of pre-Christian or pagan writers like Cicero.[126] He corrected an idea prevailing in Europe that Muslim society was especially intolerant of dissent by referring to some more or less public disputes that he had personally observed. It is quite obvious that he applied patterns of observation inherited from his experience of confessional conflict and division in Europe while describing what he observed in North Africa, narrowing the Ibādiyya minority of Djerba to "Protestants" facing the orthodox ("Catholic") Tunisian Muslims.[127] Even if the passages altogether generate more questions than they solve – what was the exact atomist teaching discussed in Tunis private "academies"? – and even if it is clear that one probably has to subtract a certain amount – but how much? – of European "coding" from the content described, it is clear that Ludwig perceived elements of the cultural Husayinid renewal. It is, as far as I see, the most detailed description concerning those topics in any European language.

At the same time, this most attentive perception of an awakening culture corresponds with Ludwig's "groping in the dark," with an interest in and eye for as yet unfound knowledge, materials not really or not easily obtainable. He was seriously searching for the remains of Arabic wisdom, considering, as many of his contemporaries did, the ruling "Turks" to be uncultivated and even destructive. He noted a story told among Tunisians that there had been an unknown king who had forbidden the study of any other science than theology. Locals ascribed the loss of knowledge to this mythical decision in the past. Ludwig had an unclear idea that there must have been some knowledge "out there" which had just not yet been found again. "[T]here [would] always remain here and there some learned men who preserved the

---

126 "Et quando dicitur illos vixisse ante Mahomedem, dicunt possibile esse quod doctrinae illius ante illum jam cognitas fuisse." ("De studiis Maurorum. Studium linguae arabicae," December 4, 1732, UB Leipzig Ms. 0662, part I, p. 110).

127 The differences within theology of predestination, the doctrines of grace and of free will, ibid., November 15, 1732, part I, 102f. Djerba was a sheikhdom governed between 1559 and 1800/967–1215 a.h. by the family of the Bin Jallūd, but was under the central government of Tunis. It was dominated by an Ibādiyya community. Cf. Merimi, "Sulaymān al-Hīlātī."

learning. Even in Algeria there would be probably here or there some men who were just so hidden that they cannot be found by foreigners." – He just had not met them.[128] Ludwig also connected the idea of knowledge loss and preservation to the mastering of languages and to the history and merging of the different peoples in North Africa. That is how the *lingua franca* had been formed as a merger and tool of communication between the inhabitants of those North African regions and the Europeans.[129] The merging of languages was perceived as similar to the merging of the population groups. The Bedouin Arabs living "under the tents" – a topos already used by Leo Africanus and by Arabic chroniclers –, were alleged to have been the "true descendants of the old Arabs" whom he held in great esteem. But those who resided in the cities were a "mixed people" that had come together from several different regions: Turks, Renegados, Spanish fugitives, various groups of Jews that had arrived in different times, during the Middle Ages and later, during the Reconquista. Ludwig stressed that "there is an impediment [sc. to describe properly "Algerian" and "Tunisian" customs] as the customs of the original inhabitants and those of the Turks that had come later to the cities have mixed and merged [*vermischet*] so much that one cannot say easily if this or that custom or manner is Moorish [i.e. Arab/Berber] or Turkish. Those who live according to a given custom do not even know themselves according to which rule they live" if one asks them.[130]

What is present in these notes is the acknowledgment of a double or triple cognitive barrier. First, it was not easy to establish contact with the inhabitants at all. Then, the everyday problem of translation and communication across the language frontiers posed a significant problem. Finally, it still remained unclear what age and origin one might attribute to what was perhaps observed or communicated by the

---

[128] "Tamen video quod in hac re ratio perditorum studiorum sit quaerenda. Manserunt tamen hinc et inde semper viri quidam eruditi qui studia illa conservarunt. Et credo quod etiam Algiriae et in aliis Regnis hinc et inde inveniantur viri quidam sed ita tecti ut non ab exteris inveniri possint." ("De studiis Maurorum in genere," November 2, 1732, UB Leipzig Ms. 0662, part II, p. 1).

[129] December 23, 1732, UB Leipzig Ms. 0662, part II, p. 126f. He was precise in characterizing the *lingua franca* as a commercial language that had progressively more Spanish elements as one approached to the coast of Morocco and more Italian the closer one came to Tunis.

[130] Susa, January 20, 1733, UB Leipzig Ms. 0662, part II, p. 141.

inhabitants as a custom or rule they followed at the time. These are methodological reflections not about errors corrupting some preexisting, but just not reachable, knowledge, but rather about gradations of loss and oblivion where it is hard to imagine, for a single observer like Ludwig, how knowledge could ever be obtained or created, where one sees no hold and no point of attack within the fog of multilevel ignorance.

After one and a half years in North Africa, not doing anything besides observing and taking notes, Ludwig came to recognize these cognitive barriers and, therefore, a specification of the realm of ignorance and its borders which he, according to his judgment, could not overcome. This stood in direct opposition to the rationalist Wolffian apodemic program of survey, research and civilizational comparison with which he had started the voyage and which he had noted on the very first pages of the notebook before departing from the southern European port of Marseille:

One has [sc. usually] a worse opinion of others than of one's own. A false love of our own [*Eigen Liebe*, German version of *amor sui*] seduces us hereto. We think of our vices as virtues . . . This is a serious error . . . which has its origin in the fact that we judge the customs of other people according to the yardstick of our own. How can one avoid that? By trying to understand the foundations of a rational teaching and politics of customs [*die Grund Sätze einer vernünftigen Sitten Lehre und Politic*] while trying to understand the customs of other countries' inhabitants and peoples according to those principles. We will then note very clearly that we have vices that other people avoid . . . and that we practice some virtues that other people miss.[131]

Ludwig's germs of an apodemic theory are a rare example within the Wolffian Leipzig circle around Gottsched where the usual deductionist rationalist philosophical program had to be brought into greater balance with empiricism because otherwise a scientific expedition made no sense.[132] The non-empiricist starting point of a traveler like Ludwig allows us to see, even more clearly than with a British or French voyager who was already an empiricist follower of Baconian or post-Cartesian stamp, how the expectation with which he started ended in

[131] "Urtheil von den Sitten der Völcker wenn solches genau sey," Marseille, December 18, 1731, UB Leipzig Ms. 0662, p. 7f.

[132] On the circle around Gottsched that promoted Wolffianism in Leipzig cf. Döring, "Sozietäten." Interest in Oriental studies was already cultivated by the professors Kehr and Clodius at that time: Preißler, "Orientalische Studien."

deception. The basic universalist rationalist program, that the perfection (*Vervollkommnung*) of each society[133] and of each individual was accomplished simply by comparing virtues and vices is destroyed during the very act of trying to do so, because that results in a growing awareness of the hidden barriers of ignorance bound with the very process of perception *prior* to any act of comparison.

The result is an unpublished and perhaps, from Ludwig's point of view, unpublishable and unfinished notebook, overflowing with loose ends. It repeatedly and explicitly uncovered the accumulated problems of ignorance on the side of the observed as well as on the side of the observer, both victims to what was, in contemporary terms, a civilizational oblivion.

A generalized vision of the cultural consequences of civilizational oblivion beyond variations of the topos of the Biblical flood arose within political and historiographical reflections only starting in the Renaissance. It seems that the empirical challenges of history writing of non-European regions was one of the most important stimulants for that.[134] The problem of the complete disappearance of civilizations and their histories was often formulated referring to a set of passages from Plato, Livy, and later Ficinian neoplatonic reasoning about the "eternity of the world," a term that had been condemned in the Middle Ages, but reinterpreted in functional and rationalist terms, most prominently with Machiavelli and some *reason-of-state*-authors.[135] This was also the underlying main problem that was encapsulated on the title page of an important work that inaugurated a new form of history, summarizing the experience of the European expansion beyond Europe during the fifteenth and sixteenth centuries, Raleigh's *History of the world* (1614).[136] This history by one of the most famous voyagers of

---

133   Wolff, *Oratio* (1726), 89: "Dirigebant adeo Sinenses actiones suas omnes ad summam sui aliorumque perfectionem tanquam ad finem ultimum."; Klemme, "Werde vollkommen!"; Ludwig had among other books a copy of Wolff, *Vernünfftige Gedancken* (first edition 1720) with him while traveling in North Africa.

134   Pagden, *The fall of natural man*; Woolf, *A global history* provides a rather comprehensive comparative overview of different forms of history writing in the world than a history of the epistemic challenges caused by globalization.

135   Zwierlein, "Forgotten Religions."

136   Cf. Sullivan, *Memory and Forgetting*; Ivic and Williams (eds.), *Forgetting*. Literary studies, however, usually apply modern theories of forgetting/oblivion (Eco, Todorov) to past sources and seldom investigate the past theory of oblivion itself.

his time, apparently mirrored the mindset of many of the early imperial British merchants and consuls, and it was also kept in many libraries of the Levant merchants,[137] and, while referring to the same Platonic theme of the eternity of the world in the preface,[138] it opposed the powers of triumphal history over oblivion in an Herodotean allegory,[139] approximated to death. The ambit of the imperial voyager, the task of describing and writing about the present, as well as research into the foreign past, and its integration into the total of history, the more one encountered new spaces, the more porous and full of holes this all seemed. This was the case in an even more profound way concerning the traditionally "known," "old" world – as was the Levant, significant portions of whose history was now recognized more and more to be as unknown as the history of wide regions in the Americas was.

One finds echoes of this idea in particular to Oriental and North African history beginning to surge around 1700, When Simon Ockley was working on the second volume of his *History of the Saracens*, using the Arabic manuscripts of the Bodleian Library in Oxford, his friend Thirlby encouraged him in his work by stating

I heartily congratulate with you and the learned world upon the success of the labours at Oxford. Fifty years of New History snatched from Oblivion and the dusty corners of a Library, will be a very considerable augmentation of the learning and knowledge of the age.[140]

Beyond the standard early Enlightenment rhetoric similar to the humanist "ad lucem proferre," it was not one author, or one text

---

[137] There were at least four copies of *The History of the World* in Lister Bigge's and Rowland Sherman's libraries in Aleppo, SP 110/73/2 (2), pp. 52 and 74, the abridged version of the *History of the world*, the *Marrow of history* was present in Woolley & Cope's library (SP 105/178, p. 409).

[138] Raleigh, *History of the world* (1614), D3^V–E2^R.

[139] "Ἡροδότου Θουρίου ἱστορίης ἀπόδεξις ἥδε, ὡς μήτε τὰ γενόμενα ἐξ ἀνθρώπων τῷ χρόνῳ ἐξίτηλα γένηται" – The contemporary English translation "to the end, that nether tract of time might ouerwhelme and bury in silence the actes of humayne kynd ..." (Herodotus, *History* (1584), f. 1^R) was closer to how the construction is usually understood today (Bakker, "The making," 3f.) and thus made understand the juxtaposition of History and Oblivion better than did the traditional Latin translation of Lorenzo Valla.

[140] Styan Thirlby to Ockley, living with his friend John Haywood, Fellow of St. John's College, ca. August 6, 1716, UL Cambridge Ms. Add. 7113, nr. 36. Cf. for that period of Ockley's life Kararah, *Simon Ockley*, 84f.

that Ockley was supposedly saving from oblivion in this passage, but a whole period and sequence of time. Fifty years of history could supposedly be acquired from studying manuscripts. This is a slight but important change that presupposed a concept of History that was starting to synchronize the different regional histories of the world as a unified history by becoming aware of "missing" knowledge about discrete elements within the broader historical narrative, blank spaces on the "map" of History. The unknown parts of History were not just hidden in unconscious nescience, instead they were specified in duration and character. One could push oblivion back and recapture History. What was not known had become specifiable, or at least quantifiable in terms of time. With this also came qualitative forms of specification.

The growing awareness of ignorance in precise moments of interaction and observation in the Levant as well as a more generalized theory and reflection by concerned figures that they were facing zones of oblivion and forgotten history that might be reconquered, leads to the question of what forces around the turn of the eighteenth century were causing this epistemic concussion, one that necessitated the transformation of nescience into realized and explicit forms of non-knowledge.

Epistemic Forces Making Unknowns Visible: Standardization and Spatialization of History

The answer to that question may be found in two interrelated parameters, the standardization and spatialization of *historia* perceptions as well as of *historia* writing.

Standardization

With the term standardization, I want to combine two phenomena that are usually not associated. The first are developments on the level of learned culture concerning the treatment of the forms of history writing. The second are trends within imperial administrations that entailed the construction of something like a bureaucratic memory of the history of the North African states, akin to what the trade empire had for the other regions where they were active.

The evolution of historical writing and the use of history have been studied extensively for the period from the Renaissance to later early modern times.[141] The sixteenth-century treatises *De arte legendi historiam* show how new conceptions of history as a field and pool of examples and causalities had developed, in addition to how techniques arose of "using" history by more systematic extraction than had been done in medieval times. All this has been well studied.[142] Structural matrices and patterns discovered in history and noticed by every reader had arisen, and most basically, but also fundamentally, the usual subjects of history had changed. European medieval history writing had known only a restricted set of worthy subjects: beyond the level of the universal history of the *mundus*, the only legitimate topics were prominent public individuals – popes, emperors, crowned and reigning potentates –, dynasties or institutions, mostly of the church, from monasteries to dioceses.[143] As is well known, the history of states and nations as historical actors themselves was a late arrival. Guicciardini wrote his *Storia d'Italia* when the peninsula consisted solely of discrete and highly distinct states, and Camden wrote his *Britannia* long before a lesser or greater Britain had appeared on any maps or in any political perceptions. These are just two examples of many that illustrate how not only the techniques of using history but also of writing it had changed fundamentally.[144] On the more sophisticated level of chronology, a century-long process of reconstructing timelines and of synchronizing such narratives for different regions, dynasties, institutions, had started. Legal humanists began to historicize the medieval legal traditions by working out the series of Roman consuls and emperors, now gaining historical depth. Even more scholars were assiduously working on Biblical chronologies.[145] The texts were coming to be seen as possessing different developmental stadia. We may understand the emergence of a humanist approach to history as if suddenly a third

---

[141] Kessler (ed.), *Theoretiker*; Kessler, "Ars historica"; Cotroneo, *I trattatisti*; Landfester, *Historia*; Seifert, *Cognitio historica*; Dubois, *La conception*, 69–158; Ménager (ed.), *L'écriture*; Grell (ed.), *Les historiographes*; Woolf, *The social circulation*.

[142] Zwierlein, *Discorso und Lex Dei*, 25–198.

[143] Guenée, "Histoires"; Goetz, *Geschichtsschreibung*.

[144] Cutinelli-Rèndina, *Guicciardini*; Parry, *Trophies*, 331–357. Camden's *Britannia* and/or Drayton's *Poly-Olbion* were present in nearly all Levant factory libraries, as in the Pera Library, in William Lateward's library 1702/03 (SP 105/178, pp. 47, 51), in Rowland Sherman's library (SP 110/73/2 (2), p. 74), Camden's *Remains*, an annex of the *Britannia*, was present in the Aleppo library, SP 105/145, p. 157.

[145] Levitin, "From sacred history."

dimension was discovered for hitherto two-dimensional items. The "discovery" of geographies and peoples from outside the Classical/ Biblical conceptions of the world played a key role here, but there was also the part of "forgotten" peoples and "tribes" from within the previously accepted boundaries of the world, and the acknowledgment that they had their own histories and own forms of computation. This opened a simply unending series of problems if one wanted to coordinate all these different histories. With all of these sundry cultures and peoples living on this one and same world, it seemed that there had to have been a way to integrate all the timelines together.[146] Many of the learned men mentioned so far were highly concerned with such problems of chronology for the region(s) and/or people they were dealing with. The chronology of the Samaritans for instance was circulating between Huntington, Ludolf, Bernard, Dodwell, Piques, that is between the Levant, Oxford, Frankfurt, the Netherlands and Paris, and its major motivation was the goal of integrating the sequence of events given into the usual European chronologies. Even for historical subjects as prominent as the Greek Church and the Byzantine Empire, Western scholars long struggled to establish coordinated historical concatenations.[147] And, as has been already stated by others, synchronization leads to recognition of empty spaces and items that do not fit. The arrangement of a continuous narrative in one column of a timeline makes the *lacunae* in another chronologically parallel column extremely difficult to miss.[148]

The texts of travelers, administrators and "history" writers concerning North Africa and the Levant that we are dealing with here do not normally belong to the genre of higher academic *historia*, so we find such chronological problems instead in the notebooks and papers of, for example, Pococke, Dodwell and the others already mentioned. A Mármol, a Laugier de Tassy or a Rycaut did not include such

---

[146] Grafton, *Scaliger*.

[147] Cf. as an example Dodwell's interleafed chronological notes in his copy of Nikephoros Kallistos Xanthopoulos' Εἴδησις ἀκριβεστάτη περὶ πάντων τὸν ἐν τῇ Κωνσταντινουπόλει Ἐπισκόπων καὶ Πατρίαρχῶν, BL Ms. Add 21078, pp. 105–115: it is obvious that he tried to contextualize and harmonize the events and succession of bishops and patriarchs with further chronological data. The manuscript that used Dodwell has, by the way, a different and more complete text than the Paris codex used by Migne in his 1865 *editio princeps* of this fourteenth-century text (PG 147, 449–458).

[148] Steiner, *Ordnung*.

chronological reasoning into his present-state-of descriptions or his histories of revolutions and dynastic fates. Gramaye demonstrated an awareness of the problem concerning the traditional genre of a *historia sacra* of "Africa" when he established a chronological list of the continent's bishops.[149] Around 1700, a drift toward states, "nations," governed regions as subject of histories can be seen. The general drift to becoming aware of emptiness and unfitting, along with the emergence of linear timelines demanded by synchronizing efforts, was the major cognitive context. The narrower context and "environment" of perception and production of those texts and of the administrative *mémoires* is found in the structure and organization of time and space as handled by the institutions themselves for which the imperial actors were working.

In the imperial administrations, forms of writing and notation developed that shaped the "horizon d'attente" of history, its sequentiality, and the expectations of its contents. I take again the example of the first volume of the *Mémoires et documents* on Morocco from the old archives of the Marine: When authors like the otherwise quite unimportant consul Guy de Villeneuve prepared for writing a "mémoire historique" about Morocco in 1777, they obtained one part of their information from printed material as has been seen at the beginning of the chapter. But they profited also from archival documents copied and sent to them. Such a list of items extracted from the archives of the ministry of the Marine for quite precise information about Morocco and the straits of Gibraltar for the period 1676 to 1704 reads as follows:

	Extrait du dépôt de Marine
Affrique	*Mémoire de M. le Comte destrées Vice amiral sur l'Employ des Vaissaux qui seront armés en Ponant en 1676 ont proposé diverses entreprises contre la flote d'Espagne et de Hollande et contre leurs Colonies tant en Amérique qu'en Affrique, n. 237 mémoires généraux*
Maroc	*Relation du voyage de M. de St. Amant à Maroc en 1683 pour la confirmacion du traité de paix avec le Roy de Maroc, liasse 477 Marine du mois de X.bre*

---

[149] Gramaye, *Africa*, 27–52.

(cont.)

Detroit	*6ᵉ juin 1701. M. le C. d'Estrées son projet pour empêcher l'entrée de la Mediterranée aux Anglais et Hollandois*
	22 Janvier / 3ᵉ May 1700, N. 2101. M. le Chevalier de la Pailletrie sur ce qu'on peut tenter contre les places des États du Roy de Maroc avec un Mémoire du S. Chazelle sur ce sujet pour la Navigation des Galères en ces Costes Nr. 293 Galères
Larache	*19 Juin 1700 M. de Pointis sur les Projets contre Larache et l'état de ce port N. 2013*
Tanger	*14 Juillet 1700 Le Sr. de Rosneves sur la tentative du bombardement de Tanger 18 dudit M. de Pointis. Je y joins un petit dessin du Mouillage du Cap Spartel jusqu'à Larache.*
Coste de Barbarie	*8ᵉ 7bre 1700 M. de Pointis sur la suite de ses entreprises à la coste de Barbarie N. 2013*
La Mamore remit les Observations au Bureau des plans le 20 9. bre 1737	*6ᵉ mars 1682 Observation de la Rade de la Mamore de Chardellon relatif a un Plan avec une lettre de M. de Chateaurenand du 9 mars même année liasse 442 Brest*
Affrique	*Du 21.e avril 1701 M. Begon sur le départ du Poty[?] pour la Coste d'Affrique N. 2085*
Salé	*23. 9.re 1701 a M. le C. Destrées sur la disposition etc. ses veues pour la guerre contre les Sallétins Reg. du Levant 389.*
Salé	*Propositions de M. de Pointis pour prendre la Ville de Salé 1702 N. 2219*
	Copie de la lettre de M. Cassard du 14. Octobre 1720.
	25. May 1702 M. de Pointis sur l'entreprise de l'Isle – le Mémoire y joint N. 2219 17 juillet refuttat en Conseil de guerre ces entreprises 17. aoust id. 19. 7bre de st.
Gibraltar	*9e X.bre 1704 sur l'entreprise et le siege de Gibraltar le S.r des Herbiers.*[150]

---

[150] AE La Courneuve MD Maroc 1, f. 163r–v. The list still continues after this point. Being just a list of titles with a technical reference system, I leave it untranslated. For several of the *mémoires* and plans mentioned, mostly later transferred to other archives, the *service hydrographique* of the navy or the *bibliothèque royale* cf. *Les sources inédites de l'Histoire du Maroc. IIe sér.,*

This list tells us that the usual practice to write a historical or political *mémoire* like that of Villeneuve was the creation of a *mémoire* from *mémoires*.[151] Here, I am interested in the impact that this administrative practice had on the perception and production of "History." What such a list makes clear is that the past historical *mémoires* and the *mémoires* of past present states concerning a given region were filed and recorded as a sequence, as strata and layers of a consecutive and ongoing chain of events within the same country, state or region. In other words, the sameness and identity of a given historical actor was produced and reified by that very form of archiving, classification and ordering. Some aspects and concerns were usually quite similar and sometimes passages from one narrative or description were copied directly into another. These "stable" parts usually described the location, the port, or the city of Tunis, Algiers, Tripoli, or Morocco. Descriptions of local political arrangements were also often similar. However, events and details relating to a given enterprise or problem were certainly specific to each of those *mémoires*. The process of writing and archiving these chronicles was conceived of as building a kind of stone wall. Each *mémoire* was like a rock, all of them were placed one after the other into the wall, one row on top of the next, year after year and decade after decade. Due to the individual focus and attention of each reporting bureaucrat, the "present state stones" were not completely identical in shape. Standardization and training of writing within the Ministry's bureaucracy smoothed the edges of those proverbial rocks, normalizing the very form of the *mémoires*. Thus, the wall's stones tended to approximate bricks. For the most part this process began rather late, but it was different for each region and city. That practice of *mémoire* writing went back to the late sixteenth century, but normally effective memory was created by the *mémoires* starting in the times of Colbert. It is therefore only with the 1660s that we – and the past administrators themselves – have a more or less uninterrupted series of *mémoires* at our disposal. In the case of Morocco, Villeneuve himself did not go far beyond the middle of the eighteenth century. Lafitte's archival inventory starts with 1533, but the sequence of *mémoires* only becomes dense and continuous with the 1680s. To some point, in the special case of Morocco, the scarcity is explained by the interruption in

tom. 6 (1700–1718), 166 n. 1, 172 n.1 and passim. Spelling is left as in the source, for normalizing of the names cf. the index.

[151] Cf. Head, "Knowing"; Brendecke et al. (eds.), *Information*; Soll, *Information master*.

the diplomatic relationship between France and Morocco between 1718/27 and the peace treaty of 1767, recovering only after several years.[152] For Tunis, Algiers or Tripoli, a historical *mémoire* in 1777 could rely on a less interrupted succession of *mémoires*. That is apparently also the reason why at the time of Villeneuve's request, the Marine archives started its investigation with the list of printed books that promised at least to root historical knowledge back to the sixteenth century. The archive itself was organized both thematically and by respective region. In the list of extractions from the archive cited above, one finds the very shelf mark pointing to the respective register.[153] Very seldom was one of those *mémoires* published during the Ancien Régime, and exceptions were typically more related to scientific voyages than to usual political affairs. But in principle, they effectively show how memory and history were now built and conceived of as a sequence of past present states, indifferently piled one on top of the other, with the expectation that the future would also be similar to the present state of affairs for a given country or state. There is no Newtonian absolute or any other abstract theory of time to be found behind this arrangement. Instead, it was the ordering principle of regional and state history, classified within an institutional archive that made the sequence of history itself linear, and shaped it not according to the genealogical line of a dynastic family, an ecclesiastical body, or an unlocalized "people," but instead for those city-states that were conceived, more or less, as agents, partners or enemies, in diplomatic state affairs.

Despite those caveats, the printed descriptions and histories of those regions prove to be quite similar to the *mémoires*, as they had the basic distinct feature in common: one "present state of … " followed another.[154] Here it was the decentralized public sphere and network of libraries that we can take as the functional equivalent to the ministry's archive in terms of memory and history building. In both cases this "brick-building" of past present states had an early modern starting

---

152 Cf. Caillé, "La représentation," 137; Caillé, *Tanger*, 10; Grillon, *Chénier*, vol. 1, 26. *Documents inédites* (1726–1732).

153 This classification system was introduced by Truguet in 1755. Cf. Taillemite, "Les archives," 36f.

154 Cf. e.g. Rycaut, *The present state* (1668); Samuel Martin, "The present state of Algiers" (1675), SP 71/2 box 1, f. 61–73; Rycaut, *The present state* (1679); Thomas Shaw, "An essay towards the present state of Algier," BL Ms. Sloane 3986, f. 48r–53r and SP 71/7 f. 5r–12v; Addison, *The present state* (1675); La Faye, *Etat* (1703).

point. Early modern bureaucrats could not find a "Mémoire historique sur Algier" in their archives for the year 1350 that would have been even only *grosso modo* similar to the *mémoires* of their own times, and an equivalent book did not exist in print either. The new understanding of History as such a sequence was a cognitive form that also shaped conscious or unconscious expectations and ideas of how the more remote past must have been in principal. And this structural concept of History's shape was incongruent or "jack-knifed" to past shapes, requiring the de- and recomposition of historical information. It became clear that the types of information that were necessary for a useful description of the present were not available for the past. This made ignorance visible.

Spatialization

I term the second epistemic force active in transforming the concept of history and, as applies to our interest here, in making ignorances visible, as the "spatialization of history." A crucial step forward for the regions and sources studied here can be recognized around 1700. When Simon Ockley was studying early Arabic history, Humphrey Prideaux wrote him that,

Your Book will not be well understood without giueing some light into the Eastern Geography I think it will be best don [sic] in the manner a Hobbs hath don in the English Thucidides that is by draweing a Map of the Saracene Empire & addeing a Geographicall table of all the Countreys Prouinces & Ternes in an Alphabeticall order addeing such a description of each as may be sufficient to giue the reader some knowledge thereof. Abul Feda will serue you for this purpose and that whose translation is called Geographia Nubiensis if you can get a true copy will further help you. Dr Pocock had a very fair one in his Library which had seuerall Maps in it drawn by the Pen when you now goe to Oxford by all you can get a sight of this Book for it is in the Bodleyan Library.[155]

Was this advice, only partially followed by Ockley, specific for the concept of history around 1700? If so, how different was it from prior ones?

---

[155] Humphrey Prideaux to Ockley, Norwich, June 2, 1706, UL Cambridge Ms. Add 7113, nr 32. This refers to the partial Latin translation of the first 1592 Arabic print of al-Idrīsī's Geography (the *Nuzhat al-mushtāq*), published by the Maronites Johannes Hesronita and Gabriel Sionita in Paris 1619. Cf. Hamilton, "Isaac Casaubon," 160f.; Tolmacheva, "Medieval Arabic geographers." Prideaux still seems to not know the author's identity, calling the text *Geographia nubiensis*, while John Selden had identified al-Idrīsī as author in his *Uxor Ebraica* (1646), cf. Toomer, *John Selden*, vol. 2, 619.

The Prideaux/Ockley correspondence concerned the somewhat special case of historical geography: ancient Western and early Islamic history that should be illustrated with maps. Indeed, Thomas Hobbes had accompanied his annotated translation of Thucydides' History (1648) with a map of Greece, among other illustrations, which the author had designed himself. In Latin, it delineated the names of places occurring in Thucydides on a map that used a longitude/latitude grid, clarifying the proportion between ancient *stadia* and miles. He also supplied a catalogue of place names with short explanations and excerpts from ancient authors like Strabo as an orientation aid for readers at the beginning of the text. Beforehand, most readers probably would not have really known where a given location named in the Peloponnesian Wars might have been actually situated, but now the whole history and each event was identified and traceable in the dimension of geographical space.[156] Historical maps of the Holy Land, for instance, had acquired a great degree of precision by Hobbes' time.[157] Such historical maps had not previously existed for many regions of less cultural significance in the West than ancient Rome, Greece or the Holy Land. A merging or link between spatial representation informed by geography after the first cartographic Ptolemean revolution of the Renaissance and history writing is at stake here. To what extent can such a "spatialization" of the concept of history also be found regarding the descriptions, historical and present-state narratives of the southern Mediterranean regions and to what prior and contemporary developments within geography and cartography was that connected?

It seems that concerning the shaping of historical narratives by Ptolemy and then more generally by longitude/latitude data occurred in a four-stage development: (a) the medieval historical writing touched upon above before the Ptolemean revolution, (b) the age of the discovery of Ptolemy and the first uncritical integration of his data into narratives, such as with Enea Silvio Piccolomini during the fifteenth century,[158] (c) the ongoing creation of descriptions and historical narratives concerning new discoveries beyond Europe, but at the same time a "de-Ptolemization" of historical narratives concerning the old world and the Mediterranean

---

[156] There seems to be no specific study on Hobbes' cartographic contribution, cf. Scott, "The peace of silence"; Dubos, *Thomas Hobbes*, 85–98; Woolf, "From hystories," 53, 60.

[157] Cf. for several famous post 1560 examples Meurer, "Ortelius."

[158] His cosmography reads like a Ptolemean palimpsest: Piccolomini, "Cosmographia"; cf. Rubiés, *Travel*, 90–96.

during the late sixteenth and the seventeenth century, (d) the new impact of the spatial perception contemporary to the "second cartographic revolution" and the Newtonian shock around 1700.

The Ptolemean revolution of geographical writing and mapmaking during the fifteenth century is widely recognized, including its impact on African cartography.[159] With that in mind, if one reconsiders the texts briefly surveyed above in light of the question of the missing Arabic Middle Ages, one recognizes that, from Leo Africanus to Gramaye and Pierre Dan and to the several manuscript reports circulating in Europe, despite their different forms, they are united by an absence of precise geographical identifications and markers as organizing principles. It is true that Leo Africanus organized his seminal work in a roughly geographical manner. His narrative follows the path of a virtual traveler beginning along the coasts of Africa, enumerating one city and place after the other, stopping here and there and then moving on. Despite a spatial organization similar to a narrative portolan map, Leo did not bother about very precise indications of distances or the localization of places.[160] While Mármol's insight into parts of the Ibero-Arabic Middle Ages was an exception, this concern also meant that he inherited very old non-geographical principles of structuring historical material according to peoples, dynasties and events in a spatially undetermined or fluid sphere of historical diegesis.[161] Gramaye divided his *Africa illustrata* into approximate regions, but otherwise his history was usually rooted in older annalistic forms of history writing. A particularly clear example is his application of the medieval tradition of ecclesiastical diocesan history to North Africa. Maps were either not included in these texts or had neither a strong semantic link nor importance to the narrative.[162] Consequently they

---

[159] Broc, *La géographie de la Renaissance*, 9–42; Milanesi, *Tolomeo sostituito*, 9–24, 47; Relaño, *Shaping of Africa*.

[160] He names Ptolemy three times but only to refer to names given by him to places, regions, the river Niger. He never takes longitude/latitude information from Ptolemy, nor from any other author, and he did not cite Abulfeda.

[161] A typical way to order historical narratives: Robinson, *Islamic historiography*, 59–79.

[162] This holds true for nearly all texts/histories concerning the Maghreb before the second half of the seventeenth century. Tomacheva: The medieval Arabic Geographers, 146 noted that no Īdrīsī map was published until 1790 and the prints of that partial translation of the most important Arabic geographer were always published without maps – as was the case for the partial Abulfeda

mostly belong to the pre-Ptolemean category (a) and not to the branch of again de-Ptolemizing old-World-narratives (c).

If one compares those North African histories of that first "long sixteenth century" to the way scholars were constructing their texts around 1700, and even to the practice of several "lower-level" *mémoire* writers, travelers and observers, one sees how different things had become. Still, the secular process of editing and correcting Ptolemy and the comparison of manuscripts went on. However, other sources had gained importance as the *Geographici graeci minores*[163] and Abulfeda. His text transmitted the Greek and Arabic tradition of geographical projection, measuring and longitude/latitude notation. Not only was it a huge list of 8,000 city names and locations as in Ptolemy's appendix, but the measurements were integrated into Abulfeda's historico-geographical narrative, noting the differences of longitude/latitude for each location according to various previous Arabic authors whose texts we sometimes even do not possess today. Kircher had used his work from manuscripts for the production of maps of Egypt,[164] and in letters exchanged between seventeenth- and early eighteenth-century Orientalists, he became a common "tip" or major issue of research as above in Prideaux's letter.

The best impression of the new character of spatialized historical perception concerning our region can be gained from the letters and work of Thomas Shaw.[165] They show the contours of the fragile and changing relationship between ignorance, partial knowledge, and specified non-knowledge as evidenced by one of the most learned geohistorical and natural historical observers of North Africa. They also display how apparently firm judgments about something observed, errors detected, and prior states of knowledge surpassed were actually based on highly unstable ground, a situation that those very same observers were only partially conscious of. Instead of calling this a "rationalist" Enlightenment approach, from the point of view of a history of

editions. There were only cartographic enterprises, separated from the geographers' texts themselves, to use their data for other maps.

[163] *Geographiae veteris scriptores Graeci minores* (1698), but the central introductory dissertation is by Dodwell, printed also separately: Dodwell, *De Geographorum aetate* (1698).

[164] Cf. "Descriptio chorographica recentior," drawn from Abulfeda, inserted into the historical map of ancient Egypt, Kircher, *Oedipus aegyptiacus* (1652), tom. I, after p. 7.

[165] On Shaw cf. the unpublished thesis Zizi, *Shaw*.

ignorance, it is more interesting to acknowledge the divergence between that unstable ground, visible through the mere process of a decade-long labor on the same problems, and the ephemeral use of the rhetoric of certain discovery and unambiguous error detection. Shaw was in North Africa from 1721 to 1732, traveling between Algiers and Cairo, but mostly based in Algiers. Returning to England, he resumed his position at Queen's College at Oxford, where he was later appointed professor of theology, and then a member of the Royal Society. He worked in England for some fifteen years on the first and second edition of his *Travels* and its *Supplement*. He was not only concerned with the sources of history as has been seen above.[166] Nevertheless, while traveling through North Africa accompanied by Pliny, Livy, Virgil, Ptolemy, the Antonine Itinerary, Raleigh's World History, Hirtius, Cellarius' *Geographia antiqua*, Sanson, Ricciolus and Guillaume Delisle's map of Africa in his travel library, he was not just searching for various famous locations or monuments named in the ancient texts. He also sought to integrate Classical and contemporary geographical studies, constantly measuring and estimating longitude/latitude coordinates with instruments[167] and comparing them with the positions or with the distances given by the authors of antiquity.[168] In a long letter describing the kingdom of Tunis in 1729, which prefigures many elements of his *Travels* printed some ten years later, he makes a recension of the "modern" as well as the "ancient geographers." He finds "errors and disagreements," "doubts and contradictions," but he was careful enough to note that he was depending on a possibly incorrect Italian copy of Ptolemy. In any case, Ptolemy's "scale of longitude" was at least "10° too far to the East." Admittedly, Shaw wrote, the Antonine Itinerary was also erroneous, but it was still "a much better conductor than Ptolemy."[169] Thus, in the place of a homogenous description one finds an inquisitive discussion of authorities.

Noting and discussing longitude/latitude measurements had become a not uncommon practice in several genres since the Ptolemean

---

[166] Cf. above n. 5.   [167] Cf. for that below n. 183, 184.
[168] Cf. for that the sequence of letters Shaw to Sloane, Algiers April 15, 1729, BL Ms. Sloane 3986, f. 45r–46v; July 9/15, 1729, ibid., f. 62r–75v (this had been printed by Sloane in the *Philosophical Transactions* 1729, pp. 177–184); October 10, 1729, ibid., f. 56r–59v; May 14, 1730, ibid., f. 61r–v; May 25, 1730, ibid., f. 60r–v.
[169] Ibid., f. 72v.

reception. It is mostly found in diaries, logbooks, or similar descriptive documents very close to the practice of navigation itself. Early sixteenth-century travelers were explicitly aware of the inaccuracy of the Ptolemean data, but noted that "those measures have to be accepted until one finds more secure ones."[170] Trying to indicate absolute localization data was far more common for "new discoveries" and non-European and non-Mediterranean regions. Here, the measurements taken were an important type of information for European readers, especially, but not only, for cartographers as mariners knew well while at sea.[171] It is often not clear which exact measure of degree was used, which prior maps the writers, pilots and captains were referring to and which instruments they utilized for their observations. Thévenot, as was mentioned above, consulted a translation of an extract from Abulfeda's *Geography* concerning the latitudes and longitudes for India, and tables of coordinates for China calculated by Jesuit missionaries.[172] But for most of the early narratives, printed for instance in Ramusio's collection, the voyage was described in terms of the subjective perception of traveling from one location to another, "x miles

---

[170] "Epistola di Massimiliano Transilvano" concerning the Magellan voyage, entry for March 31, 1520, in: Ramusio (ed.), *Delle navigationi* (1563), vol. 1, 348C. Ramusio indicated the reason for this as that he had seen "le tauole della Geografia di Tolomeo, doue si descriue l'Africa, & la India esser molto imperfette" (f. aij^V). But there are quite rare references to Ptolemy or other prior and ancient authors in the travel narratives themselves. The "Discorso sopra la navigatione di Hannone Carthaginese fatto per un pilotto Portoghese," ibid., 112B–114B, 112B–C discusses the problems with Ptolemy: "si conosce manifestamente, che li gradi di detto auttore sono stati variati da coloro, che trascrissero il libro, come nelli gradi delle isole fortunate."

[171] Usually, travelers or pilots indicated the latitudes using the degree system, Ramusio (ed.), *Delle navigationi* (1563), vol. 1, 114E, 115C, 116C, 119B/C, 180A/B, 184A/B, 185B, 348B. Antonio Pigafetta paid particular attention to the indication of measurements during his circumnavigation, 1519–1522, ibid., 352E–370B.

[172] Cf. the "Iournal du Voyage précédent des Indes Orientales, dressé à la maniere des Mariniers, Par I. Le Tellier Pilote de l'Amiral" with the usual entries of direction, distances, wind, latitude/longitude and magnetic variation (Thévenot, *Relations* (1664), vol. 2, 124–128). Thévenot included an "Instruction pour les aiguilles des Compas ou Boussoles" referring to Dutch experience with the compass concerning magnetic variation around 1650, to Blaeu's *Tabulae sinuum, tangentium, secantium* – in fact by Frans van Schooten, printed by Blaeu, Amsterdam 1627 – and to Cornelis Jansz Lastman's *Schat-Kamer, Des Grooten Zee-vaerts-kunst* (Thévenot, *Relations* (1664), vol. 3, 10f.).

away from the last." And this was done first with no, or only a minimal, awareness of the problems with the Ptolemean data. As is well known, late medieval Portolan maps were still to some extent more precise than Ptolemy. This has been called the necessary Ptolemean detour.[173] Later, during the sixteenth and seventeenth centuries, cartography absorbed the fruits of Ptolemy on the one hand and regained the self-confidence to correct his data with the modern instrumental observations on the other. At this time, narratives and descriptions of the "Old World," both historical and of modern travel, seem to have been "de-Ptolemized," meaning that one seldom finds descriptions of the Levant region that feature longitude/latitude coordinates.[174] It seems that navigation had become so unproblematic in the Mediterranean that the questions of place, position and how a location fits into the overall concept of geography and the globe had become a second-order problem. The most direct seventeenth-century precursors of Shaw who might have been candidates for a similar "spatialized" perception and description should have been the most learned, such as Nointel's secretary, the sieur de la Croix versed in Oriental languages, or the academic traveler Tournefort, whose voyage was an initiative supported by the *Académie des sciences* and funded by the French king. However the difference between theirs and Shaw's writings, only three to four decades later, is stark. De la Croix never used absolute longitude/latitude localizations for his description of Ottoman Greece, the Levant and the Barbary coast, nor did he discuss the divergences of indications by prior authorities.[175] Tournefort devoted significant place to the localization of the Greek isles he was visiting in his letters to Pontchartrain. Nevertheless, he consistently provided only relative localizations by distances to other places, not only in the place of longitude readings – where that was usual before the mid-eighteenth-century invention of the instrumental measurement of longitude –, but in lieu of latitude as well. He compared geographical indications

---

[173] Jacob, *Sovereign map*, 62.

[174] Corresponding to this is the fact that the method and techniques of nautical navigation made little progress for nearly two centuries after the 1540s, cf. Stimson, "Longitude problem," 78.

[175] The closest text comparable to Shaw's is his Journal, 1673 to 1678 (BNF NAF 10839–10841 – "Il a 800 lieues d'Etendue de l'Orient à l'occident et 700 du Septentrion au midi," BNF NAF 10840, f. 83).

by Strabo, Livy and Ptolemy, again, without referring to longitude/latitude coordinates.[176] Likewise, the learned traveler and friend of Herbelot, Jean Thévenot, never used absolute forms of localization for the Levant and employed even less precise indications for travel between the Greek islands than Tournefort had done.

For Shaw, who was the most geographically-interested and specialized observer surveying in North Africa for decades, it becomes clear that all his "fieldwork" of measuring and comparing data[177] was not conducted by relying on the latest and best editions of the Ancient authors, namely Ptolemy. Nor did he have complete editions of the Antonine Itinerary or the famous map of Roman times in Algiers at his disposal which was known as the Peutinger's table. He used the latter only second hand through citations from the books he possessed in Algiers. He wrote about this to Sloane, finishing his letter with "But I believe I must have patience till I see Christendom before I can consult these and other authors,"[178] hoping perhaps that Sloane would send those books. But Sloane took him at his word, considered the difficulties of communication between London and Algiers, and responded in 1731: "As to Antoninus Itinerary Peutingers tables & other books of that kind that are not to be had, I belieue you must satisfy your self with the extracts [sc. in Ricciolus and Cellarius that Shaw had with him] till your return."[179] He drew quite skillful ink maps of the present western province of Algiers and of the kingdom of Tunis, relying on his own measurements, which were the basis of the maps printed later in the *Travels*. He used the Arabic toponyms but added a high quantity of Latin place and monument indications with the purpose of merging present and ancient geography. Back in Oxford he transformed himself into an armchair scientist and now included excerpts from ancient Western as well as Arabic authors. Now equipped with a more recent Greek-Latin edition of Ptolemy, an excerpt by Shaw from the *kanon poleon / catalogus urbium* of Ptolemy's fourth book is of interest here, concerning the most eastern cities of the ancient province of

---

[176] Tournefort, *Relation* (1727), cf. e.g. vol. 1, 56, 98, 174, 269, 287, 297, 331 (here the reference to Ptolemy III, 15, but only concerning Ptolemy's indication that there were still three cities on the island of Chios at his time), 330.

[177] Cf. "Rien de comparable, en tout cas, au grand ouvrage de Thomas Shaw … Le seul voyageur français don't l'œuvre pourrait se comparer à celle de Shaw, est Jean André Peyssonnel" (Broc, *La géographie des philosophes*, 60).

[178] Shaw to Sloane, Algiers, December 26, 1730, BL Ms. Sloane 4051, f. 156r.

[179] Sloane to Shaw, London, February 17, 1731, BL Ms. Sloane 4068, f. 193r–v.

Mauretania Caesariensis and the most western cities of the province *Africa propria*. He noted well what was known at the time about the notation system of numbers and fractions expressed in the Greek letter-number system for latitude and longitude.[180] He also took a note from Abulfeda's Prolegomena to the *Geography* – not yet translated at that time – about the quantity of one degree in relation to the circumference of the globe.[181] Gagnier's partial translation of Abulfeda did not cover the regions that Shaw was concerned with. Obviously, the quality of past geographical sources and the complexity of his understanding of it was far more pronounced at Oxford than in Algiers.

At stake in this example is both the problem of the relationship between the epistemic value of direct observation in comparison to "armchair science," and the thorny relationship between "imperial center" and "periphery" for the production and circulation of knowledge, not to mention, more specifically, the quality of ignorances, partial knowledge and uncertainty. The main catalyst for his work of

---

[180] "N.B. I have purposely omitted to name the 3rd & 4th Columns, viz the divisions of degrees, because I know no one word which expressed the Idea they stand for. Minutes in plain they are not, for the Greeks divided not their degree into sixty, but only twelve parts. Nor do the letters in these columns stand for such a number of these twelve parts, but such a fraction of degree, as the letter properly stands for, exempli gratia (γ) in these columns stands not for 3 twelves, but for ⅓ of a degree id est 20. (δ) stands not for 4/12, but ¼ of a degree. So (γ'β) stands for ⅓ + 1/12 = 25, & (γ°) for ½ + 1/6: id est 40 & so the rest. The Character L' signifies half a degree, & L'δ = ½ + ¼: is 45 d. &c." (Bodleian Add. Ms. D 27, f. 281r). He also noted that he was using the Greek-Latin edition by Peter Bertius at that time (*Theatrum geographiae veteris* [1618]). When he was in North Africa, he had still used an "Italian edition" of Ptolemy which probably refers to one of the editions prepared by Giovanni Antonio Magini (cf. Shaw to Sloane, Algiers, July 9/15, 1729, BL Ms. Sloane 3986, f. 72r). That Italian translation contained many errors in converting Greek notation into the modern system of numerical degrees. The Bertius edition provided the Greek text at least for those sections excerpted here identical to the Erasmus *editio princeps* of 1533, so they still followed the manuscript of the Heidelberg Palatine library, which is known today in Ptolemy philology as codex "A" (Vat. Pal. Graec. 388, fifteenth century, transferred in 1623 to the Vatican). Neither the Erasmus nor the Italian or the Mercator editions provide the reader with a rule of how to decipher the Greek notation. As the numbers differ sometimes up to ten degrees in the flawed earlier decipherings, these were no small details for someone seriously interested in geographic location. Cf. Mittenhuber, *Text- und Kartentradition*, 165–169.

[181] "Ex Abulfedae Prolegom. ad Geographiam de quantitate unius gradus majoris circuli coelestis in circuita terrae – videlicet sex et quinquaginta Milliaria cum duabus tertiis Milliaris" (Bodleian Ms. Add. D. 27, f. 306v)

observation was that of specifying non-knowledge. He "abused" the position of chaplain to the Algiers factory to some extent for those more scientific than religious purposes.[182] First-hand observation had to answer the questions now being formulated, because ignorance about the precise coordinates, locations, the existence or not of geographical features, social information, natural history, and so on had become highly explicit. This was the case not only for his exchange of letters with Sloane, but also for the printed result itself. Already in the Preface, he informed his reader about the importance of measuring longitude and latitude and about his "method" for doing so. Admitting that he had not yet had a Harrison clock or a similar instrument to measure time precisely, a prerequisite for non-astronomical longitude calculations, he specified exactly what kind of quadrant and compass he used to measure latitude and magnetic variation while traveling across the land. For measuring all East-West distances, that is, longitudes, he told the reader of the printed *Travels* that he used his camels' allegedly very constant marching pace as timekeeper for travels in the hinterland,[183] while in a letter to Sloane of 1729 he described a way to estimate longitudes for the coastal sites by a combination of observing the travel time and by the observation of stars.[184] He methodically

---

[182] Ludwig, who, together with Hebenstreit, visited Shaw in Algiers, supposed that he was "a man who had come probably rather for studying the country than for preaching the evangile into this region" (UB Leipzig Ms. 0662, part II, p. 28f., September 8, 1732).

[183] "The Horses and Camels of these Countries *keep generally one constant Pace*: the latter at the Rate of two Miles and an half, the other of three Geographical Miles an Hour. Sixty of these, according to my Calculation, constitute one Degree of a great Circle. *The Space we travelled was first of all computed by Hours*, and then reduced into Miles. . . . I took the Meridian Altitude, with a Brass Quadrant of twelve Inches Radius, which was so well graduated, that I could distinguish the Division upon the Limb to at least one 12th Part of a Degree. (This operation the Arabs call *The weighing of the Sun*.) Towards the further carrying on likewise of these Geographical Observations, I Had a Pocket Compass, with the Needle well touched; the Variation whereof was, at a Mean, in the Kingdom of Algiers, (A.D. 1727) fourteen Degrees, to the Westward, and in the Kingdom of Tunis sixteen; to which in like Manner I paid a proper Regard, in laying down the Maps and Geographical Observations" (Shaw, *Travels* (1738), viii–ix).

[184] The description of the instruments used is fairly similar to the one just cited from the *Travels*, but the longitudes were measured here differently: "I made use of a small but very good Mariners Compass, and found the variation at Carwan 10° W, at Biserta something more than 12° and at Algier I find it now to be 13°30°. I carried along with me likewise a brass Quadrant of a foot

discussed the exact positions of a given place, enumerated the different positions supplied by previous authors, and in so doing offered an exposition of the relativity and uncertainty of what was known. Yet his final text is not just a pure geographical representation that only described the earth's surface without any mention of humans and their artefacts.[185] It still contains large parts on history in the narrower sense, ancient monuments, natural history and descriptions of the current constitutions of the city-states. From the point of view of the organization of the early modern epistemic system, it is therefore better to understand this in the first instance as a spatialization of *historia* writing than as the differentiation of a modern form of geographical narrative. For nearly all prior "histories" and descriptions concerning North Africa, and even the Levant in general, this seems not to be the case. None of these earlier authors ever explicitly acknowledged lacking historical episodes such as the forgotten Middle Ages discussed previously, or not having precise geographical localizations.

This change was happening at the same time as the European, particularly the British and the French, republic of letters was receiving and discussing Newton's theory about the Earth's shape as laid down in his *Principia mathematica* (1687).[186] The first "national" reaction by

Radius, and took the Latitudes of Tunis, Cairwan, Spetula, Gaffsa, Toser, Ebille, Gaps, Sfax, Susa, Lowharia and Biserta with all the exactness such an Instrument would admit. The Longitude I have calculated in this manner: The Transit of several stars through this Meridian of Algier, and as they are calculated for Paris, gives a difference of 30° nearly of a degree, making Algier so much to the East of Paris, which is 2°30° East of London. Most mariners whom I have conversed with upon this agree within 10 or 12 miles, that the distance between Algier and the Goletta (or port of Tunis) is 400 Miles: I have made this voyage four times, and the reckonings we made aboard amounted only to 390. I have made therefore the meridional distance betwixt this place and Cape Carthage 350 Miles (allowing 48 to a degree of Longitude) for as this whole Course is not upon the Same Parallel, we may very well allow 40 or 50 Miles for the oblique sailing, because the Course is in 37° N.Lat. but Algier lies in 36°48° and the Goletta in 36°40°" (Shaw to Sloane, Algiers, July 9, 1729, BL Ms. Sloane 3986, f. 75r-v). Despite Shaw's obvious strong desire for precision, it remains unclear how he "recognized" the distance aboard, cf. Andrewes, "Even Newton," 193 for the development of the discussion about instruments to be used within the London circles at Shaw's time.

185 "Es sollte stets im Auge behalten werden, dass der Gegenstand der wissenschaftlichen Geographie in erster Linie die Oberfläche der Erde für sich ist, unabhängig von ihrer Bekleidung und ihren Bewohnern." (Richthofen, *China*, vol. 1, 730).

186 Greenberg, *Earth's shape.*

the French was an attempt to prove that the English hero of science must have been wrong, Cassini II upholding the idea of the earth being a prolate spheroid. French expeditions to Peru (1735–1744) and Sweden (1734–1737), however resulted in proving Newton's theory that the earth was shaped as an oblate spheroid and its implication for the changing length of a degree according to its position on the earth.[187] The huge project of a completely new trigonometric survey of France that then commenced under Colbert and was orchestrated for decades by the Cassini dynasty of geographers, was connected to those discussions, but is also an indicator of that new focus on spatial and cartographical precision.[188] Furthermore, the as yet unsolved problem of how to instrumentally measure longitude was ongoing, most famously with the 1714 founding of the Board of Longitude. Only in 1735 did Harrison invent the first of his famous clocks capable of serving as nautical "time keepers," and it was not until after the invention of his the fourth model ("H.4") around 1750, which eventually won the Board of Longitude's prize, did instrumental longitude observations really come into wide use.[189] That Shaw employed the term "time keeper," then usually used for clocks able to keep exact time, for his putatively consistently marching camels, shows that he was acquainted with that continuing "quest for longitude" and transferred the problem from the more common discussions about nautical navigation problems to terrestrial traveling. The quest for longitude itself is just the proof that the beginning of the eighteenth century was a period when spatial perceptions were, again, deeply shaken. This is the phase (d) in the development outlined above. The drive for the spatialization of the (geo-)historical narrative was both effect and causal trigger for that coterminous change of spatial perceptions. It was "causal," because it would be mistaken to conceive of the historical process as just starting in the brain of Newton. Instead, regular geographical investigation, new forms of empiricist academic traveling, *and* the narrower field of progress within physics, mathematics and

---

[187] "Whether the French liked it or not, the issues regarding the earth's shape cut across national lines" (Greenberg, *Earth's shape*, 87). On La Condamine, who was the major actor of the Peru expedition cf. Safier, *Measuring* passim.

[188] Konvitz, *Cartography*, 1–31.

[189] Cf. Andrewes (ed.), *The quest*; Bennett, "The travels" and the results of the Cambridge Project to digitize the papers of the Board led by Simon Schaffer; for the later eighteenth century Croarken, "Tabulating"; Barnett, "Explaining themselves"; Dunn and Higgitt, *Ships*.

technology have to be seen in interaction with each other, and all of this also involved growing international competition in the sphere of the scientific and exploratory enterprises.[190] For instance, recent research about one of the major figures of the French cartographic school gathered under Louis XIV within the *Académie royale des sciences*, Guillaume Deslisle, does not mention anything about the reception of Newton's theories.[191] Delisle was the cartographer who reportedly made the first important steps to the new form of depicting empty spaces on his 1707 map of Africa that was later developed into a pattern that would be so important for the nineteenth-century's scramble for Africa."[192] In a 1720 *mémoire* to the Academy, Delisle claimed that, according to the results of his new calculations and measurements, the whole Mediterranean was in fact more than 1,000 miles shorter than had been traditionally thought:

I have found that this part of the Mediterranean where I have employed the portolan maps without help by observation, was at least as divergent from the usual maps as the other ... From Tripoli to the straits of Gibraltar I have found the divergence to be 7 degrees of 26 in total ... Following those measures, the length of the Mediterranean would be 41 degrees and 30 minutes from Alexandrette [*İskenderun*] to Gibraltar which are 860 *lieux* on that parallel, instead of 56 degrees or 1160 *lieux* which one is usually noting in the ordinary maps. This error of 300 *lieux* concerning the extension of a Sea which has been so familiar to us since time immemorial has to be considered far more than the error of 500 *lieux* from here to China that the Academy has [sc. recently] proven, as China is three times further away from here than the oriental part of the Mediterranean.[193]

This was an epistemic, not a practical problem; it did not change the daily work of a ship's captain much. But it is just this change in the conception of space that was occurring while authors like Shaw were writing history. The very position of each place – on ancient and

[190] Bourguet and Licoppe, "Voyage," 1150; Safier, *Measuring*, 90.
[191] Dawson, *L'Atlier Delisle*.   [192] Surun, "Le blanc," 122.
[193] Guillaume Delisle, "Détermination géographique de la situation et de l'étendue des différentes parties de la Terre – Mémoire présenté à l'Académie royale des sciences, November 27, 1720," Archives de l'Académie des Sciences, Mémoires, printed in: Dawson, *L'Atelier Delisle*, 246. Deslisle's 41° is essentially the length that is given in modern atlases. Fifteenth-century editions of Ptolemy had given the Mediterranean 60° of longitude, and in the Hondius-Mercator Atlases of the 1630s, it had shrunk only to between 51° to 53°, cf. Albert Van Helden, "Longitude," 88.

present maps and in "reality" – became unclear again, and the awareness of that ignorance was an element added to previous perceptions
and descriptions, all of which had originally been executed in state of
unconsciousness.

I have argued above that the developments in higher academic
chronological investigation and the evolution of "administrative timescapes" of the sequentiality of present-states from the past into the
future as normal *horizon d'attente* were incongruent with older forms
of organizing history and cut across them. In so doing, they made it
clear that there were elements "missing" that had gone unnoticed in
those older forms of narratives, timetables and lists. Concerning the
spatialization of historical narratives, the argument here functions on
two levels. The first is that ignorances about exact localization entered
into narratives via discussions of several discrepancies/options given by
authors or different forms of measurement. These discussions explored
the relativity of the known, and they therefore left open a window to
the unknown. The other point is that insisting on exact localization and
the spatial organization of texts operated alongside the continual combination of historical narratives with geographical information and
cartography. These combined processes directed attention to places
and names on maps and opened up questions about their histories.
Standardized geographical perception meant that each of those places
had to possess a history as long as humans had lived there. If a history is
organized along a sequence of events defined by the focus of, for
example the biography of a person or a family, the attention of the
reader is not particularly directed toward something not narrated. But
within the framework of a spatialized concept of history, new questions
arise: what about this or that village, city or ruin, depicted on a map,
and featured within a narrative at least as a place that one passes by
between points X and Y? What could be known about the history of a
given and otherwise unfamiliar location? Actually, the general development of Western history writing points in that direction. The revised
edition of Camden's *Britannia*, released around 1700 and supplemen-
ted by regional surveys authored by regional specialists, is a typical
example, as is the flourishing of other regional histories, starting from
the bird's eye view of the cartographer and investigating the dimension
of historical depth for locations seen on the surface. The stretching and
shrinking of measurements went unquestioned for a long time until the
Newtonian challenge jumpstarted a process that had its roots in the

Renaissance, and emphasized both the quest for accuracy as well as a sense of ignorance and uncertainty.

The result was the difference between a geohistorical narrative like Shaw's of 1738 on the one hand and one such as Gramaye's or Mármol's from around 1600 on the other, a divergence that is evident from the very first page.

Just as Newton's head is not the single starting point of physico-geographical learned discussion, Abulfeda's writing is neither to be understood as a suddenly discovered matrix of geohistorical writing that became slavishly emulated now by early Enlightenment Europeans with Oriental interests. It is true that the way the fourteenth-century Abulfeda integrated discussions unresolved among Arabic geographers about the precise location of locations in his geohistorical narrative was much more congenial to what Shaw produced than either the earlier Western descriptions or the narrative-bare Ptolemean data tables. However, the reception of Abulfeda in interested academic circles ranging from Cambridge, Oxford, Paris to Leiden, from Rome to Leipzig and elsewhere was actually the effect of a post-Baconian, even now Newtonian search for material to fill the voids of now more explicit ignorance, finding along the way matching prior forms of explicitness about unknowns. Abulfeda's textuality was in elective affinity with the drive for spatializing *historia*, it was in elective affinity with the early Enlightenment's new way of recognizing and explicitly bringing ignorance to light.

Conclusion

The findings of this chapter are, to some extent, opposed to a general tendency in recent scholarship that stresses cross-cultural entanglements, the importance and impact of cultural brokers, and the high degree of exchange between Europeans and Ottoman subjects of all ethnic and religious identifications. Yet on the other hand it can be easily reconciled with these perspectives, by distinguishing between different levels of coexistence, cooperation and communication. There was, certainly, on a practical level, considerable acquaintance and know-how that merchants, consuls, and dragomen possessed concerning exchange and communication with their Ottoman, Berber, Armenian, Greek and Jewish counterparts in the Levant, at least regarding business matters. They made contracts, bought, sold and exchanged goods, signed contracts in the original language and European translations in the consulate's chancery register. Through translators, they negotiated with the Ottoman government and the corsairs over peace treaties and concerning the handling of ships, goods and everyday legal conflicts. However, on a second "epistemic" level, it is hard to detect a deep exchange and acquaintance with large parts of their counterparts' culture and history. There was a perceptional lacuna and a long term state of nescience concerning all that what would be, in early modern terms, the *historiae* of those regions and peoples. It might be something of an overstatement, but when the French and British started to establish their consuls in the North African port cities, they were entering countries with a mindset not so different from one entering the American "New World" – with the difference being what was known from classical ancient authors. This is, by the way, why identificatory projections like the blending of the British with the Phoenician people could work so well: projections function far better on a *tabula rasa* than they would on a historical tableau highly "populated" and filled with centuries of medieval Arabic culture and history. The label of "parallel societies" would be perhaps misleading, as a good part of the society in Algiers, Tunis and Tripoli did not draw upon the elements of earlier Arabic history writing and philosophy either, as we have seen. Moreover, it is not the aim of this chapter nor should it be the intended or unintended byproduct of adopting the perspective of a history of ignorances, to return to the idea of a strong cultural or civilizational barrier between "Europeans" and "Islam." This makes no sense regarding just the manifold crossings and transformations of particular actors between the cultures as converts and renegades. If one looks, as has been done here, at what the knowledge of orientation used in the trade empire's administrations was, what was written by its agents, what was perceived by them, one finds no deep and lasting influence of – to invent counterfactually something not existing – "renegado expert writing" on the lists from the Marine archives or from Shaw's letter to Sloane with which this chapter started. The Europeans used European, and mostly even European vernacular, descriptions and histories for those purposes of politico-commercial positioning, not even the specialized productions of learned Orientalism. Oriental manuscripts traveled to Europe, yes, and were used, and to some extent were read there by specialists, but until 1750, traces are extremely scarce of any sense that this reinformed the actors of the trade empires in the Paris and London centers, as well as in the Levant, and that it gave them a more finely-grained concept of the *historia* of the people with whom they negotiated. As has been shown, this is overwhelmingly supported by the evidence of what was contained in the surviving Levantine library catalogues of consuls and merchants. After work, one read Camden or Rousseau in Aleppo, not Ibn Khaldun. The focus therefore, on everyday practices can also obfuscate the recognition of the historical presence of ignorance which is then, nevertheless an important object to be understood historically: One is dealing to some extent with a decades- and centuries-long paradox that the trade empires, while assuming the importance of general and historical knowledge about all those regions, while exercising the gesture and ritual of "empirical" description and observation, remained largely autoreferential in their text production. Or, to smoothen that statement, there is at least a much larger portion of texts produced and in circulation that had such an autoreferential character, referring to and repeating prior Western histories. This is in striking contrast to the bits and pieces of direct translations from Oriental texts that were incorporated into what was the open canon of historico-political descriptions to serve as orientation for the men of politics and commerce. These were extremely small in number and extent.

All this was not static. There was a non-knowledge cycle at work whose movement seems to have been similar in shape and chronology to those within the realm of religion and science. The political operative (non-)knowledge touched upon in the first chapter is somewhat different from that, as the inaugurated questioning into the national had no endpoint. Still, the starting point was similar in all four epistemic fields, around 1650/60: Not only the specific form of how the replacement of ignorance evolved, as the semantic structure of the Barbary inhabitants" proneness to revolutions demonstrates – it became elaborated in a definitive form around 1620/50 and it partially dissolved around 1730/50. Furthermore, the epistemic forces of standardization and spatialization analyzed here that triggered the specification and treatment of ignorance also had their main starting point around 1650/60: the practical administrative sequentialization of history in the archives – the "laying of bricks of history" to recall that metaphor – only started to function uninterruptedly during the 1660s in France. The first Western attempts to uncover the nescience about the Arabic Middle Ages in its learned form also began during the seventeenth century, if one takes

Pococke's *Specimen* of 1650 as a starting point. The spatialization of perceptions in a post-Ptolemean form was certainly a long process, but we have seen how around 1700 a new phase began, and how now, most of all, unresolved, optional and opposite ideas about localizations (re-) entered the narratives about the closest regions of the "old world," contemporary to the abovementioned Newtonian and praeter-Newtonian shifts. Ignorance became explicit, framed, and one dealt with it terminologically and technically. Spatialization then led to *historia* becoming differentiated into history and geography in a modern sense around and after 1750, which is beyond this study. However, the emergence of a sophisticated cartographical *technique* to draw empty spaces on maps and to reflect explicitly on these voids (Jean Baptiste d'Anville) just at that time is an easily recognizable point in that development of explicitly exposing ignorance and dealing with it.

4 | *Science: Mediterranean Empires and Scientific Unknowns*

As has been outlined in the introduction, the methodology chosen here resembles a universalizing of an approach more common within the narrower field of "empire and science" studies of all the epistemic types that imperial actors encountered. Scientific preoccupations certainly had their place in the lives of the French and the British in the Mediterranean – indeed, as will be seen, they devoted a good deal of their time to science. But for a number of good reasons this topic figures as the final chapter. First, science was not the central preoccupation of imperial actors in this context. Consuls and merchant nations were not maintained primarily for scientific purposes; it was only once they had been established for politico-economic reasons that they could serve other goals within the fields of religion, history and science. They certainly did apply themselves to these tasks: consuls, chancellors and travelers protected by the empires were all active in the search for specimens, animals and plants and in the observation of the geological and hydrological characteristics of the Levantine regions.[1] Second, this study should not be misunderstood as yet another work on "empire and science" – only this final chapter falls easily within that designation.[2] The emphasis is rather on the generalization of the epistemic approach: on a history of ignorance that cuts through different types of epistemes to discover their shapes, similarities and differences through parataxis and paralleling. The field of science certainly allows a particularly clear demonstration of the problems involved. Like

---

[1] The list of active botanists from Rauwolf to Tournefort, Ludwig/Hebenstreit and Adanson with specific interest in the Levant regions is long and does not need to be elaborated here. Another particular interest from Greaves, Golius and Shaw to the nineteenth century concerned the hydrological system of the Nile and the functioning of the ancient as well as the Arabic medieval Nilometers. Cf. for instance Bodleian Ms. Add. D 27, f. 297r–304v: Latin-Arabic synopsis of a treatise on the Nile and the Nilometers. Cf. Popper, *Cairo Nilometer*; Mikhail, *Nature and empire*.

[2] Cf. the literature above, Introduction, n. 1.

the second and third chapters, this section concentrates on a major example: the idea of a petrified city that was believed to exist in North Africa. The epistemic evolution and development of this idea within this chapter can be summarized in two sentences: until 1730/40, the natural sciences accepted the hypothesis that at certain moments spontaneous petrifications of organic material and creatures occurred, and stories about petrified cities and human bodies were taken as proof for that hypothesis. Subsequently, different forms of falsification – all prior to the insights gained on volcanic eruptions, human bodies and organic material following the excavations in Herculaneum and Pompeii in 1738 – led to something approaching a Popperian law of nature in the form of a negative universal "there-is-sentence": "There are no spontaneous petrifications in nature."[3] For the questions here pursued, the way between those two points, the reconstruction of the growth of the "story" of the petrified city as a specified unknown, is in itself of greater importance than that clear endpoint. The non-knowledge concerning nature is, in the end, a clearly heteroreferential one: the question is simply whether something exists or not – a city with living men and women petrified in one instant – and if one can find it, and on a second level, if therefore a natural law exists that allows such an object to come into being. In this context, it is strategically better to choose the "abnormal normal" idea – normal in early modern terms – of spontaneous petrification, although, of course, the present scientific consensus is that it is impossible. Analyzing this idea will reveal systematically more about how unknowns and ignorance were managed than would the analysis of objects like clearly existing species of the *echinus* or the *henna* plant, for example. By choosing the "impossible" object, one profits from the constant unconscious simultaneous perception of otherness that the reader cannot escape. It helps to illustrate that, despite this otherness, the epistemic maneuvers and movements can be astonishingly similar, and that therefore the parallels between science, history, religion and politics become even more compelling. In this way, it becomes possible to compare very different forms of (non-)knowledge – such as that of laws of nature and the non-knowledge cycles involved with religion, history and politics – and it will be possible to realize the unfolding of the different realms of possibilities involved, of the differentiation between them, and the

---

[3]   Cf. Popper, *Logik*, chap. 3, nr. 15, chap. 4, nr. 19–24, 39–41, 47–59.

switching of classifications and functions concerning an epistemic item such as the story of a "petrified city."

From Avicenna to the Queries of the Royal Society (1692)

From a Prodigium to Atomist Science

In 1594, the patrician Martin Baumgartner published in Nuremberg a small and rather conventional description about his travels to the Holy Land, Egypt and the North African coast. In the twelfth chapter, titled "Medals dug up about Tripoli. Arabian metamorphosis. Persian Water, its Virtue and Use," Baumgartner described some discoveries of Roman coins found in what he identified as the ruins of "ancient Tripoli," before narrating: "We were credibly inform'd, by very grave and prudent Men, that on the way as you go to Meccha, there was, or rather had been, a certain great City, in which, by the dreadful Judgment of God, not only Men and Beasts were turned into Stone, but likewise their very Utensils, and each according to its own former Shape and Figure."[4] From the 1630s to the 1670s, that brief passage was echoed in further reports of willing and unwilling travelers to the Tripolitan hinterland, an area largely unknown to Europeans at that time. The information was also integrated into early modern scientific discussions. In 1637, Athanasius Kircher traveled to Malta acting as confessor to the landgrave Friedrich of Hassia. With the support of Fabio Chigi, the future pope Alexander VII, who was at that time papal inquisitor and delegate to Malta,[5] Friedrich was due to assume the new titles of the Great Cross of the Order of the Maltese Knights of Jerusalem and nomination as coadjutor of the order's Grand Priorate of Germany. Kircher used his time on the island for intensive studies and investigations in natural history.[6] His most important work on proto-geology, the *Mundus subterraneus*, was published in 1664. In the preface, Kircher declared that this work had started right during his stay on Malta. Kircher dealt with the subject of petrification in Book

---

[4] Baumgartner, *Peregrinatio* (1594), 130; English translation in Churchill, *Collection* (1704), vol. 1, 486.

[5] Borg, *Fabio Chigi*.

[6] Bartòla, "Alessandro VII," 7–29 and 77–87; Hein, *Athanasius Kircher*; Totaro, "L'autobiographie," 113–123; Fletcher, *A Study*, 526–528; Stolzenberg, *Egyptian Oedipus*, 1–3.

VIII ("Of Stones"), Section II ("Of the transformation of juices, salts, herbs, plants, trees, animals and men into stone or about the power of petrification"). He first supplied a systematic theory of a petrification liquid ("De origine succi petrifici"), before going to recount a story, in Chapter II ("Various observations of things converted to stones"), reportedly told to Kircher by Gian Francesco d'Abela, the vice-chancellor of the Maltese Order, known as the founder of Maltese natural and geographic history.[7] According to this tale, a black Ethiopian girl, named Victoria Calsurakie, then ten years old, from the city of Cucu in Tongil province, had been apprehended by the Knights of Malta in 1634 when she and some 600 other Ethiopians were to be transported as slaves on five ships from Tripoli to Constantinople. In Malta she was now serving the order's Archdeacon Eugenio Testaferrata. When she was five years old and lived in a village called Biegioda, Victoria heard that the village Biedoblo, which was midway between Cucu and Biegioda, had been suddenly transformed into the hardest stone during an earthquake that had occurred amid a thunderous and stormy summer night. Her parents and older brother brought this news to Biegioda, where it was interpreted as a sign of divine justice against the Arabs, who had inhabited the village.[8]

The few existing comments on the passages of Kircher's *Mundus subterraneus* dealing with the theory of petrification have classified them as not particularly innovative or "distinct."[9] His dependence on Aldovrandi and Gesner, and therefore on a state of reasoning proper to the previous century, was noted:

> Kircher described the fossil ammonites as "caraway seeds" and attributed both their creation and that of kindred forms to a mysterious power possessed by nature ... which produced living forms, even those of plants, modelled in stone ... Kircher attributes the origin of "dendrites" to the fact that seeds and spores of fern had fallen into the stone when it was still soft and had then matured inside the rock, thus giving it its outward form.[10]

Robert Hooke and other Baconian experimentalists in the Royal Society took the idea that "fossils owed their formation and figuration

[7] Luttrell, "Manduca and Abela," 113; Freller, Epitome, 317–321.

[8] Kircher, *Mundus subterraneus*, tom. 2 (1664), 50.

[9] Fletcher, *A study*, 175. [10] Ibid.

to some kind of *Plastick virtue* inherent in the earth" as a paradigm to contrast with their new vision and techniques of research. Hooke, for example, used a microscope to compare the structures of fossils and of living wood.[11] To some extent, this was also a Protestant answer to what was perceived as a Catholic scientific approach. In addition to its neo-Platonicism and Orientalism, the Jesuit marks of this approach were evident in passages where Kircher welcomed prodigious "obser-vations" by missionaries, such as the appearance of a Madonna hold-ing Jesus in a rock in Chile or a petrified *Deipara* found in the Gotteswald near Lucerne in Switzerland in the year 1659. Kircher classified these phenomena as "not the work of a mortal sculptor, but made by the paintbrush of Nature by the intention of God."[12] Sudden, spontaneous and miraculous acts of God in creating stones thus belonged at the center of Kircher's beliefs and teaching, and they relied on medieval traditions from Albertus Magnus to Eusebius Norimbergius.[13] On the one hand, the petrified Tripolitine city fitted with these miraculous stories, showing in this case God's power to punish evil men on earth. On the other, unlike the cases of divinely-shaped stones and Madonnas made of rock, the instantaneous petrifi-cation of real men and women had further epistemological implications for the conception of the natural history of stones and of the earth as a whole. Kircher did not elaborate upon these implications. He referred the reader to the sixteenth-century geologists Gesner, Agricola, Rueus and Falloppio for further information, but then dismissed the subject: "that should be enough of those wonders of petrified things . . . If they are true, surely I estimate them being effectuated rather not by nature but by a higher cause."[14] The authors indicated by Kircher did not contain much detail about similar phenomena. Agricola mentioned only petrified plants, wood, the famous teeth of fish which were later to be discussed as "Glossopetrae" from Malta, and, as the largest item, bones of whales that had been "converted into stones."[15] Rueus' work does not contain much about "fossils" in the modern sense of the word

[11] Rudwick, *Meaning of fossils*, 56.
[12] Kircher, Mundus *subterraneus* (1664), tom. 2, 44b.
[13] Ibid., 48f. Albertus had received Avicenna's *De congelatione et conglutinatione* in a chapter concerning fossils: Albertus Magnus, "Mineralium libri quinque," 21.
[14] Kircher, *Mundus subterraneus* (1664), tom. 2, 50b.
[15] Agricola, *De ortu & causis* (1558), 324.

at all.[16] Falloppio went into the most detail about the process of petrification, and did extensively consider the notorious problem of how mussel shells had ended up on mountain tops. But he did not mention anything like petrified human bodies.[17] Agricola, Cardano and Gesner all noted the discovery of a stone or an assemblage of stony bones in Saalfeld/Thuringia, which was possibly similar to the form of a man's breast, but they added no specific reflection on that finding.[18] Gesner knew more about the stories of seemingly molded stones that could have taken their form "sponte naturae," but regarding petrified human bodies he only related Ovid's story about Niobe and that of Lot's wife.[19]

In contrast, Ulisse Aldovrandi opened a section of his *Musaeum Metallicum* (1648) with the statement that "every species [*omne rerum genus*], if it be from the family of animals, or plants, or fossils can transform into stony rigidity." For animals he referred to Avicenna; then he referred to the petrified hoard of Tartars that Kircher had compared to the Tripolitine example: "We also read in the Asia map of Cornelis de Jode that in Tartaria near Samogedes it happened that some cattle, herds and shepards were transformed into stone, not changed anyway in their form, as is to be seen still today."[20] This map was printed in 1593, but as a reprint of a 1562 map of Russia produced by Anthony Jenkinson, where the sudden petrification was dated to the middle of the thirteenth century.[21] Aldovrandi even dared to supply a copy of Jenkinson's illustration of the group of petrified

---

[16] Rueus, *De gemmis* (1596); Chapter XIX on Corals comes closest to the problem of petrification.

[17] Falloppio, *De medicatis aquis* (1569), 96–110.

[18] Cardano, *De subtilitate* (1582), 228; Gesner, De *omni rerum fossilium genere* (1565), 145v.

[19] Gesner, *De omni rerum fossilium genere* (1565),142r.

[20] Aldovrandi, *Musaeum metallicum* (1648), p. 823.

[21] Jenkinson, *Rvssiae Descriptio* (1562): "Haec saxa hominum iumentorum camelorum pecorumque, caeterarumque rerum formas referentia, Horda populi gregis pascentis, armentaque fuit: Quae stupenda quadam metamorphosi, repente in saxa riguit, priori forma nulla in parte diminuta. Euenit hoc prodigium annis circiter 300. retro elapsis." On Jenkinson Baron, "B.A. Rybakov." Jenkinson's travel narrative of 1557 does not contain a reference to that petrification story. Palmer, "Writing Russia," 51–58. The Tartar petrification story appears in Botero, *Relationi universali* (1595), vol. 1, f. 61v; Purchas, *Pilgrimage* (1614), 426; Van Helmont, "De Lithiasi" (1648), 14f. A similar story from America in Acosta, *Historia natural* (1590), 145f.

Figure 4.1 Horde of petrified men in the desert from Aldovrandi, *Musaeum metallicum* (1648), 823.

men in the Tartar desert (see Figure 4.1) within his scientific treatise on mineralogy.[22]

Like the other authors and Kircher, Aldovrandi did not enter into a discussion about the causes of this specific form of petrification.[23]

With Aldovrandi and Kircher, the story of the petrified groups of men had entered the most specialist and avant garde scientific treatises concerning a theory to explain the transformation of liquids and organic material into solid stone.

The localization of these prodigious observations remained somewhat arbitrary in the works of armchair natural philosophers. One of the places where such a spontaneous petrification had happened was "confirmed" to be near Tripoli/Barbary by several other reports. Thomas Bartholinus, physician to the King of Denmark, had visited Malta, like Kircher before him, in the spring of 1644. Apparently D'Abela had shown him petrified wood from the city near Tripoli, which he called Ras Sem in his museum. Bartholinus even noted that d'Abela had sent a petrified child from Ras Sem to Richelieu.[24] Another account from 1654/55 was not printed until 1726, when William Derham included it in the posthumous edition of Robert Hooke's *Philosophical Experiments*. The story came in a letter from the former English consul in Tripoli, Thomas Baker, to Mr. Waller, dated November 12, 1713. According to this letter, in 1654/55, when

---

22 Aldovrandi, *Musaeum metallicum* (1648), 823.   23 Ibid.
24 Bartholinus, *Centuria* (1654), 308 in the chapter "Homines Petrificati"; Deusing, *Foetus* (1661), 43.

the English Admiral Robert Blake had led a squadron of warships against the corsairs of the North African republics, he sent out a Mr Hebden in a small frigate to fetch a "Figure of a Man petrified" from the petrified city, which the latter "conveyed to Leghorne, and thence to England, and that it was carried to Secretary Thurlow." Later, when Baker was consul in Tripoli (1677–1685), the Turks, specifically the Commander of the Garrison of Derne, assured him that the petrified city did really exist, but because it was "wholly buried in the Sands" at that time, he could not procure a new specimen of petrified humans or animals.[25] Baker's letter was written sixty to seventy years after the supposed event, and published even later. As no trace of the expedition in the 1650s remains, one may suspect that the letter's text was already influenced by the subsequent steady growth and circulation of the story.[26] But with Baumgartner, Kircher, Bartholin and Baker, the North African localization was steadily confirmed.

Of further systematic interest within early modern scientific explanatory discourse is a letter from Peter Fitton, sent from Florence on July 2, 1656 to Kenelm Digby in Toulouse, who transmitted the letter to an anonymous friend on September 27. It was then printed in the first November issue of the weekly newspaper *Mercurius politicus*. Fitton wrote:

Sir, This is to . . . let you know of a strange Metamorphosis hapned in Barbary not long since, which is, the turning of a whole City into Stone; that is, men, beasts, trees, houses, utensils, &c. every thing remaining in the same posture (as Children at their Mothers brests, &c.) when the petrifying Vapor fell upon this place. This City is under the King of Tripoli, some 4 days journy into the Land. One Whiting the Capt. of an English ship (who hath bin a slave in these parts) coming to Florence, told the great Duke of this accident, and he himself had seen the City. The Duke desirous to know the truth, wrote to the Bassa of Tripoli about it, there having been a friendly correspondence between them these many years: The Bassa hath now answered the Dukes Letter, and assures him, that the thing is most true, and that he himself is an Ey-witness of it, going to the place purposely to see it, & that it hapned in the

---

[25] "Mr Waller's Relation of petrified Bodies of Men," in: Hooke, *Philosophical experiments* (1726), 386–388.

[26] Powell (ed.), *Letters of Blake*, 266–279; Curtis, *Blake*, 135–147; Powell, *Robert Blake*, 252–292; no reference to the pretended search during the consulate of Baker in his diary, nor in his dispatches, Pennell, *Piracy*; SP 71/22.

space of very few hours; and withall he hath sent to the great Duke divers of those things petrified, and among the rest, (Venetian Zecchines) turned into stone. Thus Mr. Fitton.[27]

Digby commented on this skeptically: "It seems strangeth to me, that an unactive body (as all dry and cold earthy ones are) should thus change Gold, the strongest Resistent in Nature. But it is true also, that little dense Atoms force their way most unresistably into all bodies, when some impellent drives them violently."[28] Evidently, Digby took the petrified city with its petrified Venetian Gold zecchines as a given and difficult *explanandum*. In his scientific works, Digby did not devote a special chapter or passage to petrification. He reasoned about the changing forms of matter in terms of the composition and dissolution of bodies belonging to the four classical elements earth, water, air and fire and thus partially adopted atomist teaching in a decidedly anti-Aristotelian way;[29] consequently, he was renowned as a reviver of atomist teaching in "our lesse partiall & more inquisitive times."[30] What did Digby mean with that brief comment on the Tripolitine petrification of gold coins? The most pertinent passage is his description of how Mercury, Antimonium and Gold supposedly interacted. "Cold" metals would normally not alter the substance of gold; but, "If Gold but touches Mercury, that sticks close to it, and whitens it so, that it scarce appears Gold, but silver only. If you cast this blanched Gold into the fire, the heat chases and drives away the Mercury." The heating of the gold by fire would not dissolve both completely again; the reiteration of the process would even lead to an "amalgation, joyn'd Mercury therwith corporally." He took this as argument that if even "cold Mercury" did "penetrate the whole body; we ought not to think it strange, that subtil atoms of fruit composed of many fiery parts wil pass with more facility and quickness" through gold.[31] Although he did not provide an example for this in his work, it becomes clear that Digby conceived of petrification as a process of composition of bodies and that it was hard for him to understand how a dry, cold and earthy

27 *Mercurius Politicus*, Nr. 334, Wednesday, October 29 to Thursday, November 6, 1656, p. 7363.
28 Ibid.
29 Digby, *Two treatises* (1644), chap. XIV, 116–129. Dobbs, "Studies," 16–25; Henry, "Kenelm Digby."
30 Boyle, *Works*, ed. Hunter et al., vol. 13, 227.
31 Digby, *Sympathetick Powder* (1657/69), 180f.

element could overcome the substance of gold (the Venetian zecchines), forming a stony amalgamation as its end product. This was even harder to understand than the combination of cold mercury and gold, because at least mercury was liquid. The doctrine of atoms entered where the processes conceived as normal by him and the normal balance of elements during processes of composition did not correspond with what nature was able to produce in such exceptional cases. The spontaneous petrification of a whole city and of all matter regardless of its elementary status could only be explained with help of atomist theory.

The background to this letter from Fitton/Digby is the Italian-English and French scientific connection that developed during the Civil Wars and to some extent prepared the ground for the institutionalization of academic circles in the Royal Society during the Restoration, as has been studied with reference to Newton's teacher Isaac Barrow and John Evelyn. And it is a connection continued by the members of the early Royal Society, such as John Ray, Philip Skippon, Francis Willughby and Nathaniel Bacon.[32] The Englishman Peter Fitton spent a long time in Italy, as agent of the English secular clergy in Rome, from 1631 to 1637, and 1640 to 1643. From 1650 to 1655, he also served as art collector,[33] and was employed at that time by the Grand Duke Ferdinand II de' Medici as keeper of the Grand Ducal medal cabinet.[34] Fitton and Digby were Blackloist English Catholics, the followers of Thomas White, who supported a national Catholic Church similar in form to French Gallicanism.[35] Blackloists had mostly left England during the Civil War and established themselves in France, at the Louvre court of Henrietta Maria, and in Italy. Digby had already been active during the 1630s in the Mediterranean, ransoming English

---

[32] On them and their visits of Kircher's museum in Rome cf. Cook, "Rome."

[33] Cf. Ferris, "Shopping-List"; Gabrieli, *Kenelm Digby*, 208–216.

[34] Fitton wrote an instruction *Of the value and rarity of medals* during the summer of 1656 for Barrow's use when the latter was searching for and collecting ancient coins and medals in Constantinople: Osmond, *Isaac Barrow*, 55–58; Feingold, "Isaac Barrow," 47–50. When Fitton describes the Tripolitine petrification as "metamorphosis" one might suspect the influence of Baumgartner's account which he might have known for professional reasons since: Baumgartner had also included a description of ancient coins in the same chapter.

[35] Collins, "Thomas Hobbes," 310ff.; Petersson, *Kenelm Digby*, 223–226; Henry, "Kenelm Digby."

captives from Algiers.[36] But he was also, as Fellow No. 34, one of the very early members of the Royal Society, elected on December 2, 1666.[37] The learned prince and later Grand Duke Ferdinand is known to have promoted science in the tradition of Galileo and to have founded only one year after the Fitton/Digby letter the *Accademia del Cimento* (1657–1662), an institution long considered a forerunner of the English Royal Society.[38] Fitton was also in close contact with Isaac Barrow during the latter's stay in Florence from February to November 1656.[39] And the Tuscan shores are renowned for stimulating scientific development: just as the ship of Captain Whiting landed there, bringing with it the story of the petrified city, ten years later in 1666, Livorno fishermen brought ashore a huge shark, which Grand Duke Ferdinand II ordered Niels Steensen to dissect. Steensen then wrote his memorable excursus about tongue stones as petrified shark teeth, which was received immediately in London in the Royal Society, marking an important milestone in the history of geology.[40]

With Digby, Fitton and Kircher, the problem of the Tripolitine petrified city was inscribed into the realm of the emerging English Baconian-Boylean scientific public sphere. Eight years later, Kircher's *Mundus subterraneus* was published. The Royal Society's secretary Henry Oldenburg saw Kircher's work simultaneously as the main target of the Society's critiques and as one of its inspirations, inserting a review of it in the *Philosophical Transactions* in 1666. He highlighted the "Story of a whole Vilage in Africa turned into Stone, with all the People thereof" as one of Kircher's statements that should prompt "severer and more minute Inquiries and Discussions."[41] Some of the early Royal Society's instructions for travelers explicitly mentioned Kircher's work to encourage empirical research on whether the claims in the *Mundus subterraneus* were true or not. This shows how central Kircher's work was at the very beginning of the newly founded Society.[42]

---

36 Hebb, *Piracy*, 162, 212, 233.
37 Hunter, *Establishing*, 60; Hunter, *Royal Society*, 138.
38 The *Accademia* began its work on June 19, 1657, but there are hints of earlier experimental work at the court, cf. Middleton, *Experimenters*, 42s.; Boschiero, *Experiment*, 111; Feingold, "Accademia del Cimento."
39 Iliffe, "Correspondents Network."
40 Rudwick, *Meaning of Fossils*, 49–56. Steensen, *De Solido*.
41 Oldenburg, "*Mundus subterraneus*" (1665/66), 116s.
42 "Directions For Observations and Experiments to be made by Masters of Ships, Pilots, and other fit Persons in their Sea-Voyages," in: *Philosophical*

Boyle, Hooke, Evelyn and others were all strongly interested in the problem of petrification in the years 1664/65, commenting for instance on the fossil wood from Acquasparta petrified by volcanic fluids that had been brought from Rome and the circles of the *Accademia dei Lincei* to London.[43] In addition to Hooke's *Micrographia* (1665) and Evelyn's *Silva* (1664), the short treatise *Observables touching petrification*, published in the first issue of the *Philosophical Transactions*, demonstrates this interest in particular: the text expressly highlighted the problem of petrified human bodies (and body parts) as a problem for future research.[44]

Robert Boyle, who wrote a great deal – but published much less – on the theory of matter and mineralogy and on petrification from the late 1650s onwards,[45] did not refer precisely to the Tripolitan city. He speculated on an explanation for spontaneous petrification, referring to what he thought was a Bavarian example from a "modern writer." Boyle cited a passage from the Bavarian historian Johannes Aventinus" *Annalium Boiorum libri VII*: according to Aventinus, "above forty Country-men, as also some Milk-maids with their Cows kill'd upon an Earthquake, had their Bodies by a terrene Spirit turned into statues" – so an Alpine scenario. Boyle believed that such spontaneous petrifications were possible and thought that they were caused in this case by the forces of the wind and sulfurous fumes that, according to Aristotelian theory, were thought to accompany earthquakes (*Metereologica* II, 7–9). He argued:

that an external agent of almost insensible bulk may turn animal Bodies into stony ones, by introducing a new texture into their parts ... in these strange petrifications, the hardning of the Bodies seems to be effected principally, if not only, as in the induration of the fluid substances of an Egg into a Chick, by altering the disposition of their parts, since the petrifying wind or steam cannot be suppos'd to have any such considerable (perhaps not any sensible)

Transactions 2 (1666/67), 433–448, 441s.; Kircher was cited in *General Heads* (1692), 11, 13.

[43] Cook, "Roman correspondence," 6, 9, 16: mention of Fitton in a letter of George Ent from 1638.

[44] "Observables touching petrification," in: *Philosophical Transactions* 1 (1665/ 66), 320s.

[45] Cf. Boyle, "History of fluidity and firmness" (1661). Cf. the "Papers on petrifaction and mineralogy," Boyle, *Works*, ed. Hunter, vol. 13, lvii–lxii, 364–425.

proportion as to bulk to the body chang'd by it, as to be thought to effect this change principally as an Ingredient.[46]

What Boyle did not mention is that his "modern writer" Aventinus was in fact relying on the famous *Book of Nature* by the fourteenth-century encyclopedist Konrad von Megenberg.[47] Megenberg had given the example of the petrified men and cattle as an observation that had been related to him by the chancellor of Duke Frederick of Austria, "maister Pitrolf." As scholarship on the history of natural hazards has well established, the story was therefore based on an infamous Carinthian earthquake in 1348, on which Konrad commented in his *Book of nature*.[48] But the theoretical aspect of Konrad's passage was taken explicitly from Avicenna, whose treatise *De congelatione et conglutinatione lapidum* had been translated in a somewhat mutilated form by Alfred of Sareshel in the circle of Gerard of Cremona during the eleventh century. Avicenna's treatise had been attached to 110 manuscripts of Aristotle's *Metereologica* and so formed the medieval Aristotelian Vulgata, which was also printed during the sixteenth century and was the standard text to explain earthquakes and their effects.[49] The complete text of Avicenna in its modern translation reads:

---

[46] Boyle, "History of fluidity and firmness" (1661), 198.

[47] Aventinus, *Sämmtliche werke*, vol. III/2 (*Annales ducum boiariae*, VII), lib. VII, cap. 20, 462.

[48] "Daz ain ist, daz dike von dem dunst, der ausget von dem ertpidem, laut vnd andrev tier ze stainn werden vnd allermaist ze saltzstainn vnd allermaist auf dem gepirg vnd da pei, da man saltz ertz grebt. Daz ist da von, daz derlai dunst vnd chraft so stark ist vnd so veberswenkig, daz si die tier also verchert. Also lerent die maister von der natur Avicenna vnd Albertus. Also sait mir auch maister Pitrolf, hertzog Fridreichs chantzlaer in Österreich, daz auf ainr hohen alben in Chärnden wol fünfzig haupt menschen vnd rinder hie vor ze stainen worden wärn, vnd daz die mayt noch vnder dem rind säzz mit einem handtschuh, reht als siv sazz, e si paidew ze stainen würden." (Konrad von Megenberg, *Buch der Natur*, cap. 33, 136). Cf. Borst, "Erdbeben," 541–544; Rohr, "Man and natural disaster."

[49] "Fiunt ergo lapides ex luto per calorem Solis, vel ex aqua coagulata virtute terrea sicca, vel ex aqua calida desiccata … Estque locus in Arabia qui colorat omnia corpora in eo existentia suo colore. Panis prope Toratem in lapidem conuersus est, remanserat tamen illi suus color. Sunt talia mira quia raro accidunt, tamen causae eorum manifestae sunt. Saepe tamen etiam lapides fiunt ex igne cum extinguitur" (Avicenna, "De congelatione" [1593], 375f.); Gerard himself had been the translator of Aristotle's *Meteorologica*, cf. Schoonheim, *Aristotle's meteorology*, xix; Lettinck, *Aristotle's meteorology*, 141–145.

In Arabia there is a tract of volcanic earth which turns to its own colour everyone who lives there and every object which falls upon it. I myself have seen a loaf of bread in the shape of a *raghîf* – baked, thin in the middle, and showing the marks of a bite – which had petrified but still retained its original colour, and on one of its sides was the impression of the lines in the oven. I found it thrown away on a mountain near Jâjarm, a town of Khurâsân, and I carried it about with me for a time.[50]

The Latin Avicenna Vulgate tradition lacked the detail of the bite mark in the bread and did not precisely locate the volcanic earth (*ḥarrah*). The topos of petrified bread with bite marks would turn up again only at the end of the seventeenth century; then, it would become embedded in a fully developed epistemological framework of experimental methodology. The period of omission occurred despite the fact that the detail had been present within eleventh-century Arabic natural philosophy; and what this all shows is that Boyle's Bavarian-Austrian example possessed a hidden Arabic-Mediterranean background.

As this topic became a central problem for the early Royal Society, its Parisian counterpart and rival, Colbert's 1666 *Académie des sciences*, took up these questions and discussed in 1668 and 1669 the Latin translation of Boyle's *Certain Physiological Essays*, a text that contained the *History of Fluidity and Firmity*. The Académiciens – Samuel Duclos, Claude Perrault, Christiaan Huygens and Edme Mariotte – devoted many sessions precisely to the problem of "coagulations transmutatives," the transformations from liquid to solid and vice versa.[51]

From Baumgartner to Kircher, Digby and Boyle, the story of a petrified city in Tripoli entered early modern geological and mineralogical discourse between 1594 and 1664 and was discussed there as a case of spontaneous petrification. This phenomenon was usually

---

[50] Avicenna, *De congelatione*, 22f.

[51] Cf. Archives de l'Académie des Sciences Paris, Procès-verbaux 4 (April 14 to December 19, 1668) and 5 (January 5–December 14, 1669): On May 4, 1669 the Academy conducted several experiments concerning the coagulation of eggs, the form of coagulation that Boyle thought to be the closest to the spontaneous petrification of men (ibid., f. 68r–72r). They continued with experiments about the coagulation of sheep blood and other liquids, from July 6 onwards; the more systematic *mémoires* by Duclos, Huygens, Perrault, Mariotte were read as first results and then as something of an answer to Boyle (f. 108–149). Cf. Boantza, "Cohesion," 80. Alice Stroup recalled the Boyle/Duclos controversy in a presentation to the Society of History of Science in 2000. It was then elaborated by Kim, *Affinity*, 48–52.

considered to be possible, but the explanations offered for the process differed. However, soon other uses for this story arose.

Science/Fiction

Only a few years after Kircher's main work, two English narratives about captivity in Barbary were published that again featured the story of the petrified city, but now with more exotic details. A certain T. S., who narrated a voyage in the Levant and his captivity from 1648 to 1652 in North Africa, claimed at the very end of his *Adventures*, published in 1670, that when he was serving his Algerian master as soldier in a siege of the fortified city of Tezrim, he saw at some distance "in a little Meadow where excellent Grass grows . . . the perfect Stature of a man Buggering his Ass."[52] Beyond the intention to provoke reactions of abhorrence in his English readers, the scandalous position served quite precise functions:

it was so lively that at a little distance I fancied they had been alive, but when I came nearer, I saw they were of a perfect Stone. I enquired wherefore the Moors or Arabs, that naturally hate all sorts of Representations, should shew their Skill by making such Beastly Figures, odious to Nature. I was informed that this was never made by man, but that some body of former years had been turned into this Representation with the Ass in the very moment of the Beastly Act. God by his power had changed the fleshly Substances of the Man and of the Ass into a firm Stone, as an eternal reproach to Mankind, and a Justification of his severe Judgments against us.

If Muslim religion forbade idolatry and cherished iconoclasm, it was even more unlikely to T. S. and to his readers that they had produced a statue representing this most disgusting act. It served therefore to produce higher credibility for the very fact of spontaneous petrification:

I did further search into the Appearances of this Report, and found the Stone to represent not only the perfect shape, but also the colour of every part of the Man and of the Beast, with the Sinews, Veins, Eyes, Mouth, in such a lively manner that no Artist with all his Colours could express it better; so that I was convinced of the Truth of this Report.[53]

When T. S. returned to Tripoli he inquired among his friends and English merchants about what he had seen and was told that about

[52] T. S., *Adventures* (1670), 238.   [53] Ibid., 238s.

five days' journey from Tripoli to the southeast, in the "Gubel" mountains, there was:

a whole Town full of these Representations; stones representing all manner
of Creatures belonging to a City, with the Houses, Inhabitants, Beasts, Trees,
Walls, and Rooms, very distinctly shap'd. Our people have entered into the
Houses, and there they have found a Child in a Cradle, of stone; a Woman in
a Bed, of stone; a man at the Door looking Lice, of stone; Camels in several
postures, of stone; a man beating a Woman, of stone; two men fighting, of
stone; Cats, Dogs, Mice, and all that belonged to the place of such perfect
stone, and so well expressing the several Shapes, Postures and Passions in
which the Inhabitants were in that time, that no Engraver could do the like.
Some may look upon this Relation as Fabulous, but let them enquire of our
Merchants, and Traders that have been in that City of Tripoly, or in the
Land, they shall find them all agree in the Confirmation of this Relation.[54]

The narrator finished the passage by judging that he saw no reason
"wherefore we should doubt of the possibility of these Relations, if we
consider the Almighty power of God that causeth all things to subsist
by his Influence, and can easily alter or change them as it seems good to
his Divine Wisdom."[55] One of the moral functions of the story – the
suggestion that Arabs and Muslims were inclined to sodomy and
homosexuality – is obvious and a frequent topic within travel and
captivity narratives, which depicted to European readers that alleged
inclination as a particular threat for captive Christians.[56] What is of
more systematic interest here is how the text employed the story, which
up until that point had circulated within "scientific" contexts, for
a passage within a travel narrative that slipped from factual descriptive
elements to fictional elements with the mixed functions of moral education and entertainment.

The text of T.S. is said to have been published posthumously by
a certain A. Roberts.[57] A footnote to the passage above refers to the
letter published by Digby/Fitton in the *Mercurius politicus* of 1656 and
shows that the source for the passage was the scientific discussion.[58]
The note also proves that Roberts, the editor, provided the final version
of the text precisely in 1670 and not earlier, and so it is unlikely that the

54   Ibid., 240–243.   55   Ibid., 244.
56   Cf. for that Colley, *Captives*, 128–130.
57   Matar, *Turks*, 77–81; Matar, *Britain*, 142.
58   "An Account of this was Printed 14 years since under the name of Sir Kenelm
     Digby." (T. S., *Adventures* [1670], 240).

passage derived from T. S.'s own captivity experience in 1654. It is impossible to discern where the "original" T. S. text ends and Roberts' rendition begins, and it is also impossible to make a clearcut distinction between true report and fiction. Gerald MacLean has noted about the *Adventures* of T. S. that some "of the incidents reported here doubtless took place, while certain descriptive passages could have been written only by, or with the advice of, someone who had visited Ottoman Algeria. There is as much of fiction as there is of fact in T.S.'s adventures." MacLean classifies the text as "historically based fiction of the kind associated with Aphra Behn and Daniel Defoe." Just as many modern authors often consciously and artfully mix descriptions from news and history well known to readers as real and then pass smoothly to complete fiction, T. S. and Roberts seem to play already quite virtuously with that border of fact and fiction. This kind of narrative belongs, therefore, not only to the context of early modern interest in the political threat of Barbary corsairs, but also to the inner-European context of the evolution of a literary system distinct and detached in its production of narratives from the realm of factual writing – "the rise of the novel."[59] Just as Fitton was a possible reader of Baumgartner, one might again ask whether T. S./Roberts was a reader of Kircher, for the marvelous setting recalls the story of the Ethiopian Victoria, and the reference to Lot and the overall interpretation as "Examples of Gods Justices" remind us of the Jesuit's account. But the story of T. S. was not framed in the scientific context of Natural History. It appeared in T.S.'s rendering in another captivity narrative of R. D.;[60] and we might suspect that if the serious Swedish scholar Johann Gabriel Sparwenfeld had read it on his voyage through North Africa in 1690/91[61] when he wrote about the petrified city to a friend in Paris, the narrative item could have wandered from "science" into

---

[59] MacLean, *Rise*, 179, 180, 184; Colley, *Captives*, 93.

[60] The passage on the petrified city was received from T. S. also in R. D., *True Relation* (1672), 6.

[61] Sparwenfeld is known as a specialist of the Russian language, but like many other early modern linguists, he searched in general for the relationships between languages and was therefore interested in the Oriental traditions in the tradition of Herbelot, whom he tried to surpass in finding new Arabic libraries and manuscripts not used for the *Bibliothèque orientale*. During his African voyage (1690/91), the major aim of which was to free 23 Swedish captives, he was mainly based in Tunis. There, he was in close contact with the French consul du Sault, cf. still fundamental Jacobowsky, *Sparwenfeld*, 153–185.

proto-fictional narratives and back to scientific research.[62] After Oldenburg's earlier exhortation to voyagers of 1666, the petrified city was included as a special problem in the first separate enlarged print of the Royal Society's cumulative Boylean *General Heads for the Natural History of a Country, Great or Small* in 1692. Michael Hunter and others have noted the "enthusiasm for such 'heads of inquiries'" in the early days of the Royal Society, judging that they "could almost be seen as a leitmotif of the society's activity in its earliest years, inspired by the Baconian imperative of data-collecting that was central to the society's rationale."[63] Already in 1661, a committee of sixteen Fellows had reflected on "proper questions to be inquired of in the remotest parts of the world."[64] The first version of Robert Boyle's *General Heads* was published in the very first issue of the Royal Society's *Philosophical Transactions* in 1666, and two further issues of the journal in 1667 were devoted almost exclusively to complementary "particular" questions specific to places such as the East Indies, Guinea, Hungary, Greenland, Turkey and Egypt. As the editors of Boyle's work have shown, the 1692 edition, which declared itself as Robert Boyle's work, was rather a compilation of articles from the first three volumes of the *Philosophical Transactions* of 1666/67, authored by fellows of the Royal Society such as Lawrence Rooke, Thomas Henshaw, Abraham Hill, Charles Howard, John Hoskins and Henry Oldenburg and

---

[62]   The indirect rendering in a letter from Piques to Ludolf with its mention of the scandalous positions of petrified men leads to the suspicion of influence by T. S.: "M Sparuenfelt est à présent à Tunis pour y chercher quelques vestiges des Vandales, mais il n'aura pas contentement il a escrit icy et m. l'Abbé de Longuerve a vu ses lettres, il n'y est pas fort content parcequ'il ne trouve pas gens qui puissent l'ayder dans ses recherches il rapporte plustost une fable qu'une histoire, d'un village fort avant dans les terres audelà de Thunis qui a esté petrifié par un torrent d'eau en sorte qu'on deterre des Gens en touttes sorte de Postures vel etiam liberis [deleted: pueris] operam dantes. C'est le Peïs de Niobe. On avoit fait Courre [sic, but should be courrir] le bruit qu'il avoit esté arresté parcequ'on l'avoit pris pour un marchand du bastion de France qui devoit une somme considérable à un des principaux de Thunis. Je chercheray l'occasion de luy faire sçauoir de vos nouvelles." (Louis Piques, Paris, de la bibliothèque Mazarine, to Ludolf, Paris, December 17, 1691, rec. January 12, 1692, UB Frankfurt/M, Nachlass Ludolf Nr. 574, f. 1053v).

[63]   Hunter, "Robert Boyle," 4; Hunter, *Establishing*; Lynch, *Solomon's Child*; Carey, "Compiling"; Collini and Vannoni, *Instructions*; on the prior more general apodemical instructions cf. Stagl, "Apodemik."

[64]   Hunter, "Robert Boyle," 14.

perhaps compiled by Denis Papin.[65] The particular heads for Turkey were originally published by John Hoskins and Henry Oldenburg in 1666,[66] but question no. 24 about the petrified city was added only in the 1692 edition. It called for future travelers to inquire "whether the Relations of a whole City's being turned into Stone be true, and if not what gave the first Rise to it, and whether it lye so near the Sea that these Bodies so metamorphosed may be easily brought into Europe." The compiler then inserted a brief digression on the subject, giving an account of someone "who was upon the Place" and stressing the credibility of the phenomenon, confirmed by a former chaplain to the factory of the Levant Company at Smyrna. This would point either to John Luke or to Edward Smith, the former being chaplain there with interruptions from 1666 to 1683 and then Fellow of Christ's College, Cambridge, and from 1685 Professor of Arabic. The latter was chaplain in Smyrna from 1685 to October 1691.[67] The chaplain's source had been an enslaved soldier in the army of the Bassa of Tripoli, who on a short leave from his military service, took time to see "this so strange Metamorphosis":

[A]t his first coming into the Place he saw a Sheep lying upon her Belly, as if it were chewing the Cud, whose Head he broke off from her Neck, with a Stone, and in the Gullet he could perceive some remainder of the chew'd Grass all petrified, which he took up, and sold afterwards to one of his Fellow-Slaves, who, having sent it to the Pope, had his Ransome returned for it: A little further they saw a Woman sitting on her Knees, with her Hands in a Trough, as if she were kneading Dough, her Mantle, that was clasp'd about her Neck being cast backward, and all turned to Stone, so hard that they could lift her and the Trough, in which the Hands were, without parting them or breaking any thing. When he asked a Priest, that was sent from the City to treat with the Commander, What way this did happen, he answer'd him, That all the Inhabitants of that Place were Sodomites, and that God rained down Fire and Brimstone from Heaven upon them; upon her [sic!] which they were all turned to Stones: And for Proof of this, he desired him to dig in the Sand, with his Hand, a Foot deep, which he found like blue Ashes; which, said the Priests, were the remainders of that Fire.[68]

---

[65] Boyle, *Works*, ed. Hunter et al., vol. 5, xli–xl. The book was entered into the Stationers' Company Register on June 3, 1692.

[66] *Philosophical Transactions* 1 (1666), 360–361, cf. Birch, *Royal Society*, vol. 2, 134.

[67] Pearson, *Biographical sketch*, 31–34; Stearns, "Fellows," 76s.

[68] *General Heads* (1692), 66–68.

One could suppose again the influence of former texts. In any case the narrator/compiler of the 1692 *General Heads* had now included an important element that could serve as *experimentum crucis* in the Baconian sense for a traveler who encountered those petrified bodies: while T. S. established his credibility through the scandalous nature of the petrified act he described, which was unlikely to be formed artificially, and by insisting that the colors of the bodies remained on the petrifications, the narrator of the *General Heads* claimed that the inner parts of the creatures were also petrified, like the chewed grass in the sheep's gullet – this a human sculptor would be unable to produce artificially. Regardless of whether this narrative was transmitted from the Levant to London or if it was merely produced in 1692 for the edition of the *General Heads* by its compiler in London, its function is clear: armed with this example and direction of inquiry, a natural historian would be able to examine objects found at the scene and produce a legitimate verification or falsification of their status as petrifications.

Before the Baumgartner story and its rather obscure Avicenna medieval antecedents, the idea of spontaneous petrification was a brief element in narrations of the miraculous, presented in the framework of a worldview that allowed for the co-presence of the extraordinary with the everyday.[69] Indeed, one could speak of a state of nescience concerning a possible conflict of different laws for distinct spheres – a sphere of natural occurrences and a sphere of divine possibilities. What happened during the seventeenth century seems to be, on the one hand, the provisory incorporation of "spontaneous petrification" into the realm of a differentiating natural science. In Boyle's work, it was conceived within an Aristotelian tradition concerning sulfuric exhalation and the effects of earthquakes; others like Digby felt the need for a supplementary atomist explanation. This process ended in the conscious and precise form of early modern specifications of unknowns, which tended to formulate, though still in narrative exemplary forms, a precise *experimentum crucis* as part of a query for scientific voyagers that would serve to determine the existence or non-existence of such an item – in this case, the inner parts of an animal's body would have to be found petrified. On the other hand, the same feature developed from an element in the miraculous narrative into an elaborated aspect of

[69] Daston and Park, *Wonders*; Platt (ed.), *Wonders*.

a narrative genre that now consciously played with the possibilities of the early modern literary system's growing autonomy. It could replace the binary distinctions of natural/divine or true/false with the distinction of factual/fictional. Next to a realm of scientific specified unknowns that possibly existed but were not yet proven, the realm of the literary fictional imaginary emerged, and with this came all the implications that have been developed within the different terminologies of modern narratology[70] – for instance, the "pact of fictionality" between author and reader, which absolves the text from a factual true/false criteria. At the same time as the two new forms developed, they interacted. The partially fictional text of T. S. borrowed its central premise from scientific discourse, and later the *General Heads* and other scientific texts borrowed from an elaborated version of T. S.'s text or its derivatives in order to formulate the *experimentum crucis* that would make establishing the truth – or not – of spontaneous petrification a task for serious scientific travelers.

Enlightened Falsifications

French Diggers in the Desert

Until that point, not one of the authors of these published works claimed to have observed the petrified city first hand. It was always integrated into scientific discourse as second-, third- or fourth-hand delegated autopsy. Kircher never went beyond Malta and never saw Egypt while writing his *Oedipus aegyptiacus*. Digby knew the southern coast of the Mediterranean from his activity as ransoming agent, but that was not related to the Fitton letter. Consul Baker had certainly lived in Tripoli but his letter, which entered the scientific context late through the edition of Robert Hooke's works, related the story in an even more manifold

---

[70] Kablitz, "Kunst des Möglichen"; Müller, "Fiktionalitätsproblem"; Lavat, "Paradoxes." The question concerns how to adapt and vary the narratological concepts of fictionality (Eco, Ricœur etc.), which are all formulated with regard to modern (or even postmodern) texts and do not pay attention to the problems that arise when looking at texts in a premodern world without a fully developed literary system detached from other functional systems of society. A further step still is the factuality/fictionality distinction in the mutual cross-fertilization of a huge variety of text genres and narrative practices, all unstable during the early modern period (cf. Zwierlein and Ressel, "Ausdifferenzierung").

and indirect way. The version incorporated in the *General Heads* was transmitted as indirect information and the chaplain of the Smyrna factory involved was not even named. Neither the London and Paris governments nor the Royal Society or the *Académie des sciences* had used the Mediterranean consular network to investigate the truth of the petrified city until that point. The French had established a consul in Tripoli, subordinate to the one in Tunis in 1577, and did not make the office autonomous until 1647.[71] The English instituted a consul in Tripoli only in 1658 – that is, after Fitton's 1655 letter to Digby.[72] In both cases, the records for an earlier period are fragmentary: in the French case, they do not start before 1642, and in the English case, are not of any considerable density before the 1660s.[73] There was no special expedition or scientific journey to the supposed petrified city until the very end of the seventeenth century. At that point, it was the French who took the lead, although beforehand the story had largely circulated in English-Italian networks. But one aspect shows an important difference from the English: all the French writers now encountered never cited any of the English sources enumerated above.

In 1690, probably in the spring, Jean-Baptiste Colbert, Marquis de Seignelay, gave orders to the then consul of Tripoli/Barbary, Claude Lemaire (?-1722), instructing him to conduct a more serious investigation into the case of the petrified city. Seignelay had a high regard for the importance of information gathering,[74] and in pursuing the legitimate politics of science, he followed in the footsteps of his father, who is known for having founded the whole array of France's scientific national academies, in particular the *Académie des sciences* in 1666. Seignelay's order is mentioned in the 1706 *Mémoire pour Messieurs de l'Académie Royalle des Sciences* by Lemaire.[75] By this point, the name

---

[71] Mézin, *Les consuls*, 738.   [72] Pennell, *Piracy*, 19.

[73] SP 71/22; AN AE B I 1088 covers the whole period from 1642 to 1695, while in later periods a volume of the consular correspondence only contains about 5–7 years on average.

[74] Cf. Dingli, *Seignelay*, 121–130.

[75] AN AE B I 1092, f. 274–278. No author of the *mémoire* is indicated and it has a false date "fin 1726." However, since Paul Lucas is mentioned as planning to travel to the petrified land, "ce qu'il me dit en partant d'icy," and since the author is writing of his fourteen years "que j'ay seroy a Tripoly en qualité de consul" (AN AE B I 1092, f. 275v, 276v), the author must be Claude Lemaire (cf. his letter of July 1706: my "14.e année de Consulat en Barbarie": AN Paris AE B I 1089, no pag.). The *mémoire* was probably inserted twenty years later,

of the city had solidified as "Ras Sem," whereas in the seventeenth-century sources there is either no name, or the names are varied or mythical.[76] The expedition to the city could be combined with the task of buying Arabic horses for which the mountains of Derne were famous and for which the Intendant de la marine in Toulon had provided a royal vessel. Lemaire and his employees had arrived at Benghazi in May 1690. Because the Bey of Derne warned him that in early summer it would be too hot for someone unacclimatized to the conditions, Lemaire sent out "un de mes Domistiques" with a caravan of eight camels and ten "Spahis arabes." Starting from Benghazi, they arrived at the "petrified land" called "Rassem, meaning in Arabic the poisoned Cap," where they found some petrified palm branches, olives and dates which had not changed color. They collected these, but then their limited supplies of water and biscuits compelled their return.[77] Lemaire took the palm and olive branches, which were heavier than marble and had been transformed into flint ("pierre à fusil"), to Versailles and showed them to Seignelay's successor Jérôme Phélypeaux de Pontchartrain and to all the scientists who were at court. Lemaire claimed that he had later asked Kalil Pasha, who traveled to Ougela, two days from Ras Sem, to bring whole petrified animals or a little child with him; but the ship transporting this curious petrified cargo sunk on the way to Tripoli.[78] A similar short note without further evidence in a description of Tripoli by François Pétis de La Croix the elder, dedicated to Pontchartrain of 1697, states that Lemaire's successor as consul of Tripoli, du Sault, traveled to Ougela and brought petrified objects back to France.[79]

After Lemaire had been reappointed consul at Tripoli, he resumed research into the petrified land. In 1705, another occasion like that of

because the former chancellor of Lemaire, Pouttion, legitimized with its help his own new attempt to head for the petrified city in that year.

[76] "Coppie de la lettre escritte a S.A.S. Mgr Ladmiral par le Sr Lemaire a Tripoly le 5 may 1706," AN AE B I 1089 (no page).

[77] AN AE B I 1092, f. 274v–276v; "Coppie de la lettre" (cf. n. 76).

[78] AN AE B I 1092, f. 274v–276v.

[79] "Du costé du midi il y a une grande étendue de pais jusqu'en Numidie comme les Provinces de Torraque, Benolete, Garyan, Feczam, et Ougela, cette dernière ville est Éloignée de Tripoli de 17 journées vers le Ouest. M. Du Sault y a eté depuis quelque temps et en a veu les arbres, hommes, oiseaux, et bestiaux petrifiés dont il a apporté en France plusieurs pièces" (Bayerische Staatsbibliothek Munich Cod. gall. 729, f. 21v).

1690 occurred when the Admiral of France Louis-Alexandre de Bourbon, Comte de Toulouse and bastard of Louis XIV, sent sieur Patrice with another vessel to Lemaire requesting again some Arabic horses from the Derne mountains. Lemaire immediately let his brother Louis replace him at the consular post of Tripoli and undertook a voyage to Benghazi and Derne between late June and autumn 1705. He was back in Tripoli at least by January 1706.[80] Patrice now brought a message from the admiral's secretary Jean-Baptiste de Trousset Valincour, who was not only concerned with naval affairs but, as the successor of his friend Racine from 1699, was also a member of the *Académie française*. Valincour asked Lemaire to gather further information about the petrified city. In a letter from July 8, 1706 to Maurepas, Lemaire sent "a *mémoire* concerning the remarkable things I have seen during the voyage made into the mountains of Derne."[81] But here as in all the letters of 1705 to 1708, Lemaire remained equivocal about whether he had really been to Ras Sem himself. It seems, on the whole, that he had always sent someone else to the city; once he specified: "I have seen all of that with the exception of human bodies and animals [sc. petrified] which I have not seen."[82] On January 20, 1707, he announced that "some persons of the Bey of Tripoli's entourage" had given an account of their voyage to Ougella and the "petrified land." He dispatched the vice-consul of Derne, Guilhaumier, to the Sheik of Ougela to possibly fetch a petrified human body "digne de la curiosité du Roy." He repeated that the olives and palms had been turned to stone and that "all those who have been there" had assured him that they had seen a "quantity of petrified human bodies in diverse attitudes as well as animals right up to a herd of sheep petrified together with their shepherd." These would all be covered and uncovered by the sands according to how the wind was blowing. He claimed to be searching for the petrified head of a child, which was said to be circulating among the servants of the Bey. He assured Maurepas that he had been "plainly informed about the truth of the petrification of the palms and olives of which I delivered some pieces to my Lord your father [sc., in 1690, to Jérôme Phélypeaux

---

80 Lemaire gave Patrice a date that had turned to stone, and claimed that he had possessed petrified bread since 1690. All that in AN AE B I 1089 (no pag.).

81 Letter of July 8: AN Paris AE B I 1089 (no pag.); the *mémoire* itself: AN Paris AE B VII 224, Lemaire, "Mémoire."

82 Lemaire, "Mémoire," 1044.

comte de Pontchartrain]."[83] However, Arabs prevented the vice-consul from actually reaching the location in question.[84] On December 10, 1707, the well-known Oriental traveler Paul Lucas arrived in Tripoli after a voyage from Cairo, which had taken him an exceptional two months because of a heavy storm.[85] Lucas lodged in the house of his good friend Lemaire and departed from Tripoli on January 25, 1708.[86] In his travel account of that voyage, printed in 1712, Lucas inserted a slightly different version of Lemaire's *mémoire* about the region of Derne, which the consul had written two years before.[87] In this version, the Arabs and Christian slaves assured Lemaire that they had seen the bodies of women and men, of animals, and of a horse still standing on all four hooves. He specified that "I myself have seen petrified bread and a date," which had brought him to the conclusion that "this catastrophe" must have happened at the peak ripeness of dates, usually September and October. This led him to believe that "this was made in one moment, or by permission of God, or by an effect of nature, which I cannot understand."[88] Although Lucas did not cite any of the sources from the circles of the Royal Society, the petrified bread, together with the dates, betrays an awareness of the new epistemology of experimental empiricism and the search for an *experimentum crucis*, in a mixture of a tradition close to Avicenna and with some elements from the discussion ranging from Kircher to the *General Heads*. It seems that Lemaire had received the English geognostic teaching at some point, as his use of the term *catastrophe* for this phenomenon is a very early incidence of the modern concept of natural catastrophes as event that

---

83 Lemaire to Pontchartrain, Tripoli, January 20, 1707, AN AE B I 1090, f. 6r–9v. The mission of the vice-consul is also touched on in Lemaire's letter of March 29, 1707, ibid., f. 24r–26v, 25v.

84 Lemaire to Pontchartrain, Tripoli, July 23, 1707, ibid., f. 38r–41v, 39r–v. The unsuccessful mission had cost 334 piastres.

85 He did not travel there himself, perhaps because his ship was struck by a severe storm when he departed from Cairo on October 17, 1707. Cf. Consul De Maillet to Pontchartrain, October 20, 1707, AN AE B I 316, f. 36r–39r, 39r: "Mr Paul Lucas . . . est parti dépuis trois jours."; Lucas, *Voyage*, tom. 2 (1712), 101, 104; Lemaire to Pontchartrain, December 22, 1707, AN Paris AE B I 1090, f. 56r (dates differ slightly);.

86 Lemaire to Pontchartrain, Tripoli, January 25, 1708, ibid., f. 60r–61v, 60r.

87 Lucas, *Voyage*, tom. II (1712), 110–132.

88 Lucas, *Voyage*, tom. II (1712), 126.

happened again and again, instead of reserving the term for the Noachian Flood only.[89]

Lemaire was frustrated. He had been hearing about this place for eighteen years, but was still unable to acquire a petrified human body or animal.[90] In 1708, a final attempt by the same vice-consul of Derne, Guilhaumier, to undertake a visit to Ras Sem was not successful; subsequently, Lemaire obtained the consulate of Aleppo and ended up finally as consul of Cairo from 1711 to his death (February 24, 1722). As a result, from 1708 onwards, the person most interested and experienced in matters concerning Ras Sem was no longer at the scene.[91]

As in the English case, the French North African investigation into Ras Sem had its background in current geological and geognostic research. The abbé Bignon was orchestrating a survey of France between 1716 and 1718, Réaumur being the leading scientist, in which the question of petrifying and calcifying fountains, was central. In the *mémoires* sent back to Paris during that vast survey, Georg Agricola, Kircher and Tournefort figure as mineralogical references.[92]

So it is not surprising that also in North Africa, efforts to investigate the realities of Ras Sem resumed in 1726 with Lemaire's former chancellor Pouttion, under the Tripoli consulship of Joseph Martin. Martin sent Pouttion to Benghazi to search for marble columns, horses and other things that might please the secrétaire d'État Maurepas.[93] Pouttion took the occasion to tell of a Moor who had traveled through the hinterland during July and had related again the story of the petrified city, now under the name of "Guibabou," where he had reportedly seen human bodies and animals and a whole ship out of water with five petrified men on board, all "d'une couleur roujatre."

[89] Robert Hooke's reflections on the catastrophic effects of earthquakes within which he also enumerated petrification, seem to match best, Huggett, *Cataclysms*, 49–51; Briese and Günther, "Katastrophe," 172–179.

[90] Lemaire to Pontchartrain, Tripoli, May 23, 1708, AN Paris AE B I 1090, f. 78r–81r, 80v.

[91] Claude Lemaire's successors as consuls of Tripoli, Pierre Poullard (from August 17, 1708 to August 1711) and his son-in-law, Pierre d'Expilly (consul 1711 to 1722) had no interest in the matter.

[92] Even small experiments in this field were conducted by throwing diverse "material" into waters considered to be petrifying in order to observe the results: Demeleunaere-Dopuyère and Sturdy (eds.), *L'Enquête*, 92f., 102, 104–106, 243, 326, 372, 594, 596, 678, 737f., 784, 793.

[93] Martin to Maurepas, AN AE B I 1092, f. 189–204, 203r–v.

Pouttion was thus aware of the story in a form close to the tradition of Avicenna and as it was told by T. S., where the remaining color of the petrified bodies was supposed to mark the difference from sculpted stone.[94] Martin and Pouttion proposed to undertake an expedition and asked the king for the necessary funds: 2000 livres for 20 camels, 25 people for excavating and 20 spahis, in addition to gifts for the Bey and the Arabs, and a tent and a horse for himself.[95] At the French court, Pouttion's initiative was appreciated and he was rewarded by nomination as vice-consul at Benghazi in 1730, but he was not put in charge of the expedition finally mounted;[96] command went instead to Pierre-Jean Pignon[97] and Claude Tourtechot alias Granger. As has been seen above in the chapter on politics, Pignon later became, as intendant de commerce of Marseille, one of Maurepas' most faithful agents for the political arithmetic of shaping the French nation in the Levant. He did not arrive at Benghazi until August 17, 1733.[98] He knew Tripoli well: he participated in the conclusion of the peace treaty between France and that regency on June 9, 1729, and then became consul of France at Cairo.[99] Writing first about those matters on July 11, 1730 from Cairo,[100] Pignon was joined there by the travelers Sr. Joinville and Granger. In contrast to Pouttion, Pignon had the attention of decision-makers in France. He received a letter from Maurepas dated November 23, 1730, giving him permission to undertake the expedition to the *pays petrifié* together with Joinville and Granger.[101]

---

[94] A possible other chronologically close French source might be "Lettre d'un Missionnaire en Egypte," in: *Nouveaux mémoires* (1717), tome 2, 73s.

[95] "Mémoire des découvertes que le Sieur Pouttion a faittes pendant dix mois de résidence qu'il a fait à Benghazi ancienne ville dépendant du Royaume de Tripoly de Barbarie, envoyé à la Cour par Monseigneur Martin consul de France audit Tripoly le 24 septembre 1726," AN AE B I 1092, f. 240–243v.

[96] Pouttion to Maurepas, Tunis, April 29, 1730; July 29, 1730; September 30, 1730; November 11 and 25, 1730 (AN AE B I 206, f. 2r, 8r–v, 14r., 17v–18r, f. 20r) and Tripoli, December 12, 1732, describing his findings of marble statues and accompanied by copies of inscriptions, AN MAR B VII 311.

[97] Pouttion to Maurepas, Benghazi, July 25, 1731, ibid., f. 25r-v: He has received the letters from Maurepas of November 15, 1730 and March 27, 1731 and he would now wait for Pignon.

[98] Pouttion to Maurepas, Benghazi, August 30, 1733, ibid., f. 78r.

[99] Cf. Pignon's *mémoire* on Egypt written for René Duguay-Trouin, who commanded the fleet that forced Algiers and Tunis in 1731 to release the French captives in AN MAR B VII 311.

[100] Pignon to Maurepas, Cairo, July 11, 1730, AN AE B I 320, f. 340v–341r.

[101] Pignon to Maurepas, Cairo, July 24 and 27, 1731, AN AE B I 321, f. 57r.

The preparation of the expedition took still some time, which Granger and Joinville used to observe the Nile and search for antiquities, while Pignon worked to identify the exact position of the city and wrote enthusiastically from Alexandria in August: "I have found the petrified land without doubt, it is around 28 degrees 2 Minutes at West South West of Cairo, 13 days of marching with camels are enough to get there ... I will go there at the end of October or beginning of November."[102] The cartography Pignon used for locating that position, two maps of North Africa by Guillaume Delisle from 1705 and 1707, are examples of a new trend in mapmaking, which marked the emergence of "modern" explicitness about uncertainty and unknowns and the developing self-discipline of the cartographer, who replaced zones that had formerly been filled from imagination with written commentaries and discussions of differing opinions within geography.[103] Back in Cairo, Pignon wrote that he had recently been informed by Maurepas' letters of May 16 that Sr. de la Condamine (1701–1774) also wanted to go with him to the petrified land, and it seems that he became anxious that he would be deprived of the glory of the probable discovery.[104] La Condamine, elected in 1730 as a member of the *Académie des sciences* as a chemist, had asked Duguay-Trouin to accompany him during his military mission against the Barbary states in 1731.[105] Maurepas and the court seem to have developed a real interest in the case and perhaps expressly prompted several researchers

---

[102] Pignon to Maurepas, Alexandria, August 16, 1731, AN AE B I 321, f. 100r

[103] "Le Géographe Delile dans sa Carte latine de la Libie de 1705 place la fontaine du soleil et le Temple de Jupiter Ammon dans le même Endroit où on voit dans la Carte françoise de 1707 de cet auteur la ville de Siwar, bien des Marchands de cette ville m'ont assuré que la fontaine y étoit et qu'on voyaoit aux Environs plusieurs ruines des Anciens Monumens, la ville petrifiée est à 3 Lieues au Nord est de Siwar" (ibid.). – The 1707 map is the *Carte de la Barbarie, de la Nigritie et de la Guinée* on which Surun, "Le blanc," 122 commented. From the two Latin maps that Delisle produced in 1705 (Dawson, *L'Atelier*, 254), Pignon must have consulted the *Theatrum historicum ad annum Christi quadringentesimum in quo tum Imperii Romani tum Barbarorum circumincolentium status ob oculos ponitur*, a map which covers the whole Western Roman Empire and, for North Africa, the *Cyrenaica* including a location "Augila" close to the place named "Ouguela."

[104] Pignon to Maurepas, Cairo, November 5, 1731, ibid., f. 118–120; Villeneuve to Maurepas, November 8, 1731, BNF Ms. fr. 7184, f. 368v–370r, 369v., also in AN AE B I 403.

[105] The partial diary of that period, BNF Ms. fr. 11333, concerns only Algiers, cf. Emerit, "Condamine."

at the same time to conduct a multipronged and reciprocally controlled series of personal inspections. Pignon tried to profit from yet another source to better prepare the expedition: he organized a translation of the geographical description of Egypt by al-Maqrīzī, an early fifteenth-century historian and geographer. Pignon employed his consular chancellor Tassin, a local *enfant de langue*, the sieur Latine, and his young brother, to translate, as he described it, "the best historian who has written about Egypt and its neighbouring lands, [that is] Macrisi in four volumes and 2400 pages," a work that he knew the Royal Library in Paris possessed. The masters of Arabic language and culture in Cairo had their own difficulties in explaining al-Maqrīzī; but Pignon claimed to have read and corrected the translation work weekly.[106] This process seemed to produce dividends: in a letter sent from August, Granger[107] and Pignon claimed to have established the precise location of the petrifications thanks to al-Maqrīzī. While Lemaire had always regarded Ras Sem near Ougela as the place in question, Pignon and Granger now headed first for Guerzé (Ghirza).

Pignon and Granger finally set out on their voyage, starting from Tripoli on January 18, 1734 and arriving at Ghirza on January 27. Unfortunately, they found only ruins in the sand, as Pignon recounted: "But to my great grief and surprise with the renegade who accompanied me and who had been with the dragoman of the consulate of Tunis in the entourage of Kalil Pasha when they passed through that city twenty-eight years ago, we have found it [= the city] reduced to ruins." It was impossible to dig deeper into the earth. From the specimens that were brought to Pignon from Ras Sem and Ougela, he "understood that one had mistaken stone pikes, which perfectly recall those that one finds in the desert of St. Macaire [= Wadi El Natrun, desert of Skete in Upper Egypt], for petrified wood," and there was nothing that resembled a human body. "If at any time there were human bodies petrified at Ghirza, that city does not exist anymore and it is impossible to dig in those ruins due to the lack of water (to be found only at a distance of fifteen leagues) as well as the danger of being attacked by

---

[106] Pignon to Maurepas, Cairo, March 24, 1732, AN AE B I 321, f. 196r–201r, 199v–200v.
[107] Who had returned to Cairo with Joinville at that time: Granger to Sauveur François Morand, Cairo, June 28, 1732, Archives de l'Académie des Sciences Paris, Pochettes [sub dato]: "Je compte de partir au mois de Septembre pour me rendre au pais petrifié ou l'on dit que les habitans y sont dévenus Pierre."

Arabs ... One may not count anymore on verifying that fact, and therefore there shall be no more talk about Petrification."[108]

Granger provided the same negative result and added a brief reflection on the subject in general:

But as the question of the petrifications is a matter of scholarship, the problem will only be decided in favor of those who believe in these petrifications by the presence of some of those petrified animals. All the reasons, even if plausible, do not seem strong enough to convince those who do not accept these petrifications – which I do not deny, if we regard the other bodies which petrify. Those who believe [sc. in the petrifications] could object to us on two grounds: first, that these petrifications might be hidden under the ruins of that city, second that we have not been where they are ... There are only two places where one indicated those petrifications [sc. on the maps], which are Ghirza and Ras Sem. Ghirza is the one which corresponds best to all the information that we have gathered relating to the subject. Kircher assures us, following Makrizi, that the city whose inhabitants who were said to be petrified is to be found nine days south of Tripoli, which applies to the place where we went – even more as there is no other city within fifty leagues. I add a piece of that pretended petrified wood to the other stones that I send to your Highness.[109]

Kircher seems not to have used al-Maqrīzī in any of his works. The standard Arabic authority for the geography of that region was *Abulfeda*, which Kircher did indeed use for drawing his *Descriptio chorographica recentior ... Aegypti*.[110] That map, the only one Kircher produced for the region, reaches no further west than Alexandria, so does not depict any place near Tripoli. Some copies of al-Maqrīzī's historico-geographical work on Egypt, the *Kitab al-mawā'iẓ wa-l-i'tibār fi dhikr al-khiṭaṭ wa-l-āthār*, had already appeared in European libraries,[111] but no use of the text has been

108 Pignon to Maurepas, Cairo, March 22, 1734, AN AE B I 322, f. 150r–151r.
109 Granger to Maurepas, Tripoli, February 22, 1734, AN Paris B VII 322 (no pag.). The passage is transcribed but divided into two pieces without reason in Granger, *Voyage*, 45, 49s.
110 Cf. above, chapter "History," note 164.
111 Cf. Holt, "Study," 451 nr. 16; a Turkish translation of Maqrīzī's work on the geography and history of Egypt had been brought to the Netherlands by the Berlin Orientalist Christian Ravius (UB Utrecht MS I B 8 a, bought 1640 in Constantinople), cf. Schmidt, "Between author," 36, n. 23. The BNF does not possess today a complete version of the work, the Ms. Arabe 1726, 1727, 1729, 1730, 2144 (Ancien fonds arabe 672–675) to which Pignon perhaps referred, are a history of the Mamluk dynasty.

documented. No printed version or translation into a European language was available, and there is also no trace of the translations that Pignon purported to have organized in Cairo.[112] I have not found either a petrified city or village mentioned by al-Maqrīzī, even though his work on Egypt contains some similar extraordinary elements.[113] One may therefore suggest that Pignon and Granger had been boasting a little, claiming a pseudoscientific basis for their calculations.

Be that as it may, this was still the first time that learned European researchers within the context of a more or less formal scientific expedition ordered and sponsored by the French court had traveled to one of the places that had been identified as the petrified city. During the seventeenth century, the only information available had come from unlearned voyagers, captives, and was transmitted by chance to Europe. It seems that, at least in the French context, the case of the petrified city was closed – "no question of petrification," as Pignon had put it. Autopsy by a multiheaded team of scientists or officials that were recognized to enjoy a similar scientific reputation was the French way of finally falsifying that hypothesis. In great contrast to the English way of organizing scientific communication, the whole enterprise between 1690 and 1734, which involved more than two generations of the leading figures at the court and in the marine, contacts in Marseille, the consuls, chancellors, dragomen, and enfants de langue of the consulates in Tripoli, Cairo, Benghazi, Seyde and many local Arabic and Ottoman collaborators, remained restricted to the sphere of unpublished letters and *mémoires*, and failed to lead to an elaborated systematic scientific publication on "spontaneous petrification" by one of the protagonists.

112 Kircher, *Oedipus Aegyptiacus*, tom. 1 (1652), f. 8; cf. also the list of authors used that Kircher gives in tom. II/2 (1653), f. Aaaar–Aaaa2v; Stolzenberg, *Egyptian Oedipus*, 156–162; Wansleben, *Nouvelle relation*, 65, 112, 117, 274 referred to a historical work on Egypt by Maqrīzī very generally in some places in his second report, but it seems not to be the *Khiṭaṭ*. Hamilton, *Copts*, 156 mentions *en passant* that Renaudot was one of the first Europeans who made use of Maqrīzī in 1713.

113 The most fitting story might be that of Ibn Ouacif Shah who built the cities in the Oasis in inner Egypt between Masr, Alexandria, Saïda, Nubia and Ethiopia, and created a magic guard statue who was able to turn strangers to stone (Al-Maqrīzī, *Description*, chap. LXXI, 691s.).

English Armchair Science

Even though the English had maintained a consul in Tripoli since 1658, there is no evidence in correspondence of the consul's personal involvement with such affairs of scientific research or culture. The Royal Society and the court remained interested in the question of the petrified city. In 1715, when the Tripolitan envoy Salé Aga was in London, an Armenian Christian member of his entourage claimed that he had seen the petrified city three times, and that he had even taken the figure of a little child with him, but had to leave it in the sands when Arabs attacked his group.[114] Thomas Shaw, who in 1720 became chaplain of the English Church at the factory in Algiers, sent a letter in 1722 to the famous natural historian and antiquarian John Woodward. With this message, the phenomenon reentered the spheres of the Royal Society. During his stay in the Maghreb, Shaw kept close contact with people such as Edmond Halley, the Secretary of the Royal Society, James Pound, Thomas Birch, and the President of the Society itself, Hans Sloane, who had succeeded Newton. Surely, Woodward, a member of the Royal Society since 1693 and its most renowned geoscientist at that time, was interested in the "Petrifyd Village, in the Kingdom of Tripoly, in Barbary"? Shaw had obtained all his information from Lemaire, whom he met in 1721 in Cairo.[115] When, in August 1728, an official envoy of the Regency of Tripoli, Cassem Aga, also named sometimes Cossum Hoja, came to London to pay homage to George II and to renew the peace treaties with England at a time when Tripoli itself was under pressure by the French fleet,[116] he

---

[114]   BL Ms. Harley 6824, f. 22r.

[115]   "The village is calld Ras Sem, or poyson'd-head. The petrifycations of Olive trees, & Olives, of Palm trees, & dates; of Children, Men, & Women, Horses, & variety of other natural & animal Beings, are frequently seen, according as the wind (which affects these sandy Deserts, in waves, similar to those of the Sea) shall have exposed, & made bare any part of these plains. The compass & Sphere of these petrifycations are 30 or 40 Miles: & where any of the olive, or Palm is found, there is allways found a ruinated Grove, or a great collection of them. The Consul did me the favour, to give me a piece of the Palm, which, as far as I can judge, by comparing it with the true one, agrees exactly, both as to the external, & internal Contexture of its fibres with it. I shall put this in a little better method, & communicate it, by you, if you think it, to the Royal-Society" (Shaw to John Woodward, Algiers, June 8, 1722, BL Add Ms. 4432, f. 51v; copy in BL Ms. Harley 6824, f. 13r–v).

[116]   In July 1728, a French squadron had bombarded the city under M. Gran' Pré, and in 1730 the city was bombarded again.

and his whole entourage were immediately interviewed about the petrified city. Cassem Aga had come with an entourage of twenty-one men, including his brother Soliman, a secretary, a steward, several servants and even his own coffee maker.[117] Jezreel Jones, who had been employed by the Royal Society and remained in close contact with it, worked as his interpreter and English secretary,[118] while on the Royal side the Greek Theocharis Dadichi served as translator.[119] First Cassem Aga himself made a declaration on the petrified city, which was translated into French by Dadichi and then communicated to Sr. Clement Cotterel, the master of ceremony, then to the Secretary of State. This report quickly went on to circulate in intellectual circles (Alexander Pope)[120] and in the Royal Society.

Report concerning the petrified city delivered by Cassem Aga, envoy of Tripoli to His Majesty of Great Britain in the year 1728

Glory to God alone.

One of my friends has asked me to write him down what I have learned at Tripoli concerning the petrified city. I have said to him that I have heard it said by different persons and particularly by a man worthy of trust who has been there in his own person. That it is a large town, of circular shape which had large and small streets, [equipped with] their shops, with a great castle built in magnificence. That here are different trees, but mostly olives and palms, all of blue stone or of the color of ashes. That there were men performing different jobs, some holding goods, others bread in their hands and everyone doing something; there are even women suckling their children, others coupled/lying together with men, and all that in stone. That he entered the castle through three different doors, where he had seen a man sleeping on his bed of stone. That there were guards at the gates holding lances in their hands. That there were, finally, in that admirable city several sorts of animals, like camels, cattle, horses, donkeys, sheep and birds, all in stone and of the color above mentioned. That's what I have learned from many persons, and mostly from the one who has seen all that.

---

[117] Cf. the list "Domesticks belonging to his Ex.cy Cossum Hoja From the States of Tripoly," rec. September 1, 1728 by the master of ceremony, Clement Cotterel, SP 71/22, f. 137.

[118] Jezreel Jones (d. 1731) was employed as clerk at the Royal Society as successor of Edmond Halley; he traveled to North Africa several times, was envoy to Morocco and also translator for other Moroccan and Barbary ambassadors, cf. Baigent in ONDB and BL Add. Ms. 61542, ff. 1–117.

[119] On Dadichi cf. above, chapter "History," note 27.

[120] George Harbins to "My Lord," December 3, 1728, BL Ms. Harley 6824, f. 16r.

This city is located two days of marching south from Ouguela; and Ouguela is distant from Tripoli seventeen days of caravan South-East.

The Arabic report, written by Cassem Aga in his own hand, is with me, and this is a faithful translation.

Theodore Charis Dadichi Interpreter.[121]

Dadichi continued to interview not only Cassem Aga himself, but his whole entourage. All had heard of the city and affirmed it to be true, but no one had been there in person. In November 1728, George Harbins[122] investigated the matter further with one of the group who claimed to be somewhat better informed and was confident of having seen petrified sheep. Harbins asked: "Why, being so near as within two days journey of so remarkable a place, he did not go to see it? He answer'd me, with a shrug & a grin, with true Turkish incuriosity, he did not think it worth while."[123]

In a letter of April 1729, Shaw doubted that the petrified specimen that Lemaire had shown to him was from Ras Sem. Instead, he believed that the olives and the palm tree were rather "petrifications ... from the Deluge, and not from any blast of wind as pretended by the Arabs." The famous petrified date, which had been Lemaire's proof that the petrification had to have taken place in autumn, during the peak ripeness of dates, was not at hand when Shaw visited; and the slice of petrified bread – the other specimen Lemaire had always claimed to be proof of the instantaneous petrification of everyday urban goods – was identified by Shaw as one of "those Echinites, called by Mr. Lloyd Discoides."[124] Later, he even specified the exact species with reference to Edward Lhuyd's *Lithophylacium Britannicum*, which he had shown to Lemaire.[125]

---

[121] "Memoir touchant la ville petrifiée delivrée par Cassem Aga, envoyé de Tripoli, aupres de Sa Majesté Britannique l'an 1728," BL Ms. Harley 6824, f. 20r–21r. Shaw, *Travels* (1757), 156s supplied a somewhat free translation.

[122] If I interpret well the "I" in Ms. Harley 6824, f. 21r: Harbins sent the copy of the Dadichi investigation to Edward Harley, 2nd Lord of Oxford on December 3, 1728, ibid., f. 16r. Harbins (1665–1744) was chaplain to the nonjuring Francis Turner, bishop of Ely. The "I" could also refer to Harley himself.

[123] November 7, 1728, BL Ms. Harley 6824, f. 21r–22v.

[124] Shaw to Hans Sloane, Algiers, April 15, 1729, rec. London June 5, BL Ms. Sloane 3396, 45f.

[125] Cf. Shaw, *Travels* (1757), 162 n. 2: "This is called Echinites clypeatus sive discum referens, pentaphylloides, Lith. Brit. class. vi. tab. 13. N° 971."

Thomas Shaw's interleafed version of that book is still extant, and is probably the exact copy he took to Cairo and showed to Lemaire (see Figure 4.2). Indeed, if one takes it out of the context of the *Echinites* fossil illustrations, the large central petrified urchin specimen, no. 971, might well look like a delicious bun.

In the first edition of his *Travels* (1738), Shaw still only referred to Ras Sem in one footnote, denouncing it as a fable: "there is nothing at this Place besides such Remains of the Deluge as are common at other Places."[126] Since the Cassem Aga interview of 1728,[127] the vice-president of the Royal Society, Martin Folkes (appointed in 1723 by Isaac Newton), had collected a large number of sources regarding the petrified city, which he then communicated to Shaw. Shaw made use of them in a lecture delivered in London before the Royal Society, probably in the winter of 1745;[128] then he published the argument of that paper as the second chapter of his 1746 supplement to his major work, the *Travels*.[129] Here, he went through all published sources that he had at hand, either unaware of or ignoring the 1734 Pignon/Granger expedition, and gave again an account of his meeting with Lemaire. Shaw himself never journeyed to the city in the desert. He had traveled along the North African coast from Algiers to Cairo and ventured to some points in the hinterland, but Ludwig, for example, was sceptical about Shaw's own eyewitness investigations.[130] Shaw's method of falsification was not by field research, as the French had done; instead he just employed the praeter-philological methods of exact analysis, checking, comparing of sources and calculations of general probability. He simply reversed the burden of proof: as every testimony that had been gathered until that point about petrified men, groups of men, cities, and so forth, had no certain basis, they had to be taken as nothing – as

126 Shaw, *Travels* (1738), 383 n. 1.
127 Note in BL London Ms. Harley 6824, f. 22v–23r.
128 The Journal-Book of the Royal Society (XIX, 435–438, 20 June 1745) has an entry on an item sent by Shaw, cf. Stearns, "Fellows," 80 n. 36.
129 Shaw, *Supplement* (1746), 10–21.
130 Ludwig noted that Shaw accepted many copies of inscriptions from ancient monuments from the surgeon of the Bey of Constantinople, a former slave, without having ever seen them. Moreover, Ludwig described Shaw not particularly kindly as "very fat" and suffering from gout; therefore "he was not so well disposed for the studying of natural things," i.e., not well disposed for physically challenging autoptic survey travels (UB Leipzig Ms. 0662, part II, 28f., September 8, 1732).

(a)

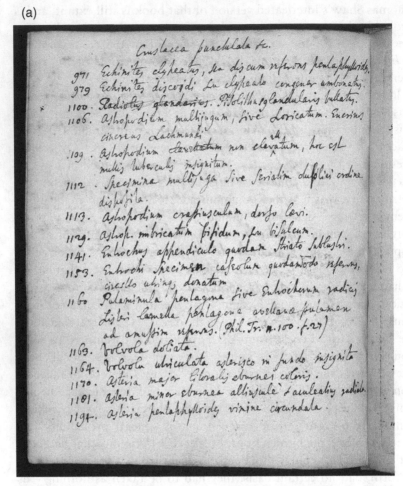

Figure 4.2 (meaning 4a on the left, 4b on the right) Thomas Shaw's travel hand book of fossils; the illustrations are from Edward Lhuyd's *Lithophylacium Britannicum*, Bodleian Ms.Lat.misc.e.76, f. 43v–44r.

fantasy, or speculation. This was no modern Popperian falsification, but a de-probabilization in the best possible way at the time.

One may note that all this happened before the first relevant news from the Herculaneum and Pompeii excavations arrived in England and France. The excavations started in 1738 and the Royal Society entered a letter referring to the discovery in the 1740 *Philosophical*

(b)

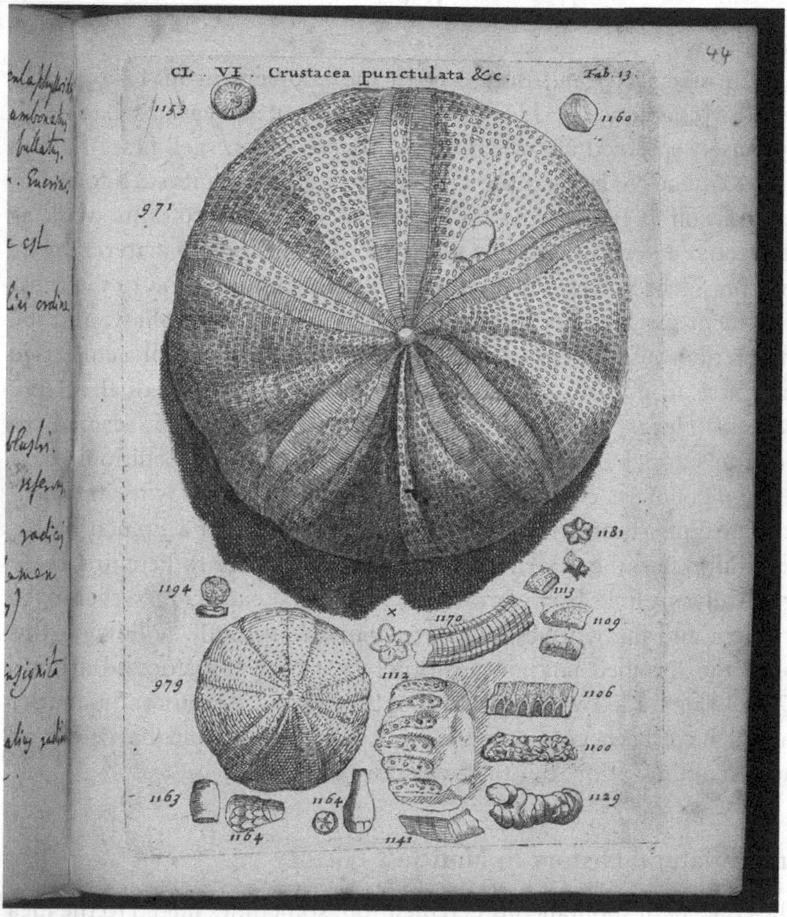

Figure 4.2 (meaning 4a on the left, 4b on the right) (cont.)

Transactions, but the first printed descriptions date from 1748, and the first author to inquire after and mention human bodies found in Stabae was Winckelmann in 1762. Later, in the 1770s, more impressions of human bodies together with skeletons – as all that remained of the victims of the 79 CE eruption – were found and described; the technique of filling the holes left by human bodies with plaster was only developed and used from 1860 onwards. These observations thus could not help the falsification of the story, but it is curious and telling at the same time that classical archaeology developed a new line to its science just as natural history excluded spontaneous petrifications from its purview.[131]

With his speech and the essay in the Supplement in 1746, Shaw closed a chapter. The idea of spontaneous petrifications had been a precisely specified unknown. The unknown concerned first of all its basic existence – the question of whether nature permits such forms of petrification in geology. As far as the Society understood its work as a collective enterprise, one can interpret the collection of material by its later president Folkes, the distribution of this information to the most competent specialist for North African affairs, Thomas Shaw, and the latter's rigorous analysis of the collected sources, as part of a collective enterprise to give an ultimate answer to query no. 24 of the 1692 *General Heads* for that region. Just as the French reacted to Newton's armchair theory of the earth with global expeditions, here the two countries chose the same two different methods for falsification: an English armchair "rationalist" analysis and a French heroic expedition into the desert. Besides Shaw's reference to Lemaire, there are hardly any links between the two "national" endeavours, even if the problem and the wording used are common to both. Following this logic, the specified unknown should have been transformed around 1730/40 into a known about non-existence and therefore, the matter should have been considered closed. Ras Sem should have lost all interest and function. But that was not the case.

From Natural History to Nation's History

The concept of spontaneous petrification, sometimes linked to the idea of petrified men from the time of the Flood as theorized by Scheuchzer – who claimed that a giant salamander skeleton was a *Homo diluvii* – became marginal or even vanished completely within natural science around 1750.[132] But when Shaw presented his successful falsification

---

131   Cf. Winckelmann, "Sendschreiben" (1762), 75; Dwyer, "Science." For the early excavations cf. Parslow, Rediscovering antiquity. For selected citations from French travel writings concerning carbonized organic material – bread, grains, olives – and concerning the impressions of human bodies in the lave since the 1760s/70s cf. Grell, *Herculaneum*, 88f., 132–134.

132   Gessner, *Tractatus* (1758), 73s., 101s. still referred to the stories, but classified them as "fabulous." In an earlier chapter, Gessner had discussed Scheuchzer's *Homo diluvii*: Scheuchzer, *Syntheo* (1726). Cf. on that Kempe, *Wissenschaft*,

as being in harmony with the dominant direction that geophysical theories were taking, there was one dissenting voice. This was William Stukeley, who had been a Fellow of the Royal Society for sixteen years longer than Shaw and had only just failed to be elected as its secretary following Halley in 1721.[133] At the moment of Shaw's lecture, Stukeley was at the height of his fame, having published a work on Stonehenge in 1740 and on "Abury" in 1743. Shaw's academic career was located in Oxford; Stukeley had graduated from Cambridge and cultivated the legacy of Isaac Newton, having been his friend in his last years. As Haycock has shown, Stukeley's "Newtonianism" also implied a strong religious conviction that Newton had only rediscovered "an ancient, but lost, wisdom" in the tradition of a *prisca theologia* combined with a *prisca scientia*.[134] Stukeley had attended the meeting of the Royal Society when Shaw's discourse on the petrified city was read. On July 10, 1747 he wrote "of RAS SEM the petrify'd city in Africa, a letter to Dr. Shaw, president of Edmund Hall, Oxford," which is more of a treatise than a letter. Stukeley altered the perspective on the problem: while the interest of scientists and travelers had been concentrated until then on the issue of petrification, Stukeley hoped to see something different in Ras Sem. He started:

My late ingenious friend, Dr. Harwood of the Commons, a lover of antiquity studys, one who was with us, when we founded the antiquarian society, London, in 1717, he was well acquainted with Sr. Christopher Wren, & has told me, that Sr. Christopher often spoke of such temples, as Stonehenge[135] being in Africa. & I have heard the same thing from Sr. Isaac Newton.

Tis the constant opinion of the oldest inhabitants of Brittain now extant, that Stonehenge came from Africa: by which we must understand, that there were such works in Africa, as Stonehenge.

What authority Sr. Christopher Wren had, or Sr. Isaac Newton had, for their assertion, I know not: or whether they did not mean this very place of Ras Sem. but I concluded, Ras sem to have been one of those temples.[136]

125–135. Gessner's was perhaps the most comprehensive treatise on the subject until that time. For a list of over ninety titles between the sixteenth and eighteenth century on petrification cf. Bourguet, *Traité* (1742), 20–28.

[133] The older monograph Piggott, *Stukeley*, is surpassed not only concerning the amount of material used but also the more nuanced contextualization of Stukeley's thought by Haycock, *Stukeley*.

[134] Haycock, *Stukeley*, 4.   [135] Ms. has "Stonehenge the being."

[136] BL Add. Ms. 51049, f. 45v–46r. Another copy of the letter/treatise is located in Bodleian Ms. Eng.misc.e. 390. Scholarship has not paid attention to this text.

He then made four points. First, he argued that the name Ras Sem, "head of poison, the poisonous head," indicated "the snake's head"; consequently, Ras Sem had been a "serpentine temple, like that of Abury: where the same name remains to this day, hak-pen: meaning in old language, the snake's head." Although the region was rich in snakes, Stukeley alleged that no snake was ever found on the grounds of the former temple. "[S]o that the memorial of a prophylactic & sacred power, in that wonderful antiquity, is still kept up." Second, Stukeley concluded that Ras Sem would be a Serpentine temple "from its form." For this, he cited the report of Cassem Aga translated by Dadichi, which said "that it was a town of circular figure." Stukeley had no problem in deducing the parallel from that very vague description:

now this is but a description of the town of Abury, with proper allowance ... the Stones of those temples, especially the ruder ones, are supposed by vulgar fancy, to be like men, women, dogs, horses etc. with some a head, some legs, arms broken off. The houses are built among these & whether they be really the foundations & ruins of houses, or that they are gate-like as at Stonehenge, & Persepolis ... [T]he people that live near these Druid temples of ours, as pertinaciously maintain the stones to have been human figures, as the Arabians do, at Ras sem.[137]

Stukeley's third point develops this last argument. Very often, he maintained, common people living near to old "patriarchal" or "Druid" temples identified the remaining stones as humans. At Rollrick Stones, i.e. Rollright Stones, Oxfordshire, "one taller stone is called the king, others in a circle are his nobles"; the remaining stones of "the weddings in Somersetshire, a large & most ancient temple of the Druids" were similarly identified as the parson, bride-groom, bride, and the guests at dinner, and one group of stones as fiddlers. William Camden had written that the neighbors of "the Hurlers in Cornwall ... have a pious belief, that they were men trans-form'd into stones." The fourth argument is drawn from a very imagi-native interpretation of the Medusa mythology, much remembered in North Africa and, according to Stukeley, perhaps also related to the Serpentine form of the temples. Those serpentine temples would have taken their form from the ancient symbol of the deity itself, the hiero-grammaton: "[G]od is a winged sphere, from which a serpent proceeds,

---

[137] BL Add. Ms. 51049, f. 49v.

the circle shows the divine number without beginning, or end. The serpent means the word, which animates, & fecundates the world. The wings indicate the serpent of God, which vivifys the world, by motion."[138] Stukeley took this interpretation of a ring (a globe), penetrated by a serpent and wings from Kircher's *Oedipus aegyptiacus*, where Kircher had identified it as one of the oldest signs of the Trinitarian Deity, as found among Zoroastrians.[139] Stukeley was not the first or the only Anglican to be interested in this sign, as has been shown in the correspondence between Baxter and Lhuyd in 1703/05.[140] In the specific English theological context, this matter was of importance to Stukeley because he wanted to prove that even theologoumena such as the trinity belonged to the oldest form of religion – thus making his argument one against Deism.

Stukeley was linking his architectonic and archeological reflections about Avebury and Stonehenge to the Mediterranean. As Defoe blended the British and the Phoenicians within his imperial concept, and as the Phoenician ur-religion was traced to the shores of Albion,[141] English proto-national historiography conflated assumed past connections to the Mediterranean with present ones. This also applies to the Stonehenge/Ras Sem association, which was rooted in the same historical tradition.[142]

---

[138] Ibid., f. 67r.

[139] "Exhibetur autem hoc epiphonema hieroglyphico schemate, qui est Globus ὀφιπτερόμορφος [a serpent-wing-formed globe] (cuius explicationem vide in Obelisco Pamphilio) in quo globus Mentem primam; serpens, mentem secundam, seu Verbum; alae, Spiritum amorem indicant" (Kircher, *Oedipus aegyptiacus*, vol. II/1 (1653), 133). It is the fifth of six gnomic symbols with which Zoroaster "nihil aliud ... innuere vult, nisi Trinitatem quandam diuinitatis in unitate existentem, vel unitatem in Trinitate subsistentem" (ibid., 132). Cf. Haycock, *Stukeley*, 211; Stolzenberg, *Egyptian Oedipus*, 203–206, 231). In the alleged *Obeliscus Pamphilius* the "globus serpentifer," called here "ὀφι-κυκλοπτερόμορφον," i.e. "serpent-circle-wing-form[ed]," was not yet interpreted as a sign of the Trinity, but as a sign of the vivificating spirit active in the cooperation of warmth and humidity during processes of generation (Kircher, *Obeliscus pamphilius* (1650), 448f. – it is not linked to Zoroaster there). The passage ascribed to Zoroaster taken from the 1607 Obsopoaeus edition contains the general idea of the triad as origin of everything, but does not contain that neologism.

[140] Cf. above chapter "Religion," n. 149.

[141] Cf. above chapters B.III.2, C.II.4.

[142] The literature on the history of Stonehenge does not take into account the early modern extra-insular search for antiquarian and proto-archaeological objects

Today, these stone structures are widely known as Neolithic remains. They date to the fourth (Stonehenge) or third (Avebury) millennium BCE and have been classified as British national monuments.[143] In medieval times, Geoffrey of Monmouth recounted the legend that the stones originally came from "the remotest parts of Africa" and that ancient giants had taken them from there to Ireland; from Ireland, Merlin then transported the stones to England to be used as gravestones for the Britons killed by the Saxons. This "chorea Gigantea" would be the best solution for the eternal monument that King Aurelius had requested.[144] In late Renaissance historiography, however, Stonehenge had no place: William Camden had purified British history from nearly everything pre-Roman, which included relegating the supposed direct genealogy of the Britons from Brutus to the realm of legends. According to Camden, if the Britons had any Trojan ancestry, it was from intermixing with the Romans after Caesar's conquest, and there was no textual tradition from the pre-Roman period, therefore nothing of interest to a humanist. Camden's unconcerned verdict on Stonehenge was that it would remain a mystery, just like the Rollright Stones.[145] When, at the instigation of King James I, Inigo Jones wrote the first book-length work on the stones, he followed Camden and interpreted them as the remains of a Roman Vitruvian temple of Coelus.[146] John Aubrey then advocated an idea mentioned but immediately rejected by Jones, that Stonehenge and Avebury might be druidic temples. Aubrey's *Monumenta Britannica* was "recognizably an archaeological treatise, effectively

of comparison: Piggott, *Ancient Britons*, 102–117; Bender, "Stonehenge"; Michell, *Megalithomania*; but cf. Vine, *Defiance*, 109–138, 149–159.

[143]   Cf. North, *Stonehenge* who dates the first work on Stonehenge back to 3500 BCE.

[144]   "Gigantes olim asportaverunt eos [sc. lapides] ex ultimis finibus Affricae et posuerunt in Hibernia dum eum inhabitarent. Erat autem causa ut balnea infra ipsos conficerent cum infirmitate gravarentur" (Monmouth, *History*, lib. VIII, nr. 128f., p173).

[145]   Parry, *Trophies*, 282; Vine, *Defiance*, 112–115.

[146]   Stukeley possessed Jones' copper-engraved reconstruction of that supposed Roman temple, Bodleian, Gough Maps 229, f. 273 – f. 272 to 280 contain clippings of all copper engravings of Stonehenge from the sixteenth to the eighteenth century. Aubrey owned an ink drawing copied from the Jones depiction cf. Bodlein Ms. Top.gen.c. 24, f. 60. Another competing vision, alleging Danish origins for Stonehenge, was brought forward by Walter Charleton in 1663, but is of no importance here.

the first of its kind to be written in England,"[147] and so transcended the
text-oriented, classicist approach of Camden; but Aubrey never finished his work. Lister and his team used portions of it for the 1695
edition of Camden's *Britannia*, and Stukeley obviously had access to
the manuscript. The passages about measuring the site and the archaeological survey in itself were an important step in instrumental archaeological fieldwork, but the most important part for the interpretation
of the site was a section titled *Mantissa – De Religione & Moribus
Druidum*.[148] Here one finds excerpts from and notes about all the
major ancient and contemporary authors that had written about the
druids, such as Edmund Dickinson and his *Delphi phoenicizantes* and
Thomas Smith's *Syntagma Druidibus*. One long excerpt is from John
Twyne's 1590 treatise on British history.[149] Aubrey did not use Aylett
Sammes' work inspired by Bochart, but Twyne seems to have been
a particularly inspiring source. His *De rebus Albionicis* (1590), which
was written at the same time as the beginning of English expansion into
the Mediterranean, has been characterized recently as "first and foremost a Roman history."[150] But a closer look shows how much non-
Roman and non-Trojan Phoenician and Oriental references are used
explicitly against the Galfridian mythos and the Britain-based vision of
Roman history. The comparative form of the first book's main argument, that the island of Britain had been linked and united with the
European continent like "a range of Mediterranean islands,"[151] should
make sense in the context of Jones' attempts to link British history with
the alleged general European-Mediterranean foundation of civilization. London, he claimed, had been built by Phoenician merchants
from Rhodes,[152] citing Ludovico Vives' Classical account of the
Phoenicians, arrival in Spain. He suggested, and many followed him,
that after their arrival in Spain, the Phoenicians sailed further to Britain
in search of tin.[153] "The Phoenicians as first merchants and explorers
brought it [the *ars magica*/discipline] with them from the tradition of
Zoroaster, to teach it in continuity."[154] Because Caesar also wrote
about the druids, Twyne associated them with the Phoenicians, under-
lining that druidic culture/*ars* did not derive from Troy or Italy but
"from Orient, Egypt" – that is, from the Babylonians and

---

[147] Parry, *Trophies*, 290s. The *Monumenta* are in Bodleian Ms. Top.gen.c. 24.
[148] Bodleian Ms Top.gen.c 24, f. 94–104.   [149] Twyne, *De rebus Albionicis*.
[150] Vine, *Defiance*, 41.   [151] Vine, *Defiance*, 39.
[152] Twyne, *De rebus Albionicis*, 40f.   [153] Ibid., 41–44, 81.   [154] Ibid., 82f.

Phoenicians.[155] All this was a pre-Trojan story; Twyne could not believe that for the entire period between the Deluge and the destruction of Troy – that is, for 1,300 years – Britain had been "void of all human species."[156] He identified the syllable *Caer*, presumably present in "Canterbury" or "Caer Lud," a pre-Roman toponym for London, and argued that it stemmed from a Phoenician, Oriental language and also appeared in "Cairo" and the Phoenician term *Carthago*, meaning "new city." Therefore, the first language to shape British place names was not the Latin imported by Brutus but an Oriental one.[157] "This was long before the rule of the Romans in Britain and also afterwards on this island in a miraculous way, [the science/magic of the druids was] brought here first of all without doubt by the Orientals of whom I have spoken, i.e. the Phoenicians and the Babylonians."[158] The belief system of those Oriental druids, Twyne alleged, would have been centered on divine nature and the immortality of the living through metempsychosis. By contemplating the sky and the motion of the stars according to the secret wisdom of the Egyptians, these druids supposedly acquired knowledge of natural phenomena, the size of the earth, and the description of the countries of the world and their particular regions.[159] All these aspects make Twyne's work an explicit opposite of a "Roman history."[160]

In addition to the passages from Twyne's book about the druidic religion, science and celestial observations, Aubry made much use of the Spinozist Burnet's *Archaeologia philosophica*. This work was an overview of the history of philosophy for different regions and times. It started with philosophy and science after the Flood, when Noah possessed all possible knowledge. Burnet believed that the druids, as Celtic philosophers, represented the first Occidental philosophical school. Refuting the idea that the druids were Pythagoreans, Burnet argued that they were much older than the Greek philosopher and that Pythagoras had learned from them, not vice versa.[161] For Aubrey it

---

[155] Ibid., 84, 86.   [156] Ibid., 92.   [157] Ibid., 108–111, 111.

[158] "Sed ut ad Magiam revertatur oratio, quam ab antiquorum Druidum hic & in Gallia institutis & religione profluxisse, ac tanquam manu traditam, persuasum habeo: Haec multo ante Romanorum in Britannia imperium, & post etiam in hac insula mire inualuit, delata primum sine dubio ab Orientalibus quos dixi, nimirum Phoenicibus & Babylonijs" (ibid., 134).

[159] Ibid., 142.

[160] So the short notice of Parry, *Trophies*, 301 holds true against Vine, *Defiance*, 37–43.

[161] Burnet, *Archaeologia* (1692), 7–9. On Burnet cf. Champion, *Pillars*, 157f.

seems to have been of importance that the supposed architects of Stonehenge were also the first and the oldest philosophers. Those colossal stones in Wiltshire might have been a place for a "Pythagoreorum collegi[um]" or "synedrion," where young men, bound together in companionship by oath and separated from the other citizens, might have discussed druidic secrets – a hypothetical account inspired by descriptions of Pythagorean schools in the work of Ammianus Marcellinus, Justinus and Polybius.[162] The remaining fragments of what should have been the preface to his *Monumenta* have become famous, as he concluded by:

comparative arguments, to give clear evidence that these monuments were Pagan Temples: which was not made out before: and also (with humble submission to better judgement) offered a probability that they were Temples of the Druids. – When a Traveller rides along by the Ruines of a Monastery, he knows by the manner of a building, sc. Schapell, Cloysters &c. that it was a Convent but of what order (sc. Benedictine, Dominican) it was he cannot tell by the bare view. So it is clear, that all the Monuments, which I have here recounted, were Temples: Now my presumption is, that the Druids being the most eminent Priests (or Order of Priests) among the Britaines 'tis odds, but that these ancient Monuments (sc. Aubury, Stonehenge, Kerrig y Druidd &c.) were Temples of the Priests of the most eminent Order, viz, Druids, and it is strongly to be presumed, that Aubury, Stoneheng &c. are as ancient as those times. This Inquiry I must confess is a gropeing in the Dark: but although I have not brought it into a clear light, yet I can affirm, that I have brought it from an utter darkness to a thin mist: and have gone further in this Essay than any one before. These Antiquities are so exceeding old, that no Body doe reach them: so that there is no way to retrieve them but by comparative antiquitie, which I have writt upon the spott, from the Monuments themselves.[163]

These sentences, as indeed Aubrey's whole notes and works, have earned him a place in the history of archaeology for the method they suggest of comparing artifacts and architectural elements in order to reconstruct history where no written testimony survives. As a result, the history of Britain was stretched beyond Biblical time; the notion of paleohistory was developed.[164] While in these methodological remarks

---

[162]   Bodleian Ms. Top.gen.c 24, f. 98r–99r.
[163]   Bodleian Ms. Top.gen.c. 24, f. 25v–26r. Cited partially by Hunter, *Aubrey*, 182; Tylden-Wright, *Aubrey*, 74; Parry, *Trophies*, 292.
[164]   On Aubrey's method cf. Hunter, *Aubrey*, 178–208; Parry, *Trophies*, 275–307.

in the preface, the Oriental dimension of the druids is not restated, it is supported throughout his collection of notes.[165]

Stukeley used Aubrey's papers and continued working on surveying the same sites and "stone circles" that had been discussed since Camden's time. He developed his interpretative framework in the footsteps of Newton and his thoughts about *prisca theologia*.[166] Like for Newton, Bochart's *Geographia sacra* became very important: he followed the specific interpretation of the ancient mythology and genealogy that Newton had developed in reading and "improving" of Bochart. Newton had placed emphasis on the figure of Phut being the grandson of Noah, whom he identified with Neptune, claiming that this name was linked to the linguistic root of python/serpent.[167] Stukeley also followed Newton's comparative religious perspective in the search for an ur-religion that would have served to remain in the harbour of the Anglican Church on the one hand and to surpass confessional rigidity on the other. While Newton had concentrated only on the Oriental traditions themselves, Stukeley was eager to show the alleged connection between Oriental traditions and British prehistory. He therefore acted as an integrator of Aubrey's and Newton's legacies, receiving from Kircher and others the theories on Egyptian and pre-Egyptian wisdom. Stukeley interpreted the buildings themselves and how they were laid out in the landscape. They were formed in the shape of a *hierogrammaton*;[168] he further calculated that

---

[165] For comparison of British walls with those of Tangier cf. Hunter, *Aubrey*, 180 n. 5.

[166] On the *prisca theologia* around 1700 in England cf. Champion, *Pillars*, 133–169.

[167] "Unde Nephtys Typhonis uxor et Neptunus Latinorum. A Phut vel Put formatur etiam Graecorum Pytho et literis inversis Typho. Nam Graeci Typhonem et Pythonem <promiscue> [Goldish: "promisive"] dicunt."
Newton, following Bochart, believed that each of the major gods derived from the Noachides went under a variety of names. Using imaginative if unlikely Hebrew derivations, he thereby explained an entire group of major Greek and Roman deities based on Noah's grandson Phut (Newton Ms. Yahuda 16, f. 40r, Goldish, *Judaism*, 53f.).

[168] "This is the representation of god or the great soul of the world among the Persian magi and the Egyptian priests and we find it here among the Western Druids doubtless tis of vastest antiquity & borrowd by them all from the postdiluvian times. 'Tis no mere wonder we have nothing left of the wisdom of these sages they kept their doctrin as secret in the east as here & Pythagoras learnt it from the Egyptians. His doctrine is so like out druids that if he learnt it not from there its plain theirs and the Egyptian was the same" (Comment of

the "oldest picture in the world," the "Antediluvian Sphere," was an astronomical constellation centered on the serpent/Draco.[169] The old druids, possessors of the ancient theological wisdom, supposedly built their temple by incorporating this divine sign into its structure.[170] Again, this represented the continuation of the research of a number of parties – Newton, a circle in Cambridge, and a group in the Royal Society – into the history of the Temple of Jerusalem and its semantic and architectural traditions. Stukeley had visited Newton in 1725 and they discussed some of Newton's drawings of the Temple, which were later distributed to the members and President of the Royal Society. "Sir Isaac rightly judged it was older than any other of the great temples mention'd in history; and was indeed the original model which they followed. He added that Sesostris in Rehoboams time, took the work-men from Jerusalem, who built his Egyptian temples, in imitation of it, one in every *Nomos*, and that from thence the Greeks borrow'd their architecture, as they had the deal of thir religious rites, thir sculpture and other arts."[171] Newton interpreted the ground plan of Solomon's temple, following the early seventeenth-century Villalpando, as a representation of the heliocentric universe.[172] Like Aubrey and Newton, Stukeley assembled pictures and materials of the sacred city of Jerusalem with Solomon's temple and the church of the Holy Sepulchre,[173] of the temple of Jupiter Ammon "built by Danaus brother of Sesostris, in imitation of Solomon's,"[174] of the "temple of the Sun at Palmyra, built in imitation of the Temple of Solomon,"[175] and of the "most magnificent Temple at Baalboth, Heliopolis in Syria."[176] It is not clear how he planned to integrate all these temples, largely conceived of as rectangular, into his conception of the oval and serpentine Dracontine temples. But it seems that he believed that the Dracontine temples were older than Solomon's temple. It was a search for an ur-temple "prototype" followed and imitated by others, such as

      a small picture of a priest with the hierogrammaton in his hand, Bodleian Ms. Gough maps 231, f. 31v).
[169] Stukeley, "Part of the Antidiluvian Sphere 10 Sept 1740 being the oldest picture in the world," Bodleian, Ms. Gough Maps 230, f. 10.
[170] Bodleian Ms. Top.gen.c. 24, f. 98.
[171] Stukeley, *Memoirs*, 17f.; Goldish, *Judaism*, 93f., 85–107.
[172] Goldish, *Judaism*, 95.
[173] Bodleian Ms. Gough Maps 231, f. 59–60, 69, 108, 127, 132–137.
[174] Bodleian Ms. Gough Maps 231, f. 121 (August 17, 1734).
[175] Ibid., f. 129r, 130r (1754).   [176] Ibid., f. 131r (1757).

the Syrians, Arabians, Greeks, Persians, and Egyptians, and for how they were finally connected to these oldest form of "Dracontian" temples.[177] Besides Stonehenge and Avebury, he counted other monumental sites in Britain among them, such as the "Hakpen or snakes head temple on Overton hill."[178] He even designed sketches of the fabulous Phut, who "builds a serpentine Temple in Thessaly" which resembles his own Avebury ground plan (see Figure 4.3).[179] At this point he had integrated Newtonian *prisca theologia*, English antiquarianism and Oriental legacies.

Next to the work of comparing ancient architectural forms, his instrumental measurements of the central part of Stonehenge were by far the best of his age and remained so for a considerable period.[180] He was also the first to use astronomical-magnetic calculations to determine the age of the monuments: using Edmond Halley's calculations about geomagnetic polar migration and variation and presupposing

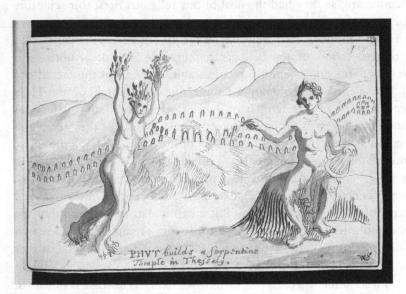

Figure 4.3 Noah's grandson Phut building a serpentine temple in Thessaly as imagined by William Stukeley (Bodleian Ms. Gough Maps 231, f. 156 r). The groundplot in the background resembles the groundplot of Avebury as given in Stukeley, *Abury*, tab. VIII after p. 14.

---

[177]   Cf. the note ibid., f. 146r.
[178]   Bodleian Ms. Gough Maps 231, f. 52r, 235r, 236r, 334r.
[179]   Ibid., f. 156r.   [180]   Cf. only North, *Stonehenge*, passim.

the builders' orientation of the stone structure according to a compass, he calculated the ages of both Stonehenge and Avebury.[181]

Within the historiography on Stukeley, some classify his approach as a weird deviation from the rational paths of Enlightenment while others defend him as standing in the older Newtonian tradition of linking science and religion;[182] there is however, perhaps room for a mediating third view, which views him more as a forerunner of a development still to come: he approached the oral "folk" tradition as a source to be appreciated, noting, for example, the memories of one Ruben Horsal who "has known this country 40 or 50 years" or of "the oldest man alive."[183] The names given to places, stones and trees by the population are recorded scrupulously as signifiers and signs of past circumstances. If this still seems difficult to distinguish from Renaissance and humanist mentalities, he also placed a great deal of value on the conservation of ancient monuments and a great "Grief which a Lover of the antiquitys of our Country" would feel if the monuments observed were exposed to danger.[184] And, while Stukeley did not cite medieval authors explicitly in his printed work, it is evident from his handwritten preparatory notes that he eschewed the Camdenian neoclassical verdict upon medieval traditions and tended to attribute value to them.[185] He reacted to Shaw's lecture about Ras Sem before the Royal Society in the same way: while Shaw rejected all

---

[181] Stukeley, *Stonehenge* (1740), 65: 460 BCE as year of construction for Stonehenge; 1800 BCE as year of construction for Avebury. "This must surely count as the very first application of natural science to the dating of any prehistoric remains." (North, *Stonehenge*, 398).

[182] Archaeologists tend to praise "the early" Stukeley and his precise observations and measurements taken in Stonehenge and Avebury, while condemning the philosophico-religious interpretative framework present in his later 1740 publication as "nonsense" (cf. Piggott, *Stukeley*, 129–131; Burl and Mortimer (eds.), *Stukeley's "Stonehenge,"* 13, 15). However, eighteenth-century intellectual historians from Hunter to Haycock have tried to reconstruct the long tradition of thought he was just continuing (Haycock, *Stukeley*, passim; Hunter, *Science*, 190; Ucko et al. (eds.), *Avebury*; for Newton himself cf. Gascoigne, "Wisdom").

[183] Bodleian Ms. Gough Maps 231, f. 25v

[184] "Had the same strange Neglect which it has hitherto lain under continu'd 20 year longer One of the most remarkeable works in the World at least in this Island must have inevitably been lost in Obscurity" (Avebury August 1721, ibid. f. 37v).

[185] Cf. for example "it seems not unlikely that the king concerned in founding Stoneheng is buryd in that calld the kings barrow or Aurelius his barrow." (Bodleian Ms. Gough Maps 231, f. 2r).

Arabic legends as "lies," suggesting that the locals merely wanted to gain money from curious Europeans, Stukeley posited that there might be some hidden truth behind the folktales about the petrified land.[186] Concerning the Oxfordshire Rollright Stones, he took up Camden's testimony that "the common people usually call them *Rolle-ritch stones*, and dreameth that they were sometimes men, by a wonderfull *Metamorphosis* turned unto hard stones."[187] Regarding Stonehenge, he did not refer to Geoffrey of Monmouth, but to "the constant opinion of the oldest inhabitants of Brittain now extant, that Stonehenge came from Africa,"[188] and throughout his treatise on Ras Sem, he cited the "country people" of England or Arabs as credible authorities.[189]

There is a disturbing co-presence of highly efficient Enlightenment instrumental science with a comparative speculative method that endeavored to bring unwritten history to light from the traditions and collective memory of "the people." This might best be understood as a proto-romantic and proto-anthropological approach to the deep unknowns of history.[190]

It has already been shown that Stukeley was not without followers.[191] The tradition of conceiving the Orient and the Mediterranean as the offspring of British imperial/national civilization and identity did not die out completely but instead changed its face, transforming into modern forms of research into humankind's common civilization and, by that,

---

186   "Which I mentioned at that time, to some of the members of the society: that the matter was partly true, partly fabulous. but if rightly understood, it would prove to be, a temple of highest antiquity" (BL Add Ms. 51049, f. 45r).
187   Camden, *Britannia* (1607), 265. Stukeley had cut out the woodcut showing the *Rollright stones* from the book, cf. Bodleian Ms. Gough Maps 229, f. 272v. – English text from the translation by Holland, *Britain* (1610), 374.
188   BL Add Ms. 51049, f. 46r.
189   "The people that live near these Druid temples of ours ... the Wilthshire people ... the country people" (ibid., f. 49v, 50r, 51r).
190   In interviewing rural populations and highly valuing their "collective memory," Stukeley, as a prehistorian, was acting like the eighteenth-century philosophers, historians and poets who traveled around the countryside to capture the old wisdom, songs, poems and traditions of "the nation," such as Macpherson at roughly the same time, before his forged publication of *Ossian* and, somewhat later, the *Sturm und Drang* poets in Germany (Lenz, Goethe) and Herder (his reception of *Ossian*), who with his concept of *Volkskultur* in the *Philosophie zu einer Geschichte der Menschheit*, sowed the first germs of a "science" of anthropology and folklore analysis, – even if those romantic roots are strongly debated today.
191   Haycock, *Stukeley*, 237–261.

what one is now used to calling "Orientalisms." Here, the example is confined to the eighteenth century and shows how two different epistemic fields interacted during the process of first recognizing, then specifying an unknown, in the most developed form, the new empiricist methodological framework provided. The falsification (in enlightened terms) of the existence of the petrified city caused its deletion from the menu of early modern natural history, but that very disproval freed the item to be reused within comparative antiquarianism and proto-national history. This effect shows how these fields of natural history and monumental human history, which had long intermingled during the sixteenth and seventeenth centuries, began separating and differentiating themselves;[192] the possibility of reactivating an item "deleted" from the realm of possibilities within one field in another shows that the disciplinary and epistemic borders between both fields had gained greater stability. The significance of the story of the petrified city changed, its function altered from a "purely" cognitive to an identificatory one. If Ras Sem was not the remains of a natural catastrophe, it might be a trace of Britain and wider civilization's and British Druidic origins. Stukeley's *idées fixes* are helpful here precisely *because* they are so overburdened with aims that cannot now be recognized by serious archaeologists as "scientific." The "abnormal normal" example illustrates the ways in which early modern science in its institutional intersection between European academies and the network of the Mediterranean empires dealt in multiple forms of ignorance on multiple levels and demonstrates how this stimulated the movement, even the very creation of the epistemic item itself (the petrification story) within and between the fields involved.

Conclusion: Scientific Unknowns and the Mediterranean

To summarize, the chapter first showed how (non-)knowledge was processed and how it was the European demand for scientific knowledge and curiosities that created the story rather than the other way around – that it was the European encounter with the story that created a demand for further investigation. The source analysis confirmed this hypothesis. Europeans often wrote about how Arabs told the story of

---

[192] For the epistemic overlap between natural and archaeological or antiquarian history cf. Hunter, *Aubrey*, 191–208 and Cooper, *Inventing*.

the petrified city, but either they cast doubts on their credibility or the content of what those "Arabs" were supposed to have said, and it resembled in an astonishing way the earlier European narratives circulating in printed form, known to those who asked the questions in North Africa. Often the Europeans did not even mention the Berbers as sources. Avicenna's name never appears in the dispatches and *mémoires*, even though the final issue between Shaw and Lemaire had been those "several Loaves of Bread petrified, whereby he [i.e. Lemaire] could infer that there must have been persons there by whom they were made and for whose use they were designed,"[193] which is a formulation extremely close to the Avicenna passage. One may speculate if the idea had spread from Avicenna's *De congelatione et conglutatione* into common knowledge among the Berber inhabitants of the Maghreb or if, conversely, the passage in Avicenna's treatise had already codified a commonplace dictum. There is no way to decide conclusively whether the origin of the story was still "autochthone" (Arab) around 1630 (Kircher's account) and around 1700 when the French started their investigation seriously. In any case, the story was fed and grew thanks to the continual inter-European competition and curiosity about the problem, and probably also as a result of the Arabic interest to profit from the European competition. The trigger for this enthusiasm seems to have been the outside demand of the epistemic forces of what we commonly refer to as the "Scientific Revolution." It is as if the inter-European scientific advance operated like two poles of a magnet, attracting the particles of the story into a large conglomeration, and repelling them as soon as their status of no value had been decided.

Interestingly, the timeline of European concern with this tale, that is, how the story ripened and expanded from the 1630s right up to when its proverbial bubble was burst by falsification in 1730/40, is again quite similar to other processes of coping with unknowns and related accumulations of knowledge. In the case of the Western interest in the Greek Orthodox Church, although situated in a completely different discursive field, the same process occurred of exposing ignorance and accumulating knowledge, while still believing that the churches might be analogous and could possibly serve as appropriate alliance partners,

---

[193] Shaw to Hans Sloane, Algiers, April 15, 1729, rec. London, June 5, BL Ms. Sloane 3396, 46.

so that the knowledge might have operative functions. The same was also true for the abrupt cessation of that interest after 1720/30, or at least, there was a similar switch of functions. After knowledge about the churches lost its operative potential, it only could serve cognitive purposes. The difference is that the "knowledge" about the petrified city, as of any other object of natural history, is typologically very different from the knowledge about a past church's rites and theology. Even more importantly, the story of the petrified city concerned something that, from our point of view, could never have existed. The isomorphism of the non-knowledge cycle of growth and disruption reveals ex-post as never having been a homologue. But this is the interest of a truly historicist approach, which seeks to parallel both cycles as they were paralleled by the historical actors during their often overlapping investigations in both fields. As the final switch and refunctionalization of the story from the field of natural history to that of national history and archaeology showed, in the end, the epistemic item ended up as part of the same search for an ur-religion.

Thanks to the abnormal character of the item searched for, the several realms of the imaginary and of the possibilities involved are highlighted, and the systematic point becomes visible that every non-knowledge item is linked to one or more of these spheres of possibility. The narrative of the petrified city grew and grew between 1630 and 1740, presenting more and more details and variations of the city: the men and women petrified during their work, sheep, locations, ships, the stability of the color, the petrification of inner organs, and so on. All that was the growth and densification of the specification of the unknown itself. This specification always had to refer to an invisible background or frame with fluid but limited borders of what was imaginable and held possible at all. The history of coping with ignorance and unknowns thus shows the relativity of all these parameters: of the way in which unknowns become specified – with empiricist Baconian or with philologico-historical methods – of the current character and limits of those spheres of possibilities in a given period, culture and place, and of the changing relationships between the epistemic item and the relevant sphere. At least three such spheres could be distinguished by 1700. First, there was the traditional imaginary space appropriate to what was only possible for God: wonders and direct causalities operated by the mysteries of divine power, an ideational space that could be populated with everything for which a natural explanation was unknown. Second was the sphere of unknowns created by Baconian Science; and third, perhaps unexpectedly, was the sphere of unknowns and imaginative possibilities belonging to literary invention. The arrival of Baconian Science had the epistemic effect of creating in the minds of researchers a more or less open empty field of declared non-knowledge. There was always a great deal not yet known. The lure of filling in this vague, blank space, delineated by the fluid borders of probability and likelihood – not everything is expected to be possible – stimulated active research and exploration. Literary imagination is perhaps not exactly a field of "unknowns," but it operates in a functionally equivalent way as narrators and readers project items into it that belong to a certain range of possibility and likeliness in reality. Despite the usual saying that literary fiction enables one to narrate "everything," narratology has always shown that fiction deviates from experienced reality only to a certain degree, like science speculates about specified unknowns as possible knowns only insofar as they are held to have a certain degree of probability. Today, one could certainly define the borders between those fields by adopting a simple form of system theory, with the in/out codes of natural/divine (the religious field of unknowns), true/false (the scientific field), or fictional/factual (the literary imagination). This would suggest a clearcut distinction that was not present in this period. The allure of those early modern processes lies rather in mists of a drifting indeterminacy and the faint and quick shifts between fields with unstable and porous borders. The history of such an epistemic item as "the petrified city" allows to see just how the necessity of classifying that specified unknown challenged a growing distinctiveness between those fields of possibilities, and how it challenged the actors to decide which reference and relationship between an item and a field of possibility was fitting.

A last point has to be made regarding the epistemic forces that led to the effects observed here within the fields of natural and then national, or civilizational, history. Again, a variation of what has been called "spatialization" – see Chapter 3 on History in this book – seems to be important within the empiricist epistemology active during the process of proto-falsification, which decided upon the deletion of the "petrified city" as object of natural history and its openness to refunctionalization within civilizational history. The most conclusive tool within the scientific field to produce that evidence of achieved falsification was, at least on the French side, simply locating the place exactly, going there, and not finding anything. Empty space or ruins in sand at precise longitude/ latitude coordinates identified on the best maps and in accordance with the best available maps (Delisle) and also the supposedly richest medieval geohistorical traditions (al-Maqrīzī) was the best answer to the question of whether the petrified city belonged to the realm of the knowables or the non-existent. In fact, in all versions of the story before, from Avicenna to Baumgartner, Kircher, Fitton/Digby, the location of the place is either not given at all, or the names varied, or there was only a vague indication that it was "near Tripoli/Ougouela/ Benghazy." The true/false code functioned with higher evidential clout if the item in question could be fixed.

The other force is the deepening of the past, the as yet unclear stretching out of the imagined possible timescale of history that leads in the end to the precarious interweaving of natural history and civilizational history. The problem of how to understand processes of congealing or even petrification were disputed among the scientists of the Royal Society and the *Académie des sciences* because both institutions were closely involved in geology and concepts of duration and the earth's history. Shaw judged the items shown by Lemaire to be, if petrifications at all, then "from the Deluge," and this points to the persistent use of a Biblical timescale; but with contemporaries like his correspondence partner Woodward, those time frames themselves were already being debated. Aubrey and Stukeley's feeling that they were "groping in the dark" for a history that had left stone monuments but no written testimonies is a similar sign of sensing a void beyond the usual known, of trying to specify the shape of the unknown and to "populate" it. Both geology and early paleohistory struggled with the relevant timescapes and durations. The impact of the empiricist approach, hardened by spatialization, challenged the emergence of a new and different semantic of "scientific truth" linked more closely to the "deeper truth" still rooted in God's providence and mythic veracity, but detached from the older field of the prodigious worlds of wonders. This was the notion of truth embedded into Aubry's and Stukeley's protoromantic method, which is thus intrinsically linked as a sidestep and reaction, to both empiricism and to the deepening and widening of historical time.

Conclusion

This study examined in parallel the different epistemic developments within imperial communication of and about Mediterranean trade between 1650 and 1750 from the perspective of how ignorance was processed. This paralleling involved dividing into separate epistemic fields what individual imperial agents handled every day in overlapping ways in the consulates and merchant colonies. It is thus the content of trade empire communication *tout court*, grasped, in the form of examples and selected cases. The question now is what emerges as common to all four epistemic domains – or, if nothing in common emerges, what can be learned for epistemic histories in general and for the function the empires had for those epistemic developments in particular? Let us recall first in a schematic form the results of the chapters for each epistemic domain:

As already mentioned in the introduction, at first glance one might think that the approach chosen here – to generalize the perspective known from the field of "science and empire" studies onto all epistemes involved and active in imperial communication – also leads to the importation of what we have learned in the past decades about the so-called "structure of scientific revolution" as the elaborated description of the shape of developments within an epistemic realm. For the fourth chapter on science, one might indeed say that the reconstruction of how query no. 24 of the Royal Society about the existence or not of a petrified city was treated, could well be understood as one of the many problems and items within the history of science that follows the basic scheme of Kuhn's revolution thesis. What started as different forms of trying to verify the existence of spontaneous petrifications, ended in different forms of falsifications – à la française and in an English way. It would then belong to a wider revolutionary shift of scientific paradigms, which, certainly, does not only concern this one example, but the general

	Politico-Economics	Religion	History	Science
Type of Non-Knowledge	Operative	Operative and epistemic	Epistemic	Epistemic
Function	(autoreferential)	Rule, mission in the Levant, identification	Cognition, orientation	Cognition; after the falsification and shift of field: identification
Entanglements	(autoreferential)	East/West alliance between Greek Church and Anglican or Jansenists; after c. 1689 this would be a tentatively antigovernmental nonjuror alliance (E); or it would lose all entangling power (F)	No specific entangling power beyond the general introduction of the Oriental reference into the West	Only after shift from field of natural to civilizational history does it become a case of East/West entanglement
Cycle	Start in the 1660s: Navigation Act/Edict of March; being	Start for both, E and F around 1650 (Claude/Arnauld	Start: one may take Pocaocke's *Specimen* (1651) as such	The publication of the first versions of the story (around 1650) coincides

(cont.)

	Politico-Economics	Religion	History	Science
	operative, the cycle has no end; in the French case, a second-degree political arithmetic process takes off (ca. 1720)	controversy, pointing back to the case of Kyrillos); endpoint: erosion of operative religious function, mostly around 1720/30	a point, the erosion of the structural prejudgment about the inclination to revolutions as a marker of an intermediate endpoint (ca. 1730–1750)	with the institutionalization of the Scientific Revolution (1660–1666); the falsifications mark an endpoint (1733–1746, F/E)
Epistemic Forces	Spatialization affects concepts of empire and government	Exhaustion of classical forms of confessional conflict and unionist solutions; deepening of History creates evidence for transconfessional basis for the religions	Spatialization, standardization and deepening of history	Spatialization affects evidence of form of proof through expedition (F); deepening of history affects concept of nature (process of petrification) as well as of replacing epistemic field (Orientalizing antiquarianism)

concept of geognosis and geology. The process leading to that would be the process of accumulation of knowledge that Kuhn always distinguished sharply from the point of the revolution of paradigms; and the whole would belong to a progress of science through revolutions of paradigms, even though Kuhn himself always strongly insisted on the relativity of "truth" and acceptance of a *communis opinio* in relationship to one given peer group. The description of the quest for the petrified city has certainly implicitly inherited many of the later historiographical developments that went far beyond Kuhn in many respects. Researchers have learned to pay much attention to the intersection and interrelatedness of epistemic realms during the overall process of scientific evolution. In contrast to that, Kuhnian narratives of the history of science, as still extant today in many cases, presuppose to some extent the identity of the epistemological framework and a functional closure and autarchy of the communication of science, an approach that seems ahistorical: one creates a "history" where Kepler, Galileo, Pauli and Einstein operate within the same field of science, thus erasing the early modern world on the one hand, and the context of the twentieth century's World Wars, mentalities and constellations on the other; only a constant emergence of new and an erosion of old peer groups and their succession is noted. Surpassing those tendencies, for a long time, much attention has been paid instead to the question of how the very different overall sociohistorical and epistemic settings led to the emergence and erosion of evidence for new and old theses – the link between theological thinking and blood circulation within heterodox trinitarian groups, for example; a similar argument would be the thesis of the elective affinity between millenarist convictions of men with Fifth-Monarchist backgrounds and the conception of economics in terms of absolute growth beyond the usual limits known to prevail within mercantilism. The question of how convictions within political philosophy influence those within science is well known, particularly in the context of the Hobbes/Boyle controversy about the vacuum. Lessons have been drawn from that example, and the conception of different spheres of possibilities – prodigious, literary and scientific ones – used here to describe the ability of the epistemic object to switch between fields is consistent with these findings.

Placed in parallel with the other epistemic fields, it becomes evident that the non-knowledge cycle within the scientific field is nevertheless specific, and it is impossible to only transfer theories or descriptions of the "structure of scientific revolution" to other fields. The reason lies first of all in the difference of the forms of non-knowledge itself. A Kuhnian and post-Kuhnian picture can only been drawn regarding the specific form of heteronomous non-knowledge that consists in nature and natural laws: there is a constant target of reference that remains stable, while only the forms of observation and interpretation change. Nature itself might also move and "evolve" but its laws do not, and within the period chosen, even natural history only slowly started to conceive of something like the extinction and emergence of species after God's creation of the world, not to speak of the stability that was assumed for objects of geology. The objects ignored within the epistemic fields of history and religion were also heteronomous according to my definition, but their status as objects to be unknown and obfuscated were to be explained by other reasons – by reference to collective oblivion, the transmission of sources, and a lack of attention for centuries. One should not underestimate the fact that the persons and most of all the methods active in both processes were identical or very similar. Claude's detailed query form of confessional articles to be asked for and checked within the Greek Orthodox churches of his time is notably similar to the detailed queries concerning problems of natural history brought forward by the Royal Society. The growing methodological reflexivity of scholars studying the Oriental religions and history, as apparent in the protoethnological reflections of Piques or in those of Ludwig and Shaw, is strikingly similar to the – comparably even quite slow and late – development of a concept of a clear *experimentum crucis* regarding this scientific problem.

Quite comparable nevertheless are the structures, forms, or shapes of those non-knowledge cycles – although not the same in other fields as within science. The usual starting point for the erosion of a paradigm was not always an anomaly observed, as Kuhn argued, or a "surprise," as the starting point for a transformation of nescience into specified non-knowledge is now termed in the sociology of ignorance. The first reception of the story of a petrified city by early modern scientists such as Kircher could be identified as such a moment, though it was less a real "anomaly" according to his standards of scientific-prodigious explanatory schemes. Rather, the aim of the century-long quest for the petrified city was in part to decide precisely whether or not it *was* an anomaly, and if so, according to what explanatory scheme – both questions were uncertain and unstable at the same time. A similar starting point could be identified for the case of religion, where the sudden use of the argument of the pretended congruency between Calvinism and Greek Orthodox teaching in 1654 within baroque France leads to the awareness of the ignorance about their precise current and past doctrines. Within the field of general knowledge and history, it is more the change in preformed assumptions observed here toward the end of the cycle, around 1730/50, that follows this form of a surprise: quite unexpectedly, the North Africans were perceived to be no longer prone to a revolutionary habitus. In the field of politics, for instance, it is the legislation and setting of the norm of the national itself in the Navigation Act, in the shipping rules for Marseille, and in every other norm ruling Mediterranean trade, that opens up the ignorance gaps between the norm and the empirical nationality of each ship, man and good encountered in a given moment. The starting point lies completely on the side of the rulers and observers; there is no stimulus from outside that starts the process. Still, the establishment and instruction of port officers in Europe and of the consuls abroad to test, check, control and diligently note the fitting of the norms of nationality with the realities of shipping introduced an empiricist form to execute norms, which became more and more typical for early modern administration, and was able to set off a comparable epistemic movement.

It is clear that with these comparisons, one touches on old questions discussed within the general methodology of the sciences – the issue of the applicability of the same rules to natural and social sciences; the question of whether the scheme of paradigm shifts and "revolutions" – although the term is now used in every discipline – is really transferable from the realm of natural sciences to other forms of knowledge production. The switch from a perceptional framework derived from Bodinist climate theory to that of more or less enlightened idealizations of foreign people as counterfactual mirrors for their own society is not the same as, for instance, the shift from a phlogiston theory of fire to oxidation, just because of the fundamental difference within the binomes of the observational framework / observed object. I have shown instances of methodological reflections by early enlightened observers on the fact that the form of observation and of asking alters the observed object itself, which would be comparable to proto-Heisenbergian reflections in the realm of nature. But still, nature or natural laws remain – at least concerning the supposedly touchable items of nature searched for by early modern researchers – as they are, while the people, histories, human artifacts and texts observed and searched for developed and changed before and at the same time as they were observed and described in those biased forms. All these differences seem to be more clear-cut today than they were for early modern contemporaries: within a determinist framework that understood the character of peoples as obeying climatic rules, the difference between a *historia* of the population of a given country and natural history is far more gradual, if it exists at all, than how the line would be drawn today.

Reconstructing the daily activities of research by consuls, chaplains and Mediterranean travelers protected by the French or the British from the same wider point of view of ignorance and the responses to it, allows the historian to be more historical or even historicist than an approach that implicitly takes for granted the fundamentality of differences between the epistemic fields and disciplines that are used today and that would distinguish in branches of "political history of empires," of "history of trade and economics," or of histories of "empire and science." This approach allows the suggestion that the simultaneity of ignoring, asking and querying within those different yet still adjacent fields was central in slowly and constantly stimulating just the awareness of those differences. In the end it becomes evident that the aim cannot be to search for an overall "structure of epistemic movements" as a kind of a generalized form of the shape of scientific revolutions, itself already highly debated. It would be misleading to search for the super formula of general knowledge revolution(s). Or it would produce only a very basic scheme: a stimulus or trigger transforms a former state of nescience into specified non-knowledge, setting off the process of non-knowledge communication. This step cannot have always had the same form, as it is not necessarily a process through which something gets accumulated – national ascriptions, for instance, are first just made and repeated, not "piled." At the endpoint, something might be considered as known, or considered as unnecessary to be known and not useful, or the object might change the epistemic type and function. The latter is important as it reminds us that – in contrast to conceptions of such movements as linear processes where, at the moment when a next step is reached, prior data, ideas and knowledge framed within older frameworks immediately loose all importance – can just be transformed from an object of natural into one of civilizational or even religious history.

Different from a history of science, a history of ignorance thus follows the paths of the epistemic objects and the paths of the "ignorance activity" itself, instead of explaining how a certain state or result in theology or science or history was reached.

Here, the biggest difference seems to be that between the operative and the epistemic forms of non-knowledge: the steady communication of national non-knowledge reifies and creates the national itself within the perception of the actors. This is where the second level of populationist management of that outcome is reached in the example of the Maurepas political arithmetic. Constantly asking for the national leads, finally, to the emergence of *the* nation. A similar emergence of a new structure cannot be observed as the result of constantly asking for purely epistemic unknowns. Only insofar as there is also an operative element, insofar as the unknowns are parts of implicit or explicit normative specifications, can similar phenomena be observed: the emergence of connections, forms of reciprocal trust if not alliances between the Western and the Eastern churches, at least for some time, and the reciprocal creation and modeling of their confessional borders can be conceived as the result of a constant asking for and testing of the theological positions. The confessionalization of the Greek Church was in fact due to that new form, developed and established within the European confessional age, of interrogation and specification of the unknown religions in the precise form of a confessional matrix, in which responses had to conform to the theologoumena reformulated as unknowns, requiring the respondent to take a position. Before, one might say, the question of the exact understanding of the Last Supper – that is, transubstantiation – within the Greek Church had been in a state of nescience for centuries. The activity of asking in such a way creates what is asked for on the side of the interrogated – again the constructive power of non-knowledge. In contrast, constantly asking for unknown parts of the history of the Berbers or for natural laws observable or not does not lead to the emerging of new social structures. Again, the development of those social, cognitive and power entanglements – as with many other social "structures" meaning counterfactual expectations – is better understood not as the effect of knowledge accumulation, but by such cycles of specified non-knowledge communication. If a society pretends to know everything about a given other society or problem or believes that it is useless to know more, the communication process itself will stop as was the case around 1720–1740 within Anglican and Jansenist circles. In such cases, the status of sufficient knowledge is or becomes close to that of nescience.

These remarks lead to the perhaps helpful question: what is the difference between such a focus on ignorance and non-knowledge, and a positive history of knowledge? First of all, no pure history of ignorance is possible. The pages of such a book would be empty. But if this book were written from the perspective of a history of knowledge accumulation, the chapter on history would be absent altogether, as it is a story of unaccumulated and unpossessed knowledge and of the replacement of content by preformed and reiterated semantic structures. The chapter on science shows how it would be completely misleading to reconstruct such an important part of the history of science as "knowledge accumulation" just because the story of the petrified city did grow and spread. The usual picture would be that, within a given frame and specification of non-knowledge, empirical data is gathered and accumulated, and the frame itself only has to be altered in the event of a surprise. The growing and accumulating here is of narratives that one might identify as the unfolding of the specified unknowns itself by speculation, since no empirical data about spontaneous petrification itself could be gathered. The process thus fulfils, in an isomorphic way, what would later become a form of methodologically empiricist research with a more strictly controlled method, but at least the part of narrative production was *not* the descriptive part of such a process – while the part of falsification might fit in that picture. It is perhaps too strong and ahistorical to call this a parasitic process – parasitic of the emerging form of "normal" science – as those processes belong indeed to the history of science. A more neutral history of ignorance shows the effect of that growth of narratives as a phenomenon on its own. What might seem ex post a process or gesture of quasi-unconscious simulation of science *was* scientific research for the actors, but the epistemic content produced was not accumulated knowledge: it was just narratives with undecided status. The chapter on religion could perhaps be framed as a history of growing knowledge about the Greek Church and the other religions and confessions. Shifting the perspective to the side of ignorance helps to reveal that the different functions within the English and French contexts started with differently rooted ignorance. It helps to see the overall inner-European functionality and to understand why the process stops after 1720/30. The first chapter can also not be understood as a process of "increasing knowledge." The reality of the national was and is an a priori unknown. To ask for it and to "brand" persons and items with a national inscription according to required specifications was an ongoing process from the shift of the 1650s/60s. One could tell that story in a traditional way as an account of shipping reforms, constructing norms, recognizing abuses and reforming the relative norms again. But then one would miss the shifts and cognitive steps behind that process and the growing consciousness concerning ignorance about the national itself. In fact, one would miss an understanding of how the national itself shaped economics in this case, and one would not be able to examine the developments in conjunction with the other epistemes discussed.

The epistemic forces identified here as spatialization, standardization and deepening of history should not be conceived as somehow magic forces outside and different from the perception and communication of the actors. They are names given to apparently active drifts. If they are used here in a partially explanatory narrative – to explain why gaps of ignorance become visible and open up – then one must also admit that this is a narrative scheme which tries to put in a explanans-explanandum-form what in fact was a multilevel epistemic and semantic process of many single moments and events, where one development – such as in cartography – might have stimulated others concerning the concept of history and vice versa. That granted, it is nevertheless important to see that those epistemic forces, as metonymic identifiers for many such micro-events of stimulation and their overall drift, are active and visible across the epistemes: spatialization affected the French imperial frames of thought during and after the Maurepas administration, it affected the forms of organizing and conceiving history (historia), and it affected the epistemological framework of scientific empiricism insofar as it gave higher evidential clout to proofs based on precise localization. The deepening of history was important for all forms of epistemic non-knowledge communication: it concerned the search for ever deeper religious roots, even beyond the Biblical chronology, it certainly concerned the narrower content of historia itself and it concerned natural history. In an abstract form this is a well-established point in the history of historiography, science and philosophy, but studying this within the practice of imperial communication, the intersection of the different epistemic forces becomes evident in a new way: they operate like forces that stretch out a tangled net or a matrix that had been bunched up, so that the holes between the knots become visible within the perception of the actors. Imperial communication, reaching out beyond the former local limits of the states and the republic of letters was central in that process, for it challenged the former frames and was challenged by the widening frames of perception at the same time.

Historicizing Ignorance, Synchronizing Empires

As has been seen, it is not a reasonable aim to formulate an abstract rule and shape of "epistemic revolutions" – perhaps not the aim of any, even deductive form of historical theory. Certainly, it is not the aim of an inductive history that follows the paths of ignorance activity within the given empirical reality of trade empire communication in the Mediterranean as this does. Nevertheless, the observed congruencies of the shapes and the synchronies of non-knowledge communication lead then back to the question about the functional link between empires and epistemes. As was stated in the introduction, the empire fades a little into the background or is liquefied and amalgamated with the epistemes, so the real heroes of this book are the voids. Focusing again more on the side of the empire, some conclusions have to be drawn about how what could be analyzed and described is historically specific for "early Enlightenment empires." That specificity can best be characterized by some comparative reflections and observations with earlier periods and links between epistemic developments and institutional settings.

From the Europeanist point of view, the chronologically closest case of comparison for imperial outreach might be the early sixteenth-century expansion to the Americas and the correlative movements within those epistemes. For the field of political economy, it seems hard to find a real equivalence to the post-1650 processing of national non-knowledge. Despite the national naval competition obviously existing between the Portuguese, Spanish, French and English, on the level of political economics, the sixteenth-century Atlantic and inter-European trade was very loosely integrated, shipping was far less nationalized, if nationalized at all, and did not have a very systematic form. Concerning the field of religion, it would be hard to find such a bidirectional entanglement of religious and theological developments in Europe with religions found and observed abroad as in the late seventeenth century and the Orient. Some American references were imported into the inner-European confessional conflict as rhetorical argument: the famous denunciation of Catholic Eucharist theology as cannibalism, or more specifically the comparison of confessional violence with pretended Brazil Tupinambá cannibalism. This was a form of rhetoric. It was a spurious element without systematic significance; no European compendium of controversial theology from Melanchthon to Bellarmino and beyond was forced to integrate sections broadly to link European confessions to non-European ones – indeed, the latter were considered pagans. And the whole dimension of searching for common religious roots is absent in the early Atlantic imperial outreach. While Acosta and others started to reflect on how to integrate the very existence of Amerindians into Biblical history, no European identificatory discourse emerged to search and find in them their civilizational roots. The field of science was already beginning to be organized in similar form to our much later post-1660 French/British constellation. Research on the *Casa de la contratación*, the Spanish royal cosmographers and the famous *relaciones geográficas de las Indias* have shown that there was already a well-ordered use of queries and other forms of specification of unknowns that the Spanish wanted to be known for their imperial administration. But what was missing is an integrative, heuristic organization and data "handling" as later started by Royal Society and the academies and their wider network of scientists; no synchronous steady and dense use of the incoming data for the elaboration and publication of scientific treatises, as occurred with Oldenburg, Boyle, Hooke, Evelyn, or Huygens, Réaumur or La Condamine. Only decades later, the *Accademia dei Lincei* published its *Tesoro messicano* relying partially on the data of the *relaciones*. At the time and in the organizational center of their gathering, there was no shaping of the non-knowledge communication, no cycle, no endpoint, no clear direction. Within the field of general knowledge and history, without doubt, the exploration and voyages of the sixteenth century produced the most significant challenge to the Eurocentric worldview. The descriptions, travel narratives and the first New World cosmographies written by Europeans may indeed be understood as an ongoing movement to overcome the erstwhile state of nescience. If one asks, as is done here for the presence of the Arabic Middle Ages in the historical consciousness of European actors in the Orient, in a similar way of the number of reflections or texts devoted to pre-Columbian America in those cosmographies, the result would be a likewise overwhelmingly presentist perception. The Europeans described what they encountered since 1492 next to short speculations on Biblical descendent of the people or their later migration, and all that mostly without any explicit awareness of a lack of anything. It would be wrong to strictly oppose an epistemically unconscious sixteenth century imperialism to an enlightened reflexive imperialism or empires of nescience to empires of specified unknowns, because the differences are gradual and not absolute. But there is a significant systematic difference: sixteenth-century inner-European epistemic developments were far less dependent on and integrated into imperial communication at all.

This might teach us to leave "empire" out of the comparative perspective, and instead try to compare the four fields within Europe during the shift of the Renaissance and Reformation to detect similarities or distinctions from that 1650–1750 period of (non-)knowledge communication. While a look at the sixteenth-century empires reveals qualitative distinctions between the epistemic fields, a look at the Renaissance/Reformation shift in general reveals complete asynchrony. One could probably reconstruct the emergence of "Renaissance" capitalism in terms of a history of ignorance: The fourteenth-/fifteenth-century Italian and the whole Mediterranean revolution of early capitalist double-entry bookkeeping, the establishment of factories, the emergence of a market and the use of the *cambio* and later other instruments and negotiable stocks are all aspects of an important epistemic shift in this sense. Asking for, and not knowing the value of, convertible currencies, or asking constantly – and by that specifying the unknown – for the present state of the balance of one's business by help of the *libro grande* of a Tuscan merchant are surely representative of a systematic shift that specifies something as organizing non-knowledge that had hitherto remained in nescience. This would be operative non-knowledge communication quite similar to the processing of the national which was to become the next step of development within that same economic field once established. One could also interpret, partially at least, the Reformation and "Confessionalization" as a process of steady non-knowledge communication about individual confessions for which a precise knowledge of elementary theological distinctions was necessary. As for the basic shape of this evolution, the starting point would be in the early sixteenth century, and the endpoint around 1650. The well-known politico-religious settlements – from the first peaces of religion in the sixteenth century to Nantes in 1598, after La Rochelle in 1629, and Westphalia in 1648 – used increasing self-awareness of the explosive feedback loop of religious "fundamentalist" demand for confessional authenticity by implementing nominalist and procedural solutions such as the establishment of spatial enclaves for religious denominations, and the use of fixed dates as frozen measures of confessional territorial settlements after all conquests, conversions and reconquests during the religious wars. The confessional is thus treated epistemically differently around 1650 than at the beginning of the Reformation. With the Renaissance, historical depth had started to expand, even if humanists were only reaching for Greek and Roman antiquity or the early church, not for other periods of history. All these are partially similar elements and forms to specify unknowns that could be observed in their extended and transformed forms in the Mediterranean trade empire communication.

The crucial distinction seems to be the asynchrony and decentralized status of those processes. Early merchant capitalism started in the 1330s at the latest, and the Renaissance/humanism at around the same time. However, the Reformation and Christian humanism are much later phenomena. And Mediterranean capitalist techniques as well as "humanism" – how one ever might define it – arrived in many countries much later, sometimes centuries later. Moreover, there is no integrative institutional setting that served as carrier for all these contents at the same time as was the case for the seventeenth-/eighteenth-century combination of governmental administration, scientific academies, cultural and church institutions, and the ambassadorial, consul and merchant networks that stretched out far beyond the borders of the state but were still part of one circuit of communication. So this extremely shortened comparative reflection suggests that the relatively high degree of synchrony that has been observed for the non-knowledge communication in all fields and for the starting and endpoints of the respective cycles or movements is indeed specific to that period from 1650 to 1750 and to the imperial framework. And so, at this point, the empire that I have obscured and partially subsumed into the four fields re-emerges as the *synchronizer* of the epistemic movements. While the epistemic movements and cycles cannot be described and understood as following one scheme and shape, for each of them incisive starting points around 1650 and endpoints around 1750 can be detected without "forcing" the sources. The very need to be explicit about the respective unknowns, and to reflect on ignorances and cope with them, is the baroque and early Enlightenment habitus that has been identified in each chapter – and these synchronized movements in the epistemic fields *are* the defining characteristic of Enlightenment empires. The synchronization is therefore an important performance of the imperial framework: it explains the mutual stimulation of one epistemic field by the other; it explains the better interchangeability of an object that switches from one field to the other if it no longer has a function; it explains how epistemic forces can have their impact at the same time across the epistemes, and it finally demonstrates, how the differentiation between the epistemic fields and even between "disciplines" in a narrower sense become visible more easily: because everything is communicated next to each other and within the same channels, the pressure to decide upon the limits of each field and about the belonging of an object of specified ignorance to the one or the other field is higher than would be the case if the one epistemic field were handled dominantly only by one type of institution – for example, the university – and the other epistemic field by another – for example, the merchant community in harbor cities. Even if the empire had no structure before and beyond the epistemic fields, as the actors and institutions were serving in its name and constantly applying the specifiers of unknowns for each field within the center and the consular network, it still produced something like a stabilizing axis shot through the epistemes.

Maybe, from the point of view of a history of science, knowledge and ignorance, this synchronization of the epistemic (non-)knowledge cycles can be understood as a step toward an unintended gradual dissolution of imperial orders as the effects of imperial communication itself, even though the high times of nineteenth-century colonial imperialism were yet to come. One might argue, that through this effect of synchronizing epistemic developments within their different fields of content and functions, a more and more global form of social order developed, where empires and nations tended to be subordinated structures in comparison to the dominant distinctions and borders of political, economic, religious and scientific fields and communication. This lies beyond the scope and period of this study, which simply confirms that enlightened empires were built on ignorance.

Bibliography

Archival Sources

Cambridge
  University Library
    Ms. Add. 7113

Frankfurt/M
  Universitätsbibliothek
    Nachlass Ludolf [Letter numbers] Nr. 573, 574, 592, 643

Harvard Map collection
  2410.1718 (Delisle, *Carte de la Barbarie* 1707), 2180.1711
  (Delisle, *Theatrum historicum*, copy Paris 1711)

La Courneuve/Paris
  Archives des affaires étrangères
    Mémoires et Documents
      Afrique 5, 7, 9
      Alger 12, 13
      Maroc 1

Leipzig
  Universitätsbibliothek
    Ms. 0662

London
  British Library
    Add. 4432, 8312, 21078–21082, 22910, 22911, 51049,
    61542
    Harley 3779, 6824
    Ms. Sloane 3396, 3495, 3986, 4051, 4053, 4068

  Kew, National Archives
    PRO

ADM 7/75–77, 630
CO 391/9
PC 2/28
SP 14/91, 105, 107, 109, 115
SP 71/1–29
SP 105/78, 145, 178
SP 110/73/2 (2–7)

Metropolitan Archives
COL_CHD_PR_07

Marseille
Chambre de Commerce
C 143–162
E 146–148
G 39–48
H 7
J 35, 46–48, 59, 902, 1563

Munich
Bayerische Staatsbibliothek
Cod. gall. 729

Nantes
Archives des affaires étrangères
Corr. Consulat Algier, A II
Tripolie de Barbarie 706/PO/01/carton 39, 45
18 PO/BO n. 40
167 PO/A/36bis

Oxford
Bodleian Library
Ashmole 1814
Ms.Add. D 27
Ms.Eng.misc.c.23
Ms.Eng.misc.e.390
Ms. Gough maps 229, 231
Ms.Lat.misc.e.76
Ms.Top.gen.c. 24
Map. Res 73
Pococke 428

      Rawlison letters 96
      Smith 28, 45, 49, 131
      Tanner 35

Paris
  Académie des sciences
    Procès-verbaux 4, 5
    Pochettes 1732

  Archives nationales
    Site Paris
      AE
        B I 206, 316, 317, 318, 319, 320, 321, 322, 403, 407,
          928, 968, 1018, 1088, 1089, 1090, 1091, 1092, 1093,
          1116, 1130
        B III 1, 2, 24, 234, 290

      MAR
        B VII 224, 311, 320, 321, 466, 473
        C VII 137, 248

    Site Pierrefitte
      K 1355

  Bibliothèque nationale (BNF)
    Ms. français 7184, 11333, 18595
    Nouvelles acquisitions françaises (NAF) 4989, 7450, 7459,
      7464–7466, 7468, 9134, 9137, 10839–10841

Unpublished PhD Theses

Ambrose, G. P. "The Levant Company mainly from 1640–1753," B.Litt. thesis Oxford 1932 (London Guildhall Library SL 66/2).

Cerbu, T. "Leone Allacci, 1587–1669: The fortunes of an early Byzantinist," PhD Harvard (1986) (Harvard University Archives 90.12423).

Kararah, A. M. A. H. "Simon Ockley. His contribution to Arabic studies and influence on Western thought," PhD Cambridge 1955 (UL Cambridge, Ph. D. Diss. 2795).

Lally, P. J. "Baldus de Ubaldis on the Liber Sextus and De regulis iuris: Text and Commentary," PhD Univ. Chicago, August 1992.

Matterson, C. H. "English trade in the Levant, 1693–1753," PhD Harvard 1936 (Harvard University Archives HU 90.3052).

Pippidi, A. M. "Knowledge of the Ottoman Empire in late seventeenth century England: Thomas Smith and some of his friends," D.Phil. Thesis, Oxford 1983.

Russell, I. S. "The later history of the Levant Company," PhD Victoria Univ. Manchester 1935 (London Guildhall Library SL 66/2).

Zizi, Z. "Thomas Shaw (1692–1751) à Tunis et Alger missionnaire de la curiosité européenne," thèse de doctorat, Université de Caen 1995 (ANRT Lille).

Published sources

Ecchellensis, A. *Chronicon orientale* (Venice: Javarina, 1729, first ed. 1651).

Abulfeda, I. *Descriptio peninsulae Arabum*, ed. J. Gagnier ([Oxford, s.n.: s.d. 1740]).

Acosta, J. *Historia natural moral delas Indias* (Sevilla: de Leon, 1590).

Acta et scripta theologorum Wirtembergensium (Wittenberg: Krafft, 1584).

Addison, L. *West Barbary, or a short narrative of the revolutions of the kingdoms of Fez and Morocco* (Oxford: at the theatre, 1671).

Addison, L. *The present state of the Jews* (London: J.C. for Crooke, 1675).

Agricola, G. *De ortu & causis subterraneorum libri V* (Basel: Froben and Episcopius, 1558).

Albertus, Magnus, "Mineralium libri quinque," in *Opera omnia*, ed. A. Borgnet, vol. 5 (Paris: Vivès, 1890), 1–116.

Aldovrandi, U. *Musaeum metallicum in libros IIII distributum* (Bologna: Ferroni, 1648).

Allacci, L. *Ioannes Henricus Hottingerus fraudis ... convictus* (Rome: Cong. Prop. Fide, 1661).

Ammirato, S. *Discorsi sopra Cornelio Tacito* (Venice: Valentino, 1607 first ed. 1594).

Angelus de Ubaldis, *Consilia seu responsa* (Lyon: Moylin, 1532).

Arnauld A. and P. Nicole, *La perpétuité de la foy de l'église catholique touchant l'eucharistie*, 3 vol. (Paris: Savraux, 1669).

Aristoteles (Pseudo-), *De mundo III = Aristoteles latinus*, vol. 11/11, 2nd ed. W. L. Lorimer and L. Minio-Paluello (Bruges: Desclée de Brouwer, 1965).

Arvieux, chevalier d', *Voyage fait par ordre du roy Louis XIV dans la Palestine* (Paris: Cailleau, 1717).

Arvieux, *Mémoires du chevalier d'*, 6 vol. (Paris: Delespine, 1735).

Aubertin, E. *De eucharistiae, sive Coenae Dominicae sacramento libri tres* (Deventer: Colomp, 1654).

Aventinus, J. (= J. Turmair), *Sämmtliche Werke*, vol. III/2: *Annales ducum boiariae VII*, ed. S. Riezler (Munich: Kaiser, 1884).

Avicenna, "De congelatione et conglutinatione lapidum," in *Artis auriferae, quam chemiam vocant* (Basel: Waldkirch, 1593), 374–382.

Avicenna, *De congelatione et conglutinatione lapidum*, ed. and transl. from Arabic E. J. Holmyard and D. C. Mandeville (Paris: Geuthner, 1927).

d'Avity de Montmartin, P. *Description générale de l'Afrique* (Paris: Sonnius, 1637).

Bacon, F. *The New Organon*, ed. L. Jardine and M. Silverthorne (Cambridge University Press, 2000).

Baldus de Ubaldis, *Opus aureum super feudis* (Lyon: Myt, 1524).

Baldus de Ubaldis, *In sextum codicis librum commentaria* (Venice: Giunta, 1599).

Bartholinus, T. *Historiarum anatomicarum rariorum centuria I et II* (The Haag: Vlacq, 1654).

Bartolus de Sassoferato, *Tractatus Tyberiadis seu de fluminibus libri tres* (s.l., s.d. = [Rome: Herolt and Riessinger, 1483]).

Baudrand, M.-A. *Geographia ordine litterarum disposita* (Paris: Michalet, 1682).

Baudrand, M.-A. *Dictionnaire géographique et historique* (Amsterdam and Utrecht: Halma and de Water: 1701).

Baumgartner, M. *Peregrinatio in Aegyptum, Arabiam, Palaestinam & Syriam* (Nuremberg: Kauffmann, 1594).

Bellarmino, R. *De controversiis christianae fidei*, tom. 3 (Cologne: Hierat, 1628).

Béranger, N. *La Régence de Tunis à la fin du XVIIe siècle*, ed. Paul Sebag (Paris: L'Harmattan, 1993).

Bertachini, G. *Tractatus de gabellis seu de vectigalibus* (Venice: de Tortis, 1489).

Bertius, T. *Theatrum geographiae veteris* (Leiden: Elzevier/Hondt, 1618).

Blake, R. *The letters*, ed. J. R. Powell (London: Navy Record Society, 1937).

Boberg, A. (praes.) and M. Ericander (def.), '*Al-leson we-torat has-somronim sive de lingua et pentateucho Samaritanorum* (Stockholm: Wernerian, 1734).

Bodin, J. *De republica libri sex* (Paris: Dupuys, 1586).

Bodin, J. *I sei libri dello stato*, ed. M. Isnardi Parente and D. Quaglioni, vol. 2 (Torino: UTET, 1988).

Bodin, J. *Methodus ad facilem historiarum cognitionem*, ed. S. Miglietti (Pisa: Ed. della Normale, 2013).

de Boisguilbert, P. *ou la naissance de l'économie politique*, 2 vol. (Paris: INED, 1966).

Bonavides, M. Fragment, inc. "De inclita Venetiarum civitate," in J. Fichard, *Vitae recentiorum iureconsultorum* (Padua: Giordano, 1565), f. 62v.

Botero, G. *Relationi universali*, vol. 1 (Vicenza: Perin, 1595).

Botero, G. "Della riputatione del prencipe," in *Aggiunte di G. B. Alla sua ragion di stato* (Venice: Ciotti, 1598), f. 40v–56r.

Bourguet, L. *Traité des petrifications* (Paris: Briasson, 1742).

Boyle, R. "The history of fluidity and firmness," in R. Boyle, *The Works*, ed. M. Hunter (London: Pickering & Chatto, 1999–2000; digital ed. Charlottesville: InteLex, 2003), vol. 2, 116–204.

Boyle, R. *The Works*, 14 vol., ed. M. Hunter et al. (London: Pickering & Chatto, 1999–2000; digital ed. Charlottesville: InteLex, 2003).

Burnet, T. *Archaeologiae philosophicae libri duo* (London: R. N. and Kettilby, 1692).

Cantillon, R. *Essai sur la nature de commerce en général* (London: Gyles, 1755).

de Capmany y de Montpalau, A. *Antiguos tratados de Paces y Alianzas* (Madrid: Imprenta real, 1786).

de Capmany y de Montpalau, A. *Memorias historicas de Barcelona*, 4 vol. (Madrid: Sancha, 1792).

Cardano, G. *De subtilitate libri XXI* (Basel: Henricpetri, 1582).

Cary, J. *An essay on the state of England in relation to its trade* (Bristol: Bonny, 1695).

Casaubon, I. *Ad epistolam Cardinalis Perronii responsio* (London: Norton, 1612).

Chalmers, G. *A collection of treaties between Great Britain and other powers*, vol. 2 (London: Stockdale, 1790).

Child, J. *A Discourse about trade* (London: Sowle, 1690).

Churchill, A. *A collection of voyages and travels*, vol. 1 (London: Awnsham, Churchill: 1704).

Cipolla, B. "De servitutibus rusticorum praediorum," in B. Cipolla, *Varii Tractatus* (Venice: Tridino, 1555), 216–441.

Claude, J. *Réponse aux deux traitez intitulez la perpetuité de la foy de l'église catholique*, 7th ed. (Charenton: Tellier, 1668).

Coke, R. *A detection of the court and state of England* (London: Bell, 1697).

Colbert, J.-B. *Lettres instructions et mémoires*, ed. P. Clément, 7 vol. (Paris: Imprimérie nationale, 1861–1870).

Corneille, T. *Dictionnaire universel géographique et historique* (Paris: Coignard, 1708).

Covel, J. *Some account of the present Greek Church* (Cambridge: Crownfield, 1722).

de la Croix, P. *Relation universelle de l'Afrique ancienne et moderne* (Lyon: Amaulry, 1688).

Crouch, H. *A complete guide to the officers of His Majesty's customs in the Out-Ports* (London: s.n., 1732).

R. D., *A true relation of the adventures* (s.l.: s.n., 1672).

Dan, P. *Histoire de Barbarie* (Paris: Rocolet, 1637).

Dapper, O. *Umbständliche und eigentliche Beschreibung von Africa* (Amsterdam: Meurs, 1670).

Dapper, O. *Description de l'Afrique* (Amsterdam: Wolfgang et al., 1686).

Davenant, C. "An essay upon the probable methods of making a people gainers in the balance of trade," in C. Davenant, *The political and commercial works*, vol. 2 (London: Horsfield et al., 1771), 164–382.

Davies, J. *The question concerning impositions, tonnage, poundage, prizage, customs,* (London: S.G. for Twyford, 1656).

Defoe, D. *A plan of the English commerce* (London: Rivington, 1728).

Deusing, A. "Foetus Mussipontani," in *Historia foetus Mussipontani* (Frankfurt: Zubrodt, 1669), 81–158.

Dickinson, E. *Delphi phoenicizantes, sive, tractatus de origine Druidum* (Oxford: Hall, 1655).

Digby, K. *Two treatises* (Paris: Blaizot, 1644).

Digby, K. *Of the sympathetick powder* (London: Williams, 1669).

Digges, D. *The defence of trade* (London: Stansby, 1615).

Documents inédites sur l'histoire du Maroc. Sources françaises, 2 vol., (1726–1732), ed. C. de La Véronne (Paris: Geuthner, 1975–1984).

Dodwell, H. *De geographorum dissertationes* (Oxford: Theatr. Sheldian., 1698).

Dodwell, H. *A discourse concerning Sanchoniathon's Phoenician history* (London: Clark, 1691).

du Perron, J. D. *Traitté du Sainct Sacrement de l'Eucharistie* (Paris: Chaudière, 1633).

du Tot, N. or P. *Réflexions politiques sur les finances et le commerce,* 2 vol. (The Haag: Vaillant and Prevost, 1738).

du Tot, N. or P. *Histoire du système de John Law,* ed. A. E. Murphy (Paris: INED, 2000).

du Val, P. *Diverses cartes et tables* (Paris: du Val, 1677).

Ductor Mercatorius: or, the young merchant's instructor (Newcastle upon Tyne: Thompson and Gooding, 1750).

Duplessis-Mornay, P. *De l'institutioñ, usage, et doctrine du sainct Sacrement de l'Eucharistie,* 6th ed. (Geneva: Chouet, 1599).

Éon, J. (i. e. M. de Saint-Jean, Ord. Carm.), *Le commerce honorable* (Nantes: Monnier, 1646).

Erpenius, T. *Historia saracenica* (Leiden: Maire and Elzevir, 1625).

Estienne, H. (ed.), *Poiesis philosophos* (Geneva: Estienne, 1573).

Evelyn, J. *Navigation and commerce* (London: T.R. for Tooke, 1674).

Falloppio, G. *De medicatis aquis atque de fossilibus tractatus* (Venice: Avanti, 1569).

Firth, C. H. and R. S. Rait (eds.), *Acts and ordinances of the interregnum, 1642–1660*, vol. 1 (London: H.M.S.O., 1911).

Fitzherbert, A. *La graunde abridgement* (London: Tottelli, 1577).

Frachetta, G. *Il prencipe* (Geneva: Chouet, 1648).

Frisch, J. *Der Schauplatz Barbarischer Schlaverey* (Altona: de Leu, 1666).

Gale, T. *The court of Gentiles, Part II: Of Barbaric and Grecanic philosophie*, 2nd ed. (Oxford: Mackock, 1676).

Galland, A. *Journal pendant son séjour à Constantinople (1672–1673)*, ed. C. Schefer (Paris: Léroux, 1881).

General heads for the natural history of a country, great or small (London: Taylor, 1692).

Gennadios, G. II Scholarios. *Homiliae de sacramento eucharistiae*, ed. E. Renaudot (Paris: Martin, 1709).

Gentili, A. *De jure belli libri tres*, ed. J. C. Rolfe and C. Phillipson, 2 vol. (Oxford: Clarendon, 1933).

Gessner, C. *De omni rerum fossilium genere . . . libri* (Zürich: Gesner, 1565).

Gessner, J. *Tractatus physicus de petrificatis* (Lyon: Haak, 1758).

Gottschling, C. *Staat von dem Königreiche Thunis in Africa* (s.l.: s.n., s.d. [1710]).

Gottschling, C. *Staat von dem König-Reiche Fez und Marocco in Africa* (s.l.: s.n., s.d. [1711]).

Gottschling, C. *Staat von dem Königreiche Algier in Africa* (s.l.: s.n., s.d. [1720]).

de Gournay, V. "Moyens proposés pour agir . . . contre les Anglais" (1755), in L. Charles et al. (eds.), *Le cercle de Vincent de Gournay. Savoirs économiques et pratiques administratives en France au milieu du XVIIIe siécle* (Paris: INED, 2011), 352–367.

Gramaye, I. B. *Asia, sive historia universalis asiaticarum gentium* (Antwerpen: Beller, 1604).

Gramaye, I. B. *Arscotum ducatus cum suis baronatibus* (Bruxelles: Mommartius, 1606).

Gramaye, I. B. *Thenae et Brabantia ultra velpam* (Bruxelles: Mommartius, 1606).

Gramaye, I. B. *Africae illustratae libri decem, in quibus Barbaria, gentesque eius ut olim, et nunc describuntur* (Tournai: Quinque, 1622).

Gramaye, I. B. *Respublica Namurcensis, Hannoniae et Lutsenburgensis* (Amsterdam: Jansson, 1634).

de Grammont, H.-D. *Correspondance des consuls d'Alger (1690–1742)* (Algiers: Jourdan, 1890).

Granger, C. *Voyage dans l'empire Ottoman 1733–1737*, ed. A. Riottot (Paris: L'Harmattan, 2006).

Grégoire, P. *De republica libri sex et viginti* (Frankfurt/M: Fischer, 1597).

Grotius, H. *Mare liberum sive de iure quod Batavis competit ad indicana commercia, dissertatio* (Leiden: Elzevir, 1609).

Grotius, H. *Appendix ad interpretationem locorum N. Testamenti quae de Antichristo agunt* (Amsterdam: Blæu, 1641).

Grotius, H. *Votum pro pace ecclesiastica* (s.l.: s.n., 1642).

Grotius, H. *De iure belli ac pacis*, ed. B. J. A. De Kanter et al. (Aalen: Scientia, 1993).

Haëdo, D. *Topographia e historia general de Argel* (Valladolid: Fernandez, 1612).

Haëdo, D. *Histoire des rois d'Alger*, ed. H.-D. de Grammont (Algiers: Jourdan, 1881).

d'Herbelot, B. *Bibliothèque orientale* (Paris: Compagnie des libraires, 1697).

Herodotus, *The famous hystory* (London: Marshe, 1584).

Histoire veritable de Tafilette (Rouen: Lucas, 1670).

Hooke, R. *Philosophical experiments and observations*, ed. W. Derham (London: Innys, 1726).

Hottinger, J. H. *Exercitationes anti-Morinianae: de Pentateucho Samaritano* (Zürich: Bodmer, 1644).

Hottinger, J. H. *Historia Orientalis* (Zurich: Bodmer, 1651).

Hottinger, J. H. *Analecta historico-theologica* (Zurich: Bodmer, 1652).

Hottinger, J. H. *Enneas dissertationum philologico-theologicarum Heidelbergensium* (Zurich: Stauffacher, 1662).

Hudson, J. (ed.), *Geographiae veteris scriptores Graeci minores* (Oxford: Theatr. Sheldian., 1698).

Huet, P.-D. *Le grand trésor historique et politique du florrissant commerce des Hollandois* (s'Gravenhage: Frik, 1713).

Jeffreys, G. *The argument of the Lord Chief Justice* (London: Taylor, 1689).

Jenkinson, A. *Russiae, Moscoviae Et Tartariae Descriptio* (London: Adams, sculp. Reinold, 1562).

Johannes Monachus, *Glosa aurea super Sexto Decretalium* (Paris: Petit, 1535).

Karmires, I. (ed.), Τὰ δογμάτικα καὶ συμβολικὰ μνημεία τῆς ὀρθοδόξου καθολικῆς ἐκκλησίας, 2nd ed., 2 vol. (Graz: Akadem. Druck-Verlagsanstalt, 1968).

Kimmel, E. J. (ed.), *Monumenta fidei ecclesiae orientalis*, pars 1 (Jena: Mauke, 1850).

Kircher, A. *Obeliscus pamphilius* (Rome: Grignani, 1650).

Kircher, A. *Oedipus aegyptiacus* (Rome: Mascardi, 1652).

Kircher, A. *Mundus subterraneus*, tom. 2 (Amsterdam: Jansson, 1664).

L'Afrique de Marmol (Paris: Jolly, 1667).

La Faye, J. B. *État des royaumes de Barbarie, Tripoly, Tunis, et Alger* (Rouen: Béhourt, 1703).

Law, J. *Œuvres complètes*, ed. P. Harsin, 3 vol. (Paris: Sirey, 1934).

Leibniz, G. W. *Consilium Aegyptiacum*, 1671–1672, in *Akademieausgabe*, vol. 4/1, 215–410.

Leo Africanus (Léon l'Africain), *Description de l'Afrique*, transl. and ed. A. Épaulard et al., 2 vol. (Paris: Librairie d'Amérique et d'Orient, 1956).

Leo Africanus, "Della descrittione dell'Africa," in G. B. Ramusio (ed.), *Delle navigationi et viaggi*, vol. 1 (1563), 1A–95F.

Lightfoot, J. *The whole works*, ed. J. R. Pitman, vol. 13 (London: Dove, 1824).

Locke, J. *An essay concerning human understanding*, ed. P. H. Nidditch (Oxford University Press, 1975).

Lucas, P. *Voyage fait dans la Grèce, l'Asie mineure, la Macédoine et l'Afrique*, tome II (Paris: Simart, 1712).

Ludolf, H. *Epistolae Samaritanae Sichemitarum* (Cizae: Hetstedt and Bielcke, 1688).

Machiavelli, N. *I primi scritti politici*, ed. C. Vivanti (= Opere, vol. 1) (Torino: Einaudi, 1997).

Malvy, A. and M. Viller (eds.), *La Confession orthodoxe de Pierre Moghila* (Rome and Paris: Pont. Inst. Orient. and Beauchesne, 1927).

Mantua Patavinus vide Bonavides, Marco.

al-Maqrīzī, *Description topographique et historique de l'Égypte*, transl. U. Bouriant (Paris: Leroux, 1895).

del Mármol Carvajal, L. *Descripcion general de Africa (1573–1599)*, ed. Instituto de Estudios Africanos, tom. 1 (Madrid: Inst. est. africanos, 1953).

Marsham, J. *Canon chronicus aegyptiacus, ebraicus, graecus* (Franeker: Strick, 1696).

Marta, G. *De iurisdictione tractatus*, vol. 1 (Avignon: Antonio, 1669).

Mastrantonis, G. (ed.), *Augsburg and Constantinople. The correspondence between the Tübingen theologians and Patriarch Jeremiah II of Constantinople on the Augsburg Confession* (Brookline: Holy Cross Orthodox Press, 1982).

Megenberg, K. v. *Buch der Natur*, vol. 2: *Kritischer Text nach den Handschriften*, ed. R. Luff and G. Steer (Berlin/New York: de Gruyter, 2003).

Melon, J.-F. *Essai politique sur le commerce* (s.l.: s.n., 1734).

de Monchrestien, A. *Traicté de l'œconomie politique*, ed. F. Billacois (Geneva: Droz, 1999).

Monmouth, G. *The history of the kings of Britain – De gestis Britonum*, ed. M. D. Reeve, transl. N. Wright (Woodbridge: Boydell, 2007).

Morin, J. *Exercitationes ecclesiasticae in utrumque Samaritanorum pentateuchum* (Paris: Vitray, 1631).

Moüette, G. *Relation de la captivité du Sr Moüette* (Paris: Cochart, 1683).

Nicole, P. *La perpétuité de la foy de l'église catholique touchant l'eucharistie*, 3rd ed. (Paris: Savreux, 1664).

Ockley, S. *The conquest of Syria, Persia, and Aegypt, by the Saracens* (London: Knaplock et al., 1708).

Ockley, S. *His account of South-West Barbary* (London: Bowyer and Clements, 1713).

Ockley, S. *The History of the Saracens*, 3rd ed. (Cambridge: Ockley and Lintot, 1757).

Oekolampad, J. *De genuina verborum domini, hoc est corpus meum, iuxta vetustissimos authores, expositione liber* (Basel: s.n., 1525).

Oekolampad, J. *Quid de Eucharistia veteres tum Graeci, tum Latini senserint, Dialogus* (s.l.: s.n., 1530).

Oldenburg, H. "Of the *Mundus subterraneus* of Athanasius Kircher," in *Philosophical Transactions* 1 (1665/66), 109–117.

Omont, H. *Missions archéologiques françaises en Orient aux XVIIe et XVIIIe siècles* (Paris: Imprimérie nationale, 1902).

Ortelius, A. *Theatrum orbis terrarum* (Antwerp: van Diest, 1570).

Papillon, T. *The East-India-trade a most profitable trade to the kingdom* (London: s.n., 1677).

Pascal, B. *Pensées* (1670), ed. M. Le Guern (Paris: Gallimard, 1977/2004).

Peregrino, A. *De iuribus et privilegiis fisci libri VIII* (Cologne: Gymnich, 1588).

Petty, W. *Britannia languens: or, a discourse of trade* (London: Baldwin, 1689).

Petty, W. "A treatise of taxes & contributions," in W. Petty, *The economic writings*, ed. C. H. Hull, vol. 1 (Cambridge University Press, 1899), 1–97.

Pezron, P.-Y. *L'Antiquité des tems rétablie et défendue, contre les Juifs & les nouveaux chronologistes* (Paris: Martin and Boudot, 1687).

Philempórios, *The scheme of the subsequent discourse. An East-India trade is highly advantageous to the true interest of England* (London: Hills, 1683).

Philopatris (J. Child), *A treatise wherein is demonstrated, I. That the East-India trade is the most national of all foreign trades* (London: T.J. for Boulton, 1681).

Piccolomini, E. S. "Cosmographia seu Rerum ubique gestarum historia locorumque descriptio," in Piccolomini, *Opera geographica et historica* (Helmstedt: Sustermann, 1699), 3–374.

Pignon, J.-P. "Sur le commerce des François au Levant," in *Journal œconomique*, September 1754, 42–104.

Plantet, E. *Correspondance des Deys d'Alger avec la cour de France, 1579–1833*, vol. 1 (Paris: Alcan, 1889).

Pococke, E. *Specimen historiae arabum* (Oxford: Hall, 1650).

Pococke, E. *Ta'rih̠ Muh̠taṣar al-duwal. Supplementum Historiae Dynastiarum* (Oxford: Hall and Davis, 1663).

de La Hestroye, Pottier "Restablissement du commerce," in *Œuvres complètes*, ed. P. Harsin, 3 vol. (Paris: Sirey, 1934), vol. 2, 67–259.

Praed, J. *An essay on the coin and commerce of the kingdom* (London: s.n., 1695).

Purchas, S. *His pilgrimage, or, relations of the world and the religions observed* (London: Stansby for Fetherstone, 1614).

Raleigh, W. *The history of the world* (London: Burre, 1614).

Ramusio, G. B. (ed.), *Delle navigationi et viaggi*, 3rd ed., vol. 1 (1563), vol. 2 (1559) (Venice: Giunti).

Richardson, D. (ed.), *The Mediterranean passes in the Public Records Office* (East Ardsley: EP Microform Ltd., 1981).

du Plessis Richelieu, A. *Testament politique*, ed. F. Hildesheimer (Paris: Champion, 1995).

du Plessis Richelieu, A. *Testament politique*, ed. L. André (Paris: Laffont, 1947).

Roberts, L. *The treasure of traffike or a discourse of forraigne trade* (London: Bourne, 1641).

Rotuli Scotiae in Turri Londinensi et in domo Capitulari Westmonasteriensi asservati, 2 vol. (London: Eyre and Strahan, 1814–15).

Rowlands, H. *Mona antiqua restaurata* (London: Knox, 1766).

Rueus, F. *De gemmis aliquot* (Frankfurt/M: Palthen and Fischer, 1596).

Rycaut, P. *The present state of the Ottoman Empire* (London: Starkey and Brome, 1668).

Rycaut, P. *The present state of the Greek and Armenian Churches* (London: Starkey, 1679).

S., T. *The adventures of an English merchant* (London: Pitt, 1670).

Sacheverell, W. *An account of the Isle of Man* (London: Hartley, 1702).

de Sacy, S. (ed.), *Correspondance des Samaritains de Naplouse* (Paris: Imprimérie royale, 1813).

de Saint-Pierre, Abbé C.-I. "Utilité des dénombremens," in *Ouvrajes [sic] de politique*, vol. 4 (Rotterdam: Beman, 1733), 255–267.

Sambstagius, J. *Oratio historico-politica de causis mutationum & interituum rerumpublicarum* (Jena: Weidner, 1616).

Sammarco, O. *Delle mutationi de' regni* (Napoli: Scoriggio, 1628).

Sammes, A. *Britannia antiqua illustrata* (London: Roycroft, 1676).

Sanson, N. *L'Affrique, en plusieurs cartes nouvelles, et exactes* (Paris: Sanson, 1656).

Sanuto, L. *Geografia dell'Africa* (1588), facsimile ed. R. A. Skelton (Amsterdam: Theatrum orbis terrarum, 1965).

Sarpi P. (pseud. F. de Ingenuis), *De iurisdictione serenissimae reipublicae Venetae in mare Adriaticum* (s.l. [Eleutheropolis]: s.n., 1619).

Scheuchzer, J. J. *Syntheo homo diluvii testis et theoskopos publicae syzetései expositus* (Zurich: Byrgklin, 1726).

Seaman, W. *Grammatica linguae Turcicae, in quinque partes distributa* (Oxford: Hall, 1670).

Selden, J. *Of the dominion, or, ownership of the Sea*, transl. M. Nedham (London: Du-Gard, 1652).

Settala, L. *Della ragion di stato libri sette* (Milan: Bidelli, 1627).

Severos, G. *Fides ecclesiae Orientalis*, ed. R. Simon (Paris: Meturas, 1671).

Shaw, T. *Travels, or observations relating to several parts of Barbary and the Levant* (London: Millar and Sandby, 1738; 2nd ed. 1757).

Shaw, T. *A Supplement to a book entituled travels, or observations* (Oxford: at the theatre, 1746).

Sheeres, H. *A discourse touching Tanger* (London: s.n., 1680).

Simon, R. *Histoire critique de la créance & des coutumes des Nations du Levant* (Frankfurt/M: Arnaud, 1684).

Simon, R. *La créance de l'église orientale* (Paris: Moette, 1687).

Smith, T. *Syntagma de Druidum moribus ac institutis* (London: Roycroft, 1664).

Smith, T. *Ignatii epistolae genuinae annotationibus illustratae* (Oxford: Theatr. Sheldian, 1709).

Smith T. (ed.), *Roberti Huntingtoni epistolae* (London: Bowyer and Churchill, 1704).

Steensen, N. *De solido intra solidum naturaliter contento dissertationis prodromus* (Florence: Stella, 1669).

Straccha, B. "Tractatus de navigatione," in B. Straccha, *Tractatus de mercatura* (Lyon: Giunta, 1558), 275–287.

Straccha, B. *Tractatus de mercatura* (Lyon: Giunta, 1558).

Stukeley, W. *Stonehenge. A Temple restor'd to the British Druids* (London: Innys and Manby, 1740).

Stukeley, W. *Abury, a temple of the British Druids, with some others, described* (London: Innys and Manby, 1743).

Stukeley, W. *Memoirs of Sir Isaac Newton's life*, ed. A. H. White (London: Taylor and Francis, 1936).

Suarez, R. "Consilium de usu maris," in B. Straccha, *Tractatus de mercatura* (Lyon: Giunta, 1558), 619–629.

Synodus bethlehemitica adversus calvinistas haereticos (Paris: E. Martin, 1676).

de Tassy, L. *Histoire du royaume d'Alger. Un diplomate français à Alger en 1724* (Paris: Loysel, 1992).

Taylor, J. *The real presence* (London: Flesher, 1654).

de Testa, I. (ed.), *Recueil des traités de la Porte Ottomane*, tome I (Paris: Amyot, 1864).

The allegations of the Turky Company and others against the East-India-Company (s.l.: s.n., 1681).

Thévenot, M. (ed.), *Relations de divers voyages curieux*, 4 vol. (Paris: Langlois et al., 1663–1672).

Thomas Aquinas, *Opera omnia*, ed. Roberto Busa, www.corpusthomisticum .org.

Thorne, S. E. M. E. Hager, M. MacVeagh Thorne, C. Donahue (ed.), *Year books of Richard II – 6 Richard II, 1382–1383* (Cambridge/Mass.: Harvard University Press, 1996).

de Thou, J.-A. *Historiarum sui temporis libri 136*, 3 vol. (Geneva: La Rouière, 1620).

Tollot, *Nouveau voyage fait au Levant, és années 1731 & 1732* (Paris: Durand, 1742).

Torres, D. *Relacion del origen y suceso de los Xerifes (1586)*, ed. M. García-Arenal (Madrid: siglo XXI, 1980).

de la Tour, S. *Histoire de Mouley Mahamet* (Geneva: s.n., 1749).

de Tournefort, P. *Relation d'un voyage du Levant*, 2 vol. (Lyon: Bryyset, 1727).

Trevers, J. *An essay to the restoring of our decayed trade* (London: Giles et al., 1677).

Troilo, F. F. v. *Orientalische Reise-Beschreibung* (Dresden: Bergens, 1676).

Twyne, J. *De rebus Albionicis, Britannicis atque Anglicis, commentariorum libri duo* (London: Bollifantus, 1590).

van Helmont, J. B. "Doctrina inaudita, de causis, modo fiendi, contentis, radice, & resolutione Lithiasis," in *Opuscula medica inaudita* (Amsterdam: Elzevir, 1648) (no pag.).

Venture de Paradis, J.-M. *Tunis et Alger au XVIIIe siècle*, ed. J. Cuoq (Tunis: Sindbad, 1983).

Vermigli, P. M. *Locorum communium theologicorum*, tom. 1 (Basel: Perna, 1580).

Veron de Forbonnais, F. *Elemens du commerce,* 2nd ed. (Leiden and Paris: Briasson et al., 1754).

Villault de Bellefond, N. *Relation des costes d'Afrique, appellées Guinée* (Paris: Thierry, 1669).

Violet, T. *The advancement of merchandize* (London: Du-Gard, 1651).

Walker, D. P. *The ancient theology. Studies in Christian Platonism from the 15th to the 18th century* (London: Duckworth, 1972).

Wansleben, J. M. *Nouvelle relation en forme de journal, d'un voyage fait en Égypte* (Paris: Compagnie des libraires associés, 1698).

Wells, E. *A specimen of an essay* (Oxford: Wells, 1720).

Welwood, W. *De dominio maris* (Cosmopolis: Fontisilvius, 1615).

Whiston, J. *A discourse of the decay of trade* (London: Crouch, 1693).

Winckelmann, J. J. "Sendschreiben von den Herculanischen Entdeckungen," ed. Marianne Gross et al. (= Winckelmann, *Schriften und Nachlass*, vol. 2: *Herkulanische Schriften Winckelmanns*, part I) (Mainz: Zabern, 1997).

Wolff, C. *Oratio de Sinarum philosophia practica* (Frankfurt/M: Andreae and Hort, 1726).

Wolff, C. *Vernünfftige Gedancken von Gott, der Welt und der Seele des Menschen, auch allen Dingen überhaupt* (first edition 1720 = *Gesammelte Werke* I/2 and I/3 [ed. 1751], ed. Jean Ecole [Hildesheim: Olms, 1983]).

Published Secondary Sources

Abdel-Halim, M. *Antoine Galland. Sa vie et son œuvre* (Paris: Nizet, 1964).

Abdesselem, A. *Les historiens tunisiens. Des XVIIe, XVIIIe et XIXe siècles* (Paris: Klincksieck, 1973).

Abun-Nasr, J. M. *A history of the Maghrib in the Islamic period* (Cambridge University Press, 1987).

Acquaviva, G. "Giulio Pace: la giurisdizione di Venezia sul Mare Adriatico," in G. Acquaviva and T. Scovazzi (eds.), *Il dominio di Venezia sul mare adriatico* (Milan: Giuffrè, 2007), 39–72.

Acquaviva, G. and T. Scovazzi (eds.), *Il dominio di Venezia sul mare adriatico* (Milan: Giuffrè, 2007).

Anderson, S. P. *An English consul in Turkey. Paul Rycaut at Smyrna, 1667–1678* (Oxford University Press, 1989).

Andrewes, W. J. H. "Even Newton could be wrong: The story of Harrison's first three sea clocks," in W. J. H. Andrewes (ed.), *The quest for longitude* (Cambridge/Mass.: Harvard University Press, 1996), 189–234.

Andrewes, W. J. H. (ed.), *The quest for longitude* (Cambridge/Mass.: Harvard University Press, 1996).

Appleby, J. O. *Economic thought and ideology in seventeenth century England* (Princeton University Press, 1978).

Arabadzoglu, G. "Σχέσεις Ὀρθοδόξων καὶ Ἀγγλικανῶν κατὰ τὰς ἀρχάς τοῦ ιη αἰωνός (Συμπλήρωσις τῆς μίκρας Ἀγγλογραικίας)," in *Orthodoxia* 27 (1952), 123–126.

Arbel, B. "Venice's maritime empire in the early modern period," in E. R. Dursteler (ed.), *A companion to Venetian history, 1400–1797* (Leiden and Boston: Brill, 2013), 125–254.

Armitage, D. *The ideological origins of the British Empire* (Cambridge University Press, 2000).

Armitage, D. *Foundations of modern international thought* (Cambridge University Press, 2012).

Armitage, D. and M. J. Braddick (eds.), *The British Atlantic world, 1500–1800*, 2nd ed. (New York: Palgrave Macmillan, 2009).

Asher, R. E. *National myths in Renaissance France. Francus, Samothes and the Druids* (Edinburgh University Press, 1993).

Aslanian, S. D. *From the Indian Ocean to the Mediterranean. The global trade networks of Armenian merchants from New Julfa* (Berkeley University Press, 2011).

Backscheider, P. R. *Daniel Defoe. His life* (Baltimore/London: John Hopkins University Press, 1989).

Backus, I. (ed.), *The reception of the church fathers in the West*, 2 vol. (Boston and Leiden: Brill, 2001).

Bacqué-Grammont, J.-L. et al., *Représentants permanents de la France en Turquie (1536–1991) et de la Turquie en France (1797–1991)* (Istanbul and Paris: ISIS, 1991).

de Baere, B. *La pensée cosmogonique de Buffon. Percer la nuit des temps* (Paris: Champion, 2004).

Baffioni, G. and P. Mattiangeli, *Annio da Viterbo, documenti e ricerche* (Rome: CNR, 1981).

Bakker, E. J. "The making of history: Herodotus' *Histories apodexis*," in E. J. Bakker, I. J. F. de Jong and H. v. Wees (eds.), *Brill's companion to Herodotus* (Leiden and Boston: Brill, 2002), 3–32.

Bargaoui, S. and H. Remaoun (eds.), *Savoirs historiques au Maghreb. Construction et usages* (Tunis-Oran: CERES, 2006).

Barnby, H. "The sack of Baltimore," in *Journal of the Cork Historical and Archaeological society* 74 (1969), 101–129.

Barnett, K. "'Explaining, themselves: The Barrington papers, the board of longitude, and the fate of John Harrison," in *Notes & Records of the Royal Society* 65 (2011), 145–162.

Barni, G. "Bartolo da Sassoferrato ed il problema del limite della giurisdizione sul mare," in *Rivista di storia del diritto italiano* 24 (1951), 185–195.

Baroja, J. C. *Las falsificaciones de la historia en relación con la de España* (Barcelona: Seix Barral, 1992).

Baron, S. "B.A. Rybakov on the Jenkinson Map of 1562," in L. Hughes (ed.), *New perspectives on Muscovite history* (New York: St. Martin's, 1993), 3–13.

Barrera-Osorio, A. *Experiencing nature: The Spanish American empire and the early scientific revolution* (Austin: Texas University Press, 2006).

Bartòla, A. "Alessandro VII e Athanasius Kircher S.J. Ricerche ed appunti sulla loro corrispondenza erudita e sulla storia di alcuni codici chigiani," in *Miscellanea bibliothecae Apostolicae Vaticanae* 3 (1989), 7–106.

Battaglia, S. *Grande dizionario della lingua italiana*, 21 vol. (Torino: UTET, 1961–2009).

Beach, A. R. "Satirizing English Tangier in Samuel Pepys's diary and Tangier papers," in G. V. Stanivukovic (ed.), *Re-mapping the Mediterranean world in early modern English writing* (Basingstoke: Palgrave Macmillan, 2007), 227–244.

Behrends, O. "Die allen Lebewesen gemeinsamen Sachen (*res communes omnium*) nach den Glossatoren und dem klassischen römischen Recht," in D. Medicus et al. (ed.), *Festschrift für Hermann Lange zum 70. Geburtstag* (Stuttgart et al.: Kohlhammer, 1992), 3–33.

Beinart, W. and L. Hughes (eds.), *Environment and empire* (Oxford University Press, 2007).

Bell, D. A. *The cult of the nation in France. Inventing nationalism, 1680–1800* (Cambridge/Mass.: Harvard University Press, 2001).

Ben Mansour, A. H. *Alger, XVIe-XVIIe siècle. Journal de Jean-Baptiste Gramaye 'evêque d'Afrique'* (Paris: Cerf, 1998).

Bender, B. "Stonehenge – contested landscapes (medieval to present day)," in B. Bender (ed.), *Landscape: politics and perspectives* (Providence: Berg, 1993), 245–279.

Bennett, G. V. "Patristic tradition in Anglican thought, 1660–1900," in *Tradition in Luthertum und Anglikanismus: Oecumenica 1971–2* (Gütersloh: Gütersloher Verlagshaus, 1972), 63–87.

Bennett, J. "The travels and trials of Mr Harrison's timekeeper," in M.-N. Bourguet, C. Licoppe, and H. O. Sibum (eds.), *Instruments, travel and science. Itineraries of precision from the seventeenth to the twentieth century* (London and New York: Routledge, 2002), 75–95.

Benton, L. and B. Straumann, "Acquiring empire by law: From Roman doctrine to early modern European practice," in *Law and History Review* 28 (2010), 1–38.

Bérenger, J. "La politique ottomane de la France dans les années 1680," in *Acta historica Academiae scientiarum* 33 (1987), 193–201.

Beretta. M. et al. (eds.), *The Accademia del Cimento and its European context* (Sagamore Beach: Watson, 2009).

Bergasse, L. and G. Rambert, *Histoire du commerce de Marseille, vol. 4 (1599–1789)* (Paris: Plon, 1954).

Berthier, A. "Turquérie ou turcologie? L'effort de traduction des jeunes de langues au XVIIe siècle. D'après la collection de manuscrits conservée à la bibliothèque nationale de France," in F. Hitzel (ed.), *Istanbul et les langues orientales* (Paris: L'Harmattan, 1997), 283–317.

Bianchin, L. "Conversiones rerumpublicarum. Zum Geschichtsbild der barocken Staatslehre," in C. Zwierlein and A. Meyer (eds.), *Machiavellismus in Deutschland. Chiffre von Kontingenz, Herrschaft und Empirismus in der Neuzeit* (Munich: Oldenbourg, 2010), 79–94.

Bireley, R. *The Counter-Reformation prince: anti-Machiavellianism or catholic statecraft in early modern Europe* (Chapel Hill: Univ. of North Carolina Press, 1990).

Bizer, E. *Studien zur Geschichte des Abendmahlsstreits im 16. Jahrhundert*, 2nd ed. (Gütersloh: Gütersloher Verlagshaus, 1962).

Blair, A. *Restaging Jean Bodin: The Universae naturae theatrum (1596) in its cultural context* (Princeton University Press, 1990).

Bleichmar, D. (eds.), *Science in the Spanish and Portuguese empires, 1500–1800* (Stanford University Press, 2009).

Blitz, H.-M. *'Aus Liebe zum Vaterland'. Die deutsche Nation im 18. Jahrhundert* (Hamburg: Hamb. Edition, 2000).

Boantza, V. D. "From cohesion to *pesanteur*. The origins of the 1669 debate on the causes of gravity," in M. Dascal and V. D. Boantza (eds.), *Controversies within the Scientific Revolution* (Amsterdam and Philadelphia: Benjamins, 2011), 77–100.

Bodenstein, W. "Ortelius' Maps of Africa," in van den Broecke et al. (eds.), *Abraham Ortelius*, 185–207.

Bolzoni, L. *La stanza della memoria: modelli letterari e iconografici nell'età della stampa* (Torino: Einaudi, 1995).

Borg, V. *Fabio Chigi. Apostolic Delegate in Malta (1634–1639). An Edition of his Official Correspondence* (Città del Vaticano: BAV, 1967).

Borschberg, P. "Hugo Grotius' Theory of Trans-Oceanic Trade Regulation: Revisiting *Mare Liberum* (1609)," www.iilj.org/publications/documents/2005-14-HT-Borschberg-Rev.-web.pdf.

Borst, A. *Der Turmbau von Babel. Geschichte der Meinungen über Ursprung und Vielfalt der Sprachen und Völker*, 4 vol. (Stuttgart: Hiersemann, 1957–1963).

Borst, A. "Das Erdbeben von 1348. Ein historischer Beitrag zur Katastrophenforschung," in *Historische Zeitschrift* 233 (1981), 529–569.

Boschiero, L. *Experiment and natural philosophy in seventeenth-century Tuscany. The history of the Accademia del Cimento* (Dordrecht: Springer, 2007).

Bosworth, C. E. *The new Islamic dynasties* (New York: Columbia University Press, 1996).

Boubaker, S. "Réseaux et techniques de rachat des captifs de la course à Tunis au XVIIe siècle," in W. Kaiser (ed.), *Le Commerce des captifs. Les intermédiaires dans l'échange et le rachat des prisonniers en Méditerranée, XVe-XVIIIe siècle* (Rome: École française de Rome, 2008), 25–46.

Bourguet, M.-N. and C. Licoppe, "Voyage, Mesures, et Instruments: Une nouvelle expérience du monde au siècle des lumières," in *Annales: Histoire, Sciences Sociales* 52, 5 (1997), 1115.

Braddick, M. J. "The English government, war, trade, and settlement, 1625–1688," in N. Canny (ed.), *The origins of empire. British overseas enterprise to the close of the seventeenth century* (Oxford University Press, 1998), 286–308.

Braddick, M. J. "Civility and authority," in D. Armitage and M. J. Braddick (eds.), *The British Atlantic world, 1500–1800*, 2nd ed. (New York: Palgrave Macmillan, 2009), 113–132.

Brendecke, A. et al. (eds.), *Information in der Frühen Neuzeit. Status, Bestände, Strategien* (Berlin: de Gruyter, 2008).

Brenner, R. *Merchants and revolution. Commercial change, political conflict, and London's overseas traders, 1550–1653* (Princeton University Press, 1993).

Brian, É. *La mesure de l'état. Administrateurs et géomètres au XVIIIe siècle* (Paris: A. Michel, 1994).

Brian, É. and C. Demeulenaere-Douyère (eds.), *Histoire et mémoire de l'Académie des sciences. Guide de recherches* (Paris et al.: Tec & Doc, 1996).

Briese, O. and T. Günther, "Katastrophe. Terminologische Vergangenheit, Gegenwart und Zukunft," in *Archiv für Begriffsgeschichte* 51 (2009), 155–196.

Brito Vieira, M. "*Mare liberum* vs. *Mare clausum*: Grotius, Freitas, and Selden's debate on dominion over the seas," in *Journal of the History of Ideas* 64 (2003), 361–377.

Broc, N. *La géographie des philosophes. Géographes et voyageurs français au XVIIIe siècle* (Paris: Ophrys, 1974).

Broc, N. *La géographie de la Renaissance (1420–1620)* (Paris: BNF, 1980).

Brodman, J. W. "Municipal ransoming law on the medieval Spanish frontier," in *Speculum* 60 (1985), 318–330.

Brodman, J. W. *Ransoming captives in crusader Spain. The order of Merced on the Christian-Islamic frontier* (Philadelphia: Penn State University Press, 1986).

van den Broecke, M. et al. (eds.): *Abraham Ortelius and the first Atlas. Essays commemorating the quadricentennial of his death, 1598–1998* (Turdijk: HES, 1998).

Bühler, K. *Sprachtheorie. Die Darstellungsfunktion der Sprache*, 3rd ed. (first 1934) (Stuttgart: Lucius & Lucius, 1999).

Bunes Ibarra, M. A. *La imagen de los musulmanes y del Norte de Africa en la España de los siglos XVI y XVII. Los caracteres de una hostilidad* (Madrid: C.S.I.R., 1989).

Burke, P. *The fabrication of Louis XIV* (New Haven: Yale University Press, 1992).

Burke, P. *A social history of knowledge*, 2 vol. (Cambridge/Malden: Scholars, 2000/12).

Burl, A. and N. Mortimer (ed.), *Stukeley's "Stonehenge." An unpublished manuscript 1721–1724* (New Haven and London: Yale, 2005).

Caillé, J. "La représentation diplomatique de la France au Maroc," in *Revue d'histoire diplomatique* 63 (1949), 104–171.

Caillé, J. "Le consul Jean-Baptiste Estelle et le commerce de la France au Maroc à la fin du XVIIe siècle," in *Revue Française d'Histoire d'Outre-Mer* 46 (1959), 7–48.

Caillé, J. *Le Consulat de Tanger (dès origines à 1830)* (Paris: Pedone, 1967).

Calafat, G. "Ottoman North Africa and ius publicum europaeum: The case of the treaties of peace and trade (1600–1750)," in A. Alimento (ed.), *War, trade and neutrality. Europe and the Mediterranean in the seventeenth and eighteenth centuries* (Milan: Franco Angeli, 2011), 171–187.

Calafat, G. *Une mer jalousée, juridictions et ports francs en Méditerranée au XVIIe siècle* (Paris: forthcoming, 2017).

Calhoun, C. (ed.), *Habermas and the public sphere* (Cambridge/Mass.: MIT Press, 1992).

Campbell, M. "'Of people either too few or too many'. The conflict of opinion on population and its relation to emigration," in W. A. Aiken and B. D. Henning (eds.), *Conflict in Stuart England. Essays in honour of Wallace Notestein* (London: New York University Press, 1960), 169–201.

Cañizares-Esguerra, J. *How to write the history of the New World. Histories, epistemologies, and identities in the Eighteenth-Century Atlantic World* (Stanford University Press, 2001).

de Caprio, V. *La tradizione e il trauma. Idee del Rinascimento romano* (Manziana: Vecchiarelli, 1991).

Caplan, J. and J. Torpey (eds.), *Documenting individual identity. The development of state practices in the modern world* (Princeton University Press, 2001).

Carbonnier-Burkard, M. "Les préambules des édits de pacification (1562–1598)," in M. Grandjean and B. Roussel (eds.), *Coexister dans l'intolérance. L'édit de Nantes (1598)* (Geneva: Labor et Fides, 1998), 75–92.

Carey, D. "Compiling nature's history: Travellers and travel narratives in the early Royal Society," in *Annals of Science* 54 (1997), 269–92.

Carrière, C. *Négociants Marseillais au XVIIIe siècle. Contribution à l'étude des économies maritimes*, 2 vol. (Marseille: Inst. hist. de Provence, 1973).

Cavaillé, J.-P. *Theatrum mundi: Notes sur la théatralité du monde baroque* (Fiesole: EUI, 1987).

Cavaillé, J.-P. *Dis/simulations. Jules-César Vanini, François La Mothe Le Vayer, Gabriel Naudé, Louis Machon et Torquato Accetto. Religion, morale et politique au XVIIe siècle* (Paris: Champion, 2002).

Cavallar, O. "River of law: Bartolus's *Tiberiadis (De alluvione)*," in J. A. Marino and T. Kuehn (eds.), *A Renaissance of conflicts. Visions and revisions of law and society in Italy and Spain* (Toronto: CRRS, 2004), 31–129.

Champion, J. A. *The Pillars of priestcraft shaken. The church of England and its enemies, 1660–1730* (Cambridge University Press, 1992).

Charbonnel, N. and M. Morbito, "Les rivages de la mer: droit romain et glossateurs," in *Revue historique de droit français et étranger* 65 (1987), 23–44.

Charles, L. et al. (eds.), *Le cercle de Vincent de Gournay. Savoirs économiques et pratiques administratives en France au milieu du XVIIIe siécle* (Paris: INED, 2011).

Cheney, P. *Revolutionary commerce. Globalization and the French monarchy* (Cambridge/Mass.: Harvard University Press, 2010).

Chérif, M. H. *Pouvoir et société dans la Tunisie de H'usayn Bin'Alî (1705–1740)*, 2nd ed., 2 vol. (Tunis: Centre de Pub. Univ., 2008).

Christian, L. G. *Theatrum mundi: The history of an idea* (New York: Garland, 1987).

Christian-Muslim Relations. A bibliographical history (Brill online resources).

Church, W. F. *Richelieu and reason of state* (Princeton University Press, 1972).

Cifoletti, G. *La lingua franca mediterranea* (Padua: Unipress, 1989).

Cipollone, G. *La casa della Santa Trinità di Marsiglia (1202–1547). Prima fondazione sul mare dell'ordine trinitario* (Città del Vaticano: BAV, 1981).

Cipollone, G. "Contributi attorno all'attività redentiva dell'ordine trinitario svolta nei secoli XIII e XIV presso alcune fondazioni costiere del Mediterraneo occidentale e del Portogallo," in *Captivis libertas. Congresso dell'apostolato redentivo-misericordioso dell'Ordine trinitario* (Rome: Centro Trinitario, 1983).

Cipollone, G. *Cristianità – Islam. Cattività e liberazione in nome di Dio. Il tempo di Innocenzo III dopo 'il 1187'* (Rome: Ed. Pont. Univ. Gregoriana, 1992).

Cipollone, G. (ed.), *La liberazione dei 'captivi' tra cristianità e Islam. Oltre la crociata e il ğihād: Tolleranza e servizio umanitario* (Città del Vaticano: BAV, 2000).

Clark, H. C. *Compass of society: commerce and absolutism in old-regime France* (Lanham: Lexington, 2007).

Cochrane, E. *Historians and historiography in the Renaissance* (Chicago University Press, 1981).

Cole, C. W. *French Mercantilist doctrines before Colbert* (New York: R. R. Smith, 1931).

Cole, C. W. *Colbert and a century of French Mercantilism*, 2 vol. (New York: Columbia University Press, 1939).

Colley, L. *Britons. Forging the nation* (New Haven: Yale University Press, 1992).

Colley, L. *Captives. Britain, empire and the world, 1600–1850* (London: Cape, 2004).

Colli, V. "Le opere di Baldo: Dal codice d'autore all'edizione a stampa," in C. Frova et al. (eds.), *VI Centenario della morte di Baldo degli Ubaldi 1400–2000* (Perugia: Università degli studi, 2005), 25–85.

Collingridge, D. *The social control of technology* (New York: St. Martin's, 1980).

Collini, S. and A. Vannoni, *Les instructions scientifiques pour les voyageurs XVIIe-XIXe siècle* (Paris: L'Harmattan, 2005).

Collins, J. R. "Thomas Hobbes and the Blackloist conspiracy of 1649," in *The Historical Journal* 45, 2 (2002), 305–331.

Conrad, L. I. "On the Arabic chronicle of Bar Hebraeus," in *Parole de l'Orient* 19 (1994), 319–378.

Constantine, S. *Community and identity. The making of modern Gibraltar since 1704* (Manchester University Press, 2009).

Cook, H. J. *Matters of exchange: commerce, medicine, and science in the Dutch Golden Age* (New Haven: Yale University Press, 2007).

Cooper, A. *Inventing the indigenous. Local knowledge and natural history in early modern Europe* (Cambridge University Press, 2007).

Cornwall, R. D. "Divine right monarchy. Henry Dodwell's critique of the Reformation and defense of the deprived Nonjurors Bishops," in *Anglican & episcopal history* 68 (1999), 37–66.

Cornwall, R. D. "The search for the primitive church: The use of early church fathers in the High Church Anglican tradition, 1680–1745," in *Anglican and Episcopal History* 59, 3 (1990), 303–329.

Cornwall, R. D. *Visible and apostolic. The constitution of the church in High Church Anglican and Non-Juror thought* (Newark: Univ. of Delaware Press, 1993).

Cotroneo, G. *I trattatisti dell'Ars historica* (Naples: Giannini, 1971).

Coudert, A. P. *The language of Adam. Die Sprache Adams* (Wiesbaden: Harrassowitz, 1999).

Crahay, R. "Réflexions sur le faux historique: le cas d'Annius de Viterbe," in *Académie Royale de Belgique. Bulletin de la Classe des Lettres et des Sciences Morales et Politiques* 69, 4/5 (1982), 241–267.

Crahay, R. "Pays, peuples et sociétés dans la *République* de Jean Bodin, contemporain de Lodovico Guicciardini," in P. Jodogne (ed.), *Lodovico Guicciardini (1521–1589)* (Louvain: Peeters, 1991), 249–272.

Cremer, A. *Der Adel in der Verfassung des Ancien Régime. Die Châtellenie d'Epernay und die Souveraineté de Charleville im 17. Jahrhundert* (Bonn: Bouvier, 1981).

Croarken, M. "Tabulating the heavens: Computing the *Nautical Almanac* in 18th-century England," in *IEEE Annals of the History of Computing* 2003, 48–61.

Cuming, G. J. "Eastern liturgies and Anglican divines 1510–1662," in D. Baker (ed.), *Orthodox churches and the West* (Oxford University Press, 1976), 231–238.

Curtis, C. D. *Blake, General-at-Sea* (Taunton: Wessex Press, 1934).

Curtis, M. *European thinkers on Oriental despotism in the Middle East and India* (Cambridge University Press, 2009).

Curtius, E. R. "Topica: Theatrum mundi," in *Romanische Forschungen* 55 (1941), 165–183.

Cutinelli-Rèndina, E. *Guicciardini* (Rome: Salerno, 2009).

Dagnaud, G. "L'administration centrale de la Marine sous l'Ancien Régime," in *Revue maritime* 193 (1912), 321–340, 712–736; 194 (1912), 22–47, 298–314, 623–649.

Darnton, R. *The forbidden best-sellers of pre-revolutionary France* (New York: Norton, 1995).

Daston, L. and K. Park, *Wonders and the order of nature: 1150–1750* (New York: Zone, 1998).

Daston, L. and M. Stolleis, *Natural law and laws of nature in early modern Europe. Jurisprudence, theology, moral and natural philosophy* (Aldershot: Ashgate, 2008).

Daston, L. *Classical probability in the Enlightenment* (Princeton University Press, 1988).

Daussy, H. *Les Huguenots et le roi. Le combat politique de Philippe Duplessis-Mornay (1572–1600)* (Geneva: Droz, 2002).

Davey, C. *Pioneer for Unity. Metrophanes Kritopoulos (1589–1639) and relations between the orthodox, Roman Catholic and Reformed churches* (Warrington: Hutson, 1987).

Davey, C. "Metrophanes Kritopoulos and his studies at Balliol College from 1617 to 1622," in P. Doll (ed.), *Anglicanism and orthodoxy. 300 years after the 'Greek College' in Oxford*, Oxford et al.: Lang, 2006), 57–77.

Davis, R. *Aleppo and Devonshire square: English traders in the Levant in the eighteenth century* (London/New York: Routledge, 1967).

Davis, R. "English foreign trade, 1660–1700," in *The Economic History Review* 7 (1954), 150–166.

Davis, R. "English foreign trade, 1700–1774," in *The Economic History Review* 15 (1962), 285–303.

Davis, R. *The rise of the English shipping industry in the seventeenth and eighteenth centuries* (first ed. 1962) (St. John's: Maritime Studies Research, 2012).

Dawson, N.-M. *L'Atlier Delisle. L'Amérique du Nord sur la table à dessin* (Sillery / Paris: Septentrion, 2000).

Debbasch, Y. *La nation française en Tunisie (1577–1835)* (Paris: Sirey, 1957).

del Treppo, M. *I mercanti catalani e l'espansione della Corona d'Aragona nel secolo XV* (Naples: L'arte tipografica, 1972).

Delattre, A. *Souvenirs de la croisade de Saint-Louis trouvés à Carthage 1876–1894* (Lyon: Mougin-Rusand, 1888).

Delbourgo, J. and N. Dew (eds.), *Science and empire in the Atlantic World* (New York: Routledge, 2008).

Demeleunaere-Dopuyère, C. and D. J. Sturdy (eds.), *L'Enquête du Régent 1716–1718. Sciences, Techniques et Politique dans la France pré-industrielle* (Turnhout: Brepols, 2008).

Demonet, M.-L. *Les voix du signe: Nature et Origine du langage à la Renaissance (1480–1580)* (Paris: Persée, 1992).

Denis, V. *Une Histoire de l'identité. France, 1715–1815* (Seyssel: Champ Vallon, 2008).

Deringer, W. "Finding the money: Public accounting, political arithmetic, and probability in the 1690s," in *Journal of British Studies* 52, 3 (2013), 638–668.

Desgraves, L. *Répertoire des ouvrages de controverse entre catholiques et protestants en France: (1598 – 1685)*, 2 vol. (Geneva: Droz, 1984/85).

Deslandres, P. *L'Ordre des trinitaires pour le rachat des captifs*, 2 vol. (Toulouse and Paris: Privat and Plon, 1903).

Dessert, D. *Argent, pouvoir et société au Grand Siècle* (Paris: Fayard, 1984).

Dessert, D. *La Royale. Vaisseaux et marins du Roi-Soleil* (Paris: Fayard, 1996).

Dessert, D. *Colbert ou le serpent vénimeux* (Brussels: Complexe, 2000).

Dew, N. *Orientalism in Louis XIV's France* (Oxford University Press, 2009).

Díaz Borrás, A. "Notas sobre los primeros tiempos de la atención valenciana a la redención de cautivos cristianos (1323–1399)," in *Estudis Castellonencs* 3 (1986), 337–354.

Díaz Borrás, A. *El miedo al Mediterráneo: La caridad popular Valenciana y la redención de cautivos bajo poder musulmán, 1323–1539* (Barcelona: CSIC, 2001).

Dickey, L. "Power, commerce and natural law in Daniel Defoe's political writings 1698–1707," in J. Robertson, *A Union for Empire. Political Thought and the British Union of 1707* (Cambridge University Press, 1995), 63–96.

Dingli, L. *Colbert, marquis de Seignelay. Le fils flamboyant* (Paris: Perrin, 1997).

Dobbs, B. J. "Studies in the natural philosophy of Sir Kenelm Digby," in *Ambix* 18 (1971), 1–25.

Dockès-Lallement, N. "Les républiques sous l'influence des nombres: le hasard et la nécessité chez Jean Bodin," in *L'œuvre de* Jean Bodin (Paris: Champion, 2004), 127–149.

Doll, P. "The idea of primitive church in High Church ecclesiology from Samuel Johnson to J. H. Hobart," in *Anglican and Episcopal History* 65, 1 (1996), 6–43.

Doll, P. (ed.), *Anglicanism and Orthodoxy. 300 years after the "Greek College" in Oxford* (Oxford et al.: Lang, 2006).

Döring, D. "Die Leipziger gelehrten Sozietäten in der ersten Hälfte des 18. Jahrhunderts und das Auftreten Johann Christoph Gottscheds," in E. Donnert, *Europa in der Frühen Neuzeit FS Günter Mühlpfordt, vol. 5: Aufklärung in Europa* (Cologne et al.: Böhlau, 1999), 17–42.

Döring, D. "Die sächsische Afrikaexpedition von 1731 bis 1733," in P. Pretsch and V. Steck (eds.), *Eine Afrikareise im Auftrag des Stadtgründers. Das Tagebuch des Karlsruher Hofgärtners Christian Thran 1731–1733* (Karlsruhe: Info Verlag, 2008), 43–56.

Downie, J. A. "Defoe, imperialism, and the travel books reconsidered," in R. D. Lund (ed.), *Critical essays on Daniel Defoe* (London et al.: Prentice Hall, 1997), 78–96.

Drapeyron, L. "Un projet français de conquête de l'Empire ottoman au XVIe et au XVIIe siècle," in *Revue des Deux Mondes* 126 (1876), 122–147.

Drayton, R. *Nature's government. Science, imperial Britain, and the 'improvement' of the world* (New Haven and London: Yale University Press, 2000).

Drønen, T. S. "Scientific revolution and religious conversion: A closer look at Thomas Kuhn's theory of paradigm-shift," in *Method & Theory in the Study of Religion* 18 (2006), 232–253.

Dubois, C.-G. *La conception de l'histoire en France au XVIe siècle (1560–1610)* (Paris: Nizet, 1977).

Dubos, N. *Thomas Hobbes et l'histoire. Système et récits à l'âge classique* (Paris: PU Sorbonne, 2014).

Dubost, J.-F. and P. Sahlins, *Et si on faisait payer les étrangers? Louis XIV, les immigrés et quelques autres* (Paris: Flammarion, 1999).

Dufourcq, C. E. *L'expansió catalana a la Mediterrania occidental. Segles XIII i XIV* (Barcelona: Vicens-Vives, 1969).

Dufourcq, C. E. "Les relations de la Peninsule Ibérique et de l'Afrique du Nord au XIVe siècle," in *Anuario de Estudios Medievales* 7 (1970/71), 39–63.

Dufourcq, C. E. "Un imperialisme médieval face au Maghreb: la naissance et l'essor de l'empire catalan d'après des travaux récents," in *Cahiers de Tunisie* 20 (1972), 101–124.

Dufourcq, C. E. "La place du Maghreb dans l'expansion de la Couronne d'Aragon: la route maghrebine par rapport à celle des îles et des épices," in M. Galley (ed.), *Deuxième Congrès International d'Études des Cultures de la Méditerranée Occidentale* (Alger: Société nat. d'édition, 1978), 217–279.

Dunn, R. and R. Higgitt, *Ships, clocks, and stars. The quest for longitude* (New York: HarperCollins, 2014).

Dupilet, A. *La régence absolue. Philippe d'Orléans et la polysynodie (1715–1718)* (Seyssel: Champ Vallon, 2011).

Dupront, A. *Le mythe de croisade*, 4 vol. (Paris: Gallimard, 1997).

Dwyer, E. "Science or morbid curiosity? The casts of Giuseppe Fiorelli and the last days of Romantic Pompeii," in V. C. Gardner Coates (ed.), *Antiquity recovered: The legacy of Pompeii and Herculaneum* (Los Angeles: Getty, 2007), 171–188.

Eldem, E. *French trade in Istanbul in the eighteenth century* (Leiden and Boston: Brill, 1999).

Emerit, M. "Un Mémoire sur Alger par Pétis de la Croix (1695)," in *Annales de l'Institut d'études orientales* 11 (1953), 6–25.

Encyclopedia of Islam, 2nd ed. (*Brill online sources*).

Engerman, S. L. "Mercantilism and overseas trade, 1700–1800," in R. Floud and D. McCloskey (eds.), *The Economic History of Britain since 1700*, 2nd ed., vol. 1: *1700–1860*, (Cambridge University Press, 1994), 182–204.

Esmonin, E. "Observations sur le Testament politique de Richelieu," in *Études sur la France des XVIIe et XVIIIe siècles* (Paris: PUF, 1964), 219–232.

Famin, C. *Histoire de la rivalité et du protectorat des églises chrétiennes en Orient* (Paris: Firmin Didot, 1853).

Faure, E. *La banqueroute de Law* (Paris: Gallimard, 1977).

Feingold, M. "Isaac Barrow: Divine, scholar, mathematician," in *Before Newton: The life and times of Isaac Barrow* (Cambridge University Press, 1990), 1–104.

Feingold, M. "Oriental Studies," in N. Tyacke (ed.), *The history of the university of Oxford, vol. IV: Seventeenth-century Oxford* (Oxford University Press, 1997), 449–504.

Feingold, M. "The Accademia del Cimento and the Royal Society," in M. Beretta et al. (eds.), *The Accademia del Cimento and its European context* (Sagamore Beach: Watson, 2009), 229–242.

Féraud-Giraud, L.-J.-D. *De la juridiction française dans les échelles du Levant et de Barbarie*, 2nd ed., 2 vol. (Paris: Durand and Pedone Lauriel, 1871).

Ferrer i Mallol, M. T. "Genoese merchants in Catalan lands," in F. A. Congdon (ed.), *Latin expansion in the medieval Western Mediterranean* (Aldershot: Ashgate, 2013), 69–88.

Ferris, J. P. "A connoisseur's shopping-list, 1647," in *Journal of the Warburg and Courtauld Institutes* 38 (1975), 339–341.

Finkelstein, A. *Harmony and the balance. An intellectual history of the seventeenth-century economic thought* (Ann Arbor: Univ. of Michigan Press, 2000).

Firth, K. R. *The apocalyptic tradition in Reformation Britain, 1530–1654* (Oxford University Press, 1979).

Fleischer, C. "Royal authority, dynastic cyclicsm and 'Ibn Khaldūnism' in sixteenth-century Ottoman letters," in *Journal of Asian and African Studies* 18 (1983), 198–210.

Fletcher, J. E. *A study of the life and works of Athanasius Kircher, "Germanus Incredibilis"* (Leiden and Boston: Brill, 2011).

Ford, J. D. "William Welwod's treatises on maritime Law," in *The Journal of Legal History* 34, 2 (2013), 172–210.

Forey, A. J. "The military orders and the ransoming of captives from Islam (12th to early 14th centuries)," in *Studia monastica* 33 (1991), 250–279.

Fox, H. S. A. "Exploitation of the landless by Lords and tenants in early medieval England," in Z. Razi and R. Smith (eds.), *Medieval society and the Manor Court* (Oxford University Press, 1996), 518–568.

Frangakis-Syrett, E. *The commerce of Smyrna in the eighteenth century, 1700–1820* (Athens: s.n., 1992).

Freller, T. *The Epitome of Europe. Das Bild Maltas und des Ordensstaats der Johanniter in der Reiseliteratur der Frühen Neuzeit* (Frankfurt/M et al.: Lang, 2002).

French, C. J. "'Crowded with traders and a great commerce': London's domination of English overseas trade, 1700–1775," in *London Journal* 17, 1 (1992), 27–35.

French, C. J. "London's overseas trade with Europe 1700–1775," in *Journal of European Economic History* 23, 3 (1994), 475–501.

Friedman, Y. *Encounter between enemies. Captivity and ransom in the Latin kingdom of Jerusalem* (Boston and Leiden: Brill, 2002).

Frigo, D. (ed.), *Politics and diplomacy in early modern Italy. The structure of diplomatic practice, 1450–1800* (Cambridge University Press, 2000).

Fukasawa, K. *Toilerie et commerce du Levant d'Alep à Marseille* (Paris: CNRS, 1987).

Gabrieli, V. *Sir Kenelm Digby. Un inglese italianato nell'età della Controriforma* (Rome: Ed. Storia e lett., 1957).

Garner, G. "L'enquête Orry de 1745 et les villes de la France septentrionale: valeur et finalité d'une statistique administrative," in *Revue du Nord* 79 (1997), 357–379.

Gascoigne, J. *Science in the service of empire: Joseph Banks, The British state and the uses of science in the age of revolution* (Cambridge University Press, 1998).

Gascoigne, J. "'The wisdom of the Egyptians' and the secularisation of history in the age of Newton," in *Science, philosophy and religion in the age of Enlightenment* (Aldershot: Ashgate, 2010), I, 171–212.

Gauci, P. *The politics of trade. The overseas merchant in state and society, 1660–1720* (Oxford University Press, 2001).

Gauci, P. *Emporium of the world. The merchants of London 1660–1800* (London: Hambledon, 2007).

Georgi, C. R. A. *Die Confessio Dosithei (Jerusalem 1672). Geschichte, Inhalt und Bedeutung* (Munich: Reinhardt, 1940).

Ghalem, M. "Historiographie algérienne du XVIIIe siècle: savoir historique et mode de légitimation politique," in S. Bargaoui and H. Remaoun (eds.), *Savoirs historiques au Maghreb. Construction et usages* (Tunis-Oran: CERES, 2006), 133–145.

Gøbel, E. "The Danish *Algerian* sea passes, 1747–1838: An example of extraterritorial production of *Human Security*," in *Historical Social Research* 35, 4 (2010), 164–189.

Godwin, J. *Athanasius Kircher's Theatre of the World* (London: Thames and Hudson, 2009).

Goetz, H.-W. "Die Anfänge der historischen Methoden-Reflexion in der italienischen Renaissance und ihre Aufnahme in der Geschichtsschreibung des deutschen Humanismus," in *Archiv für Kulturgeschichte* 56, 1 (1974), 25–48.

Goetz, H.-W. *Geschichtsschreibung und Geschichtsbewußtsein im hohen Mittelalter* (Berlin: Akademie, 1999).

Goffman, D. *The Ottoman Empire and early modern Europe* (Cambridge University Press, 2002).

Goldish, M. *Judaism in the theology of Sir Isaac Newton* (Dordrecht and Boston: Kluwer, 1998).

González Arévalo, R. "Rapporti commerciali tra Firenze e il Regno di Granada nel XV secolo," in L. Tanzini and S. Tognetti, *'Mercatura è arte'* - *Uomini d'affari toscani in Europa e nel Mediterraneo tardomedievale* (Rome: Viella, 2012), 179-203.

Grafton, A. *Joseph Scaliger. A study in the history of classical scholarship*, 2 vol. (Oxford University Press, 1983-1993).

Grafton, A. "Invention of traditions and traditions of invention in Renaissance Europe: The strange case of Annius of Viterbo," in A. Grafton and A. Blair (eds.), *The Transmission of Culture in Early Modern Europe* (Philadelphia: Pennstate University Press 1990), 8-38.

Grafton, A. "Isaac Casaubon on Hermes Trismegistus," in *Journal of the Warburg and Courtland Institute* 46 (1983), 78-93.

Grandchamp, P. *La France en Tunisie au XVIIe siécle*, 10 vol. (Tunis: Imprimérie rapide, 1920-1933).

Grangaud, I. *La ville imprenable. Une histoire sociale de Constantine au 18e siècle* (Paris: EHESS, 2002).

Gray, D. *Chaplain to Mr Speaker: The religious life of the House of Commons* (London: H.M.S.O., 1991).

Gray, T. "Turkish piracy and early Stuart Devon," in *Report and Transactions of the Devonshire Association for the Advancement of Science, Literature and the Arts* 121 (1989), 159-171.

Greenberg, J. L. *The problem of the earth's shape from Newton to Clairaut. The rise of mathematical science in eighteenth-century Paris and the fall of 'normal' science* (Cambridge University Press, 1995).

Greene, M. *A shared world: Christians and Muslims in the early modern Mediterranean* (Princeton University Press, 2000).

Greene, M. *Catholic pirates and Greek merchants: A maritime history of the Mediterranean* (Princeton University Press, 2010).

Grell, C. *Herculaneum et Pompéi dans les récits des voyageurs français du XVIIIe siècle* (Rome et al.: École française, 1982).

Grell, C. (ed.), *Les historiographes en Europe de la fin du Moyen Âge à la Révolution* (Paris: PU Sorbonne, 2006).

Grillon, P. *Un chargé d'affaires au Maroc. La correspondance du consul Louis Chénier (1767-1782)*, 2 vol. (Paris: SEVPEN, 1970).

Gross, M. *Ignorance and surprise. Science, society and ecological design* (Cambridge, MA: MIT, 2010).

Guenée, B. "Histoires, annales, chroniques. Essai sur les genres historiques au Moyen Âge," in *Annales ESC* 28 (1973), 997-1016.

Hacking, I. *The emergence of probability. A philosophical study of Early Ideas about Probability, Induction and statistical inference* (Cambridge University Press, 1976).

Haguet, L. "J.-B. d'Anville as Armchair Mapmaker: The Impact of Production Contexts on His Work," in *Imago mundi* 63 (2011), 88–105.

Hald, A. *A history of probability and statistics and their application before 1750* (New York: Wiley, 1990).

Hamilton, A. "The English interest in the Arabic-speaking Christians," in G. A. Russell (ed.), *The "Arabick" interest of the natural philosophers in seventeenth-century England* (Leiden and Boston: Brill, 1994), 30–54.

Hamilton, A. et al. (eds.), *The republic of letters and the Levant* (Leiden and Boston: Brill, 2005).

Hamilton, A. *The Copts and the West, 1439–1822. The European discovery of the Egyptian Church* (Oxford University Press, 2006).

Hamilton, A. "Isaac Casaubon the Arabist: 'Video longum esse iter'," in *Journal of the Warburg and Courtauld Institutes* 72 (2009), 143–168.

Hamilton, A. "From East to West: Jansenists, Orientalists, and the Eucharistic controversy," in W. Otten et al. (eds.), *How the West was won: Essays on literary imagination, the canon, and the Christian Middle Ages for Burcht Pranger* (Leiden and Boston: Brill, 2010), 83–100.

Haran, A. Y. *Le lys et le globe. Messianisme dynastique et rêve impérial en France aux XVIe et XVIIe siècles* (Seyssel: Champ Vallon, 2000).

Harding, A. "The medieval briefes of protection and the development of the Common Law," in *Juridical Review* n.s. 2 (1966), 115–149.

Harding, N. B. "North African piracy, the Hanoverian carrying trade, and the British state, 1728–1828," in *The Historical Journal* 43, 1 (2000), 25–47.

Harmsen, T. "High-principled antiquarian publishing: The correspondence of Thomas Hearne (1678–1735) and Thomas Smith (1638–1710)," in *Lias* 23 (1996), 1–29.

Harper, L. A. *The English navigation laws. A seventeenth-century experiment in social engineering* (1939, repr. New York: Octagon, 1973).

Harrison, M. "Science and the British Empire," in *ISIS* 96 (2005), 56–63.

Hart, A. R. *A history of the king's serjeants at law in Ireland: Honour rather than advantage?* (Dublin: Four Courts Press, 2000).

Hauser, H. *La pensée et l'action économiques du Cardinal de Richelieu* (Paris: puf, 1944).

Haycock, D. B. *William Stukeley. Science, religion and archaeology in eighteenth-century England* (Woodbridge: Boydell, 2002).

Head, R. "Knowing like a state: The transformation of political knowledge in Swiss Archives, 1450–1770," in *Journal of Modern History* 75 (2003), 745–782.

Hebb, D. D. *Piracy and the English government, 1616–1642* (Aldershot: Ashgate, 1994).

Heckscher, E. F. *Mercantilism* (first ed. Swedish 1931), transl. M. Shapiro, introd. L. Magnusson, 2 vol. (London and New York: Routledge, 1994).

Heers, J. "Gênes et l'Afrique du Nord vers 1450: les voyages 'per costeriam'," in *Anuario de Estudios Medievales* 21 (1991), 233–246.

Heers, J. *The Barbary Corsairs. Warfare in the Mediterranean, 1480–1580* (London: Greenhill, 2003).

den Heijer J., "Coptic historiography in the Fāṭimid, Ayyūbid and early Mamlūk periods," in *Medieval Encounters* 2 (1996), 67–98.

Hein, O. *Athanasius Kircher S.J. in Malta* (Berlin: Akademie, 1997).

van Helden, A. "Longitude and the satellites of Jupiter," in W. J. H. Andrewes (ed.), *The Quest*, 86–100.

Henry, J. "Sir Kenelm Digby, recusant philosopher," in G. A. J Rogers et al. (eds.), *Insiders and outsiders in seventeenth-century philosophy* (New York: Routledge, 2010), 43–75.

Hering, G. *Ökumenisches Patriarchat und europäische Politik 1620–1638* (Wiesbaden: Steiner, 1968).

Hering, G. "Orthodoxie und Protestantismus," in Idem, *Nostos: gesammelte Schriften zur südosteuropäischen Geschichte* (Frankfurt/M et al.: Lang, 1995), 73–130.

Hering, G. *Οἰκουμένικο Πατριαρχεῖο καὶ Εὐρωπαϊκή Πολιτική 1620–1638*, transl. D. Kourtobik (Athens: Morphotiko Hidryma Ethnikes Trapezes, 2003).

Hershenzon, D. "Plaintes et menaces: captivité et violences religieuses en Méditerranée au XVIIe siècle," in J. Dakhlia and W. Kaiser (eds.), *Les musulmans dans l'histoire de l'Europe. II. Passages et contacts en Méditerranée* (Paris: A. Michel, 2013), 441–460.

Heyberger, B. *Les chrétiens du Proche-Orient au temps de la Réforme catholique* (Rome: École française de Rome, 1994).

Heyberger, B. "'Pro nunc nihil est respondendum'. Recherche d'information et prise de décision à la Propagande: l'exemple du Levant (XVIIIe siècle)," in *Mélanges de l'École française de Rome (Italie et Méditerranée)* 109/2 (1997), 539–554.

Heyberger, B. "Abraham Ecchellensis dans la République des Lettres," in *Orientalisme, Science et Controverse: Abraham Ecchellensis (1605–1664)* (Turnhout: Brepols, 2010), 9–52.

Heyberger, B. "L'Islam et les Arabes chez un érudit Maronite au service de l'Église catholique," in *Al-Qantara* 28 (2010), 481–512.

Heywood, C. "Ideology and the profit motive in the Algerine corso," in C. Vassallo, M. d'Angelo (eds.), *Anglo-Saxons in the Mediterranean.*

Commerce, Politics and Ideas (XVII-XX Centuries) (Msida: Malta University Press, 2007), 16–42.

Hilaire de Barenton, E. M. B. La France catholique en Orient durant les trois derniers siècles (Paris: Œuvre de St. François and Poussielgue, 1902).

Hirsch, E. C. Der berühmte Herr Leibniz. Eine Biographie (Munich: Beck, 2000).

Hitzel, F. (ed.), Istanbul et les langues orientales (Paris: L'Hamattan, 1997).

Holt, P. M. "The study of Arabic historians in seventeenth century England: The background and the work of Edward Pococke," in Bulletin of the School of Oriental and African Studies 19 (1957), 444–455.

Homsy, B. Les capitulations & la protection des chrétiens au Proche-Orient aux XVIe, XVIIe et XVIIIe siècles (Harissa: St. Paul, 1956).

Hont, I. Jealousy of trade. International competition and the nation-state in historical perspective (Cambridge, MA.: Harvard University Press, 2005).

Hoon, E. E. The organization of the English customs system 1696–1786 (New York and London: Appleton-Century, 1938).

Hoppit, J. A land of liberty? England 1689–1727 (Oxford University Press, 2000).

Hoquet, T. Buffon: histoire naturelle et philosophie (Paris: Champion, 2005).

Horden, P. and N. Purcell, The corrupting sea: A study of Mediterranean history (Oxford University Press, 2000).

Houssaye Michienzi, I. Datini, Majorque et le Maghreb (14e-15e siècles). Réseaux, espaces méditerranéens et stratégies marchandes (Leiden and Boston: Brill, 2013).

Hunter, M. John Aubrey and the realm of learning (New York: Science History, 1975).

Hunter, M. Science and society in Restoration England (Cambridge University Press, 1981).

Hunter, M. Establishing the new science. The experience of the early Royal Society (Woodbridge: Boydell, 1989).

Hunter, M. The Royal society and its fellows (Chalfont St. Giles, Bucks: BSHS, 1994).

Hunter, M. "Robert Boyle and the early Royal Society: A reciprocal exchange in the making of Baconian science," in The British Journal for the History of Science 40, 1 (2007), 1–23.

Iliffe, R. "Making correspondents network. Henry Oldenburg, philosophical commerce, and Italian science, 1660–1672," in M. Beretta et al. (eds.), The Accademia del Cimento and its European context (Sagamore Beach: Watson, 2009), 211–228.

Isenmann, M. (ed.), Merkantilismus. Wiederaufnahme einer Debatte (Stuttgart: F. Steiner, 2014).

INED. *Économie et population. Les doctrines françaises avant 1800* (Paris: INED, 1956).

Israel, J. I. *Dutch primacy in world trade, 1585–1740* (Oxford University Press, 1989).

Israel, J. I. *Conflicts of empires: Spain, the Low Countries and the struggle for world supremacy, 1585–1713* (London et al.: Rio Grande, 1997).

van Ittersum, M. J. "*Mare liberum* versus the propriety of the seas? – The debate between Hugo Grotius (1583–1645) and William Welwood (1552–1624)," in *Edinburgh Law Review* 10 (2006), 239–276.

van Ittersum, M. J. *Profit and principle: Hugo Grotius: Natural rights theories and the rise of Dutch power in the East Indies, 1595–1615* (Boston and Leiden: Brill, 2006).

Ivic, C. and G. Williams (eds.), *Forgetting in early modern English literature and culture. Lethe's legacies* (London and New York: Routledge, 2004).

Jacob, C. *The sovereign map. Theoretical approaches in cartography throughout history* (Chicago University Press, 2006).

Jacobowsky, C. V. *J. G. Sparwenfeld. Bidrag till en biografi* (Stockholm: Lindberg, 1932).

Jaspert, N. and S. Kolditz (eds.), *Seeraub im Mittelmeerraum. Piraterie, Korsarentum und maritime Gewalt von der Antike bis zur Neuzeit* (Munich: Fink, 2013).

Joachimsen, P. *Geschichtsauffassung und Geschichtsschreibung in Deutschland unter dem Einfluss des Humanismus* (Leipzig: Teubner, 1910).

Jones, A. "Decompiling Dapper: A preliminary search for evidence," in *History in Africa* 17 (1990), 171–209.

Jouhaud, C. *Mazarinades: la Fronde des mots* (Paris: Aubier, 1985).

Jugie, M. "Le mot transsubstantiation chez les Grecs avant 1629 et après 1629," in *Échos d'Orient* 10 (1907), 5–12, 65–77.

Julliany, J. *Essai sur le Commerce de Marseille*, 2nd ed., vol. 1 (Marseille and Paris: Barile and Lib. du Commerce, 1842).

Kablitz, A. "Kunst des Möglichen: Prolegomena zu einer Theorie der Fiktion," in *Poetica* 35 (2003), 251–273.

Kaiser, W. "Vérifier les histoires, localiser les personnes. L'identification comme processus de communication en Méditerranée (XVIe-XVIIe siècles)," in W. Kaiser and C. Moatti (eds.), *Gens de passage en Méditerranée de l'Antiquité à l'époque moderne. Procédures de contrôle et d'identification* (Paris: Maisonneuve & Larose, 2007), 369–381.

Kaiser, W. and C. Moatti (eds.), *Gens de passage en Méditerranée de l'Antiquité à l'époque moderne. Procédures de contrôle et d'identification* (Paris: Maisonneuve & Larose, 2007).

Kammerling Smith, D. "Le discours économique du Bureau du commerce, 1700–1750," in L. Charles et al. (eds.), *Le cercle de Vincent de Gournay. Savoirs économiques et pratiques administratives en France au milieu du XVIIIe siécle* (Paris: INED, 2011), 31–61.

Kappler, É. *Les conférences théologiques entre catholiques et protestants en France au XVIIe siècle* (Paris: Champion, 2011).

Karmires, J. N. *Μητροφάνης ὁ Κριτόπουλος καὶ ἡ ἀνέκδοτος ἀλληλογραφία αὐτοῦ* (Athens: Theologike bibliotheke, 1937).

Kaser, M. and R. Knütel, *Römisches Privatrecht*, 19th ed. (Munich: Beck, 2008).

Katzer, A. *Araber in deutschen Augen. Das Araberbild der Deutschen vom 16. bis zum 19. Jh.* (Paderborn: Schöningh, 2008).

Kaufmann, T. *Die Abendmahlstheologie der Straßburger Reformatoren bis 1528* (Tübingen: Mohr & Siebeck, 1992).

Kempe, M. *Wissenschaft, Theologie, Aufklärung. Johann Jakob Scheuchzer (1672–1733) und die Sintfluttheorie* (Epfendorf: Bib. Academica, 2003).

Kempe, M. "Beyond the law. The image of piracy in the legal writings of Hugo Grotius," in *Grotiana* 26–28 (2005–2007), 379–395.

Kempe, M. *Fluch der Weltmeere. Piraterie, Völkerrecht und internationale Beziehungen* (Frankfurt/M: Campus, 2010).

Kepler, J. S. "Fiscal aspects of the English carrying trade during the thirty years war," in *The Economic History Review* 25, 2 (1972), 261–283.

Keppel, T. *The life of Augustus Viscount Keppel*, 2 vol. (London: Colburn, 1842).

Kessler, E. (ed.), *Theoretiker humanistischer Geschichtsschreibung* (Munich: Fink, 1979).

Kessler, E. "Ars historica," in *Historisches Wörterbuch der Rhetorik* 1 (1992), col. 1046–1048.

Kidd, C. *British identities before nationalism. Ethnicity and nationhood in the Atlantic World, 1600–1800* (Cambridge University Press, 1999).

Kim, M. G. *Affinity, that elusive dream: A genealogy of the chemical revolution* (Cambridge, MA: MIT, 2003).

Kinkel, S. "The king's pirates? Naval enforcement of imperial authority, 1740–46," in *The William and Mary Quarterly* 71, 1 (2014), 3–34.

Kitromilides, P. M. "Orthodoxy and the West: Reformation to Enlightenment," in M. Angold (ed.), *The Cambridge history of Christianity, vol. 5: Eastern Christianity* (Cambridge University Press, 2006), 187–209.

Klemme, H. F. "Werde vollkommen! Christian Wolff's Vollkommenheitsethik in systematischer Perspektive," in J. Stolzenberg and O.-P. Rudolph (eds.), *Christian Wolff und die europäische Aufklärung*, vol. 3 (Hildesheim et al.: Olms, 2007), 163–180.

Köhler, W. *Zwingli und Luther: Ihr Streit über das Abendmahl nach seinen politischen und religiösen Beziehungen*, 2 vol. (first ed. 1924, repr. New York and London: Johnson, 1971).

Konvitz, J. W. *Cartography in France 1660–1848. Science, engineering, and statecraft* (Chicago University Press, 1987).

Koselleck, R. "Geschichte, Historie V–VII," in O. Brunner et al. (eds.), *Geschichtliche Grundbegriffe. Historisches Lexikon zur politisch-sozialen Sprache in Deutschland*, vol. 2 (Stuttgart: Klett, 1975), 647–717.

Koselleck, R. *Futures past: on the semantics of historical time* (Cambridge University Press, 1985).

Kosto, A. "Ignorance about the traveler: Documenting safe conduct in the European Middle Ages," in C. Zwierlein (ed.), *The dark side of knowledge. Histories of ignorance, 1400–1800* (Leiden/Boston: Brill, 2016), 269–295.

Krüger, L., L. J. Daston, and M. Heidelberger (eds.), *The probabilistic revolution*, vol. 1: *Ideas in history* (Cambridge, MA: MIT, 1987).

Kühlmann, W. *Gelehrtenrepublik und Fürstenstaat. Entwicklung und Kritik des deutschen Späthumanismus in der Literatur des Barockzeitalters* (Tübingen: Niemeyer, 1982).

Kumar, D. (ed.), *Science and empire: Essays in Indian context, 1700–1947* (Delhi: Anamika Prakashan, 1991).

Laboulais-Lesage, I. (ed.), *Combler les blancs de la carte. Modalités et enjeux de la construction des savoirs géographiques (XVIIe-XXe siècles)* (PU Strasbourg, 2004).

Lacey, H. "Protection and immunity in later medieval England," in T. B. Lambert and D. Rollason (eds.), *Peace and protection in the Middle Ages* (Durham and Toronto: CMRS and Pont. Inst. Med. Stud., 2009).

Ladjili, J. "La Paroisse de Tunis au XVIIIème siècle d'après les registres de catholicité," in *Revue de l'Institut des belles lettres arabes* 134 (1974), 227–277.

Lammeyer, J. *Das französische Protektorat über die Christen im Orient, historisch, rechtlich und politisch gewürdigt. Ein Beitrag zur Geschichte der diplomatischen Beziehungen der Hohen Pforte* (Bona-Leipzig: Noske, 1919).

Landfester, R. *Historia magistra vitae. Untersuchungen zur humanistischen Geschichtstheorie des 14. bis 16. Jahrhunderts* (Geneva: Droz, 1972).

Langelüddecke, H. "'I finde all men & my officers all soe unwilling': The collection of ship money, 1635–1640," in *Journal of British Studies* 46, 3 (2007), 509–542.

Larquié, C. "Le rachat des chrétiens en Terre d'Islam au XVIIème siècle (1660–1665)," in *Revue d'histoire Diplomatique* 94 (1980), 297–351.

Larquié, C. "L'Église et le commerce des hommes en Méditerranée: l'exemple des rachats de chrétiens au XVIIe siècle," in *Mélanges de la Casa de Velasquez* 22 (1986), 305–324.

Larrère, C. *L'invention de l'économie au XVIIIe siècle. Du droit naturel à la physiocratie* (Paris: PUF, 1992).

Laurens, H. *Aux sources de l'Orientalisme. La Bibliothèque orientale de Barthélemi d'Herbelot* (Paris: Maisonneuve et Larose, 1978).

Laurent, V. "L'âge d'or des Missions latines en Orient (XVIIe-XVIIIe siècle)," in *L'Unité de l'Eglise: organe du Mouvement pour le Retour des Dissidents à l'Unité Catholique* 12 (1934), 217–224, 251–255, 281–288.

Lavat, F. "Paradoxes et fictions. Les nouveaux mondes possibles à la Renaissance," in *Usages et théories de la fiction. Les théories contemporaines à l'épreuve des textes anciens* (PU Rennes, 2004), 87–111.

Le Blévec, D. "Le contexte parisien et provençal de la règle des Trinitaires," in Cipollone (ed.), *Cristianità – Islam*, 119–129.

Le Thiec, G. "L'Empire ottoman, modèle de monarchie seigneuriale dans l'œuvre de Jean Bodin," in *L'œuvre de Jean Bodin* (Paris: Champion, 2004), 127–150.

Lecuyer, B.-P. "Une quasi-expérimentation sur les rumeurs au XVIIIe siècle: l'enquête proto-scientifique du contrôleur général Orry (1745)," in R. Boudon et al. (eds.), *Science et théorie de l'opinion publique. Hommage à Jean Stoetzel* (Paris: Retz, 1981), 170–187.

Leighton, C. D. A. "Ancienneté among the non-jurors: A study of Henry Dodwell," in *History of European Ideas* 31 (2005), 1–16.

Leiner, F. C. *The end of Barbary terror. America's 1815 war against the pirates of North Africa* (Oxford University Press, 2006).

Leiser, G. "A figurative meeting of minds between seventeenth century Istanbul and Paris: The world views of Kātib Çelebī and Barthélemi d'Herbelot," in *Journal of Oriental and African Studies* 19 (2010), 73–84.

Lenci, M. "Le confraternità del riscatto nella Toscana di età moderna: il caso di Firenze," in *Archivio Storico Italiano* 167, 2 (2009), 269–297.

Lenci, M. "Toscani schiavi nel Maghreb e nell'Impero Ottomano (1565–1816): una prima valutazione quantitativa," in N. Jaspert and S. Kolditz (eds.), *Seeraub im Mittelmeerraum. Piraterie, Korsarentum und maritime Gewalt von der Antike bis zur Neuzeit* (Munich: Fink, 2013), 407–430.

Leng, T. *Benjamin Worsley (1618–1677). Trade, interest and the spirit in revolutionary England* (Woodbridge: Boydell, 2008).

Lepenies, W. *Das Ende der Naturgeschichte. Verzeitlichung und Enthistorisierung in der Wissenschaftsgeschichte des 18. und 19. Jahrhunderts* (Munich and Vienna: Hanser, 1976).

Lespagnol, A. *Messieurs de Saint-Malo. Une élite négociante au temps de Louis XIV*, 2 vol. (PU Rennes, 1997).

Lestringant, F. *Le Huguenot et le sauvage*, 3rd ed. (Geneva: Droz, 2004).

Lettinck, P. *Aristotle's meteorology and its reception in the Arab world* (Boston and Leiden: Brill, 1999).

Letwin, W. *The origins of scientific economics. English economic thought 1660–1776* (Westport: Greenwood, 1963).

Lévi-Provençal, E. *Les historiens des Chorfa. Essai sur la littérature historique et biographique au Maroc du XVIe au XXe siècle* (Paris: Larose, 1922).

Levine, J. M. *The battle of the books. History and literature in the Augustan Age* (Ithaca/London: Cornell, 1991).

Levitin, D. "From sacred history to the history of religion: Paganism, Judaism, and Christianity in European historiography from Reformation to 'Enlightenment'," in *The Historical Journal* 55, 4 (2012), 1117–1160.

Libois, C. *La compagnie de Jésus au "Levant." La province du Proche-Orient* (Beyrut: Lib.orientale, 2009).

Libois, C. (ed.), *Monumenta Proximi-Orientis,* vol. 5: *Égypte (1591–1699)* (Rome: IHSJ, 2002).

Lieberman, D. "The mixed constitution and the common law," in M. Goldie and R. Wokloer (eds.), *The Cambridge history of eighteenth-century political thought* (Cambridge University Press, 2006), 317–346.

Linn, J. B. and W. H. Egle, *Pennsylvania Archives*, 2nd ser., vol. 2 (Harrisburg: Meyers, 1876).

Littleton, C. G. D. "Ancient languages and new science. The Levant in the intellectual life of Robert Boyle," in A. Hamilton et al. (eds.), *The republic of letters and the Levant* (Leiden and Boston: Brill, 2005), 151–172.

Lo Basso, L. *A vela e a remi. Navigazione, guerra e schiavitù nel Mediterraneo (secc. XVI–XVIII)* (Ventimiglia: Philobiblon, 2004).

Loop, Jan, *Johann Heinrich Hottinger: Arabic and Islamic studies in the seventeenth century* (Oxford University Press, 2013).

López Pérez, M. D. *La corona de Aragón y el Magreb en el siglo XIV (1331–1410)* (Barcelona: CSIC, 1995).

Lucchini, E. *La merce umana: Schiavitù e riscatto dei liguri nel Seicento* (Rome: Bonacci, 1990).

Luhmann, N. "Ökologie des Nichtwissens," in *Beobachtungen der Moderne* (Opladen: westdt. Verlag, 1992).

Lüsebrinck, H.-J. et al., "Kulturtransfer im Epochenumbruch – Entwicklung und Inhalte der französisch-deutschen Übersetzungsbibliothek 1770–1815 im Überblick," in *Kulturtransfer im Epochenumbruch. Frankreich – Deutschland 1770 bis 1815*, 2 vol. (Leipzig: UV, 1997), vol. 1, 29–86.

Luttrell, A. T. "Girolamo Manduca and Gian Francesco Abela: Tradition and invention in Maltese historiography," in *Melita Historica* 7, 2 (1977), 105–132.

Lydon, J. G. *Fish and flour for gold, 1600–1800: Southern Europe in the colonial balance of payments*, Philadelphia 2008 (www.librarycompany .org/economics/lydon.htm).

Lynch, W. T. *Solomon's child: Method in the early Royal Society of London* (Stanford University Press, 2001).

MacLean, G. *The rise of Oriental travel. English visitors to the Ottoman Empire, 1580–1720* (Houndmills et al.: Palgrave Macmillan, 2004).

MacLean, G. (ed.), *Re-Orienting the Renaissance. Cultural exchanges with the East* (Houndmills et al.: Palgrave Macmillan, 2005).

MacLeod, R. (ed.), *Nature and empire: Science and the colonial enterprise, special issue of Osiris* 15 (2000).

Magnusson, L. *Mercantilism: The shaping of an economic language* (London and New York: Routledge, 1994).

Martin, T. "Une arithmétique politique française?" in T. Martin (ed.), *Arithmétique politique dans la France du XVIIIe siècle* (Paris: INED, 2003), 1–13.

Martin, T. (ed.), *Arithmétique politique dans la France du XVIIIe siècle* (Paris: INED, 2003).

Maser, M. *Die Historia Arabum des Rodrigo Jiménez de Rade* (Berlin: Lit, 2006).

Masonen, P. *The "Negroland revisited". Discovery and invention of the Sudanese Middle Ages* (Helsinki: Finnish Acad. of Sciences and Letters, 2000).

Masson, P. *Histoire du Commerce français dans le Levant au XVIIe siècle* (Paris: Hachette, 1896).

Masson, P. *Histoire des établissements et du commerce français dans l'Afrique barbaresque (1560–1793)* (Paris: Hachette, 1903).

Masson, P. *Histoire du commerce français dans le Levant au XVIIIe siècle* (Paris: Hachette, 1911).

Matar, N. *Turks, Moors and Englishmen in the age of discovery* (New York: Columbia, 1999).

Matar, N. *Britain and Barbary, 1589–1689* (Gainesville: University Press Florida, 2005).

Mather, J. *Pashas. Traders and travellers in the Islamic world* (New Haven/ London: Yale University Press, 2009).

Mattingly, G. *Renaissance diplomacy* (London: Cape, 1955).

Mayoral, J. V. "Five decades of structure: A retrospective view," in *Theoria. An International Journal for Theory, History and Foundations of Science* 27 (2012), 261–280.

Mazzacane, A. "Lo stato e il dominio nei giuristi veneti durante il 'secolo della terraferma'," in *Storia della cultura Veneta. Dal primo Quattrocento al Concilio di Trento*, vol. 1 (Vicenza: Neri Pozza, 1980), 577–631.

McClaughlin, T. "Une lettre de Melchisédec Thévenot," in *Revue d'histoire des sciences* 27, 2 (1974), 123–126.

McClellan, J. E. III and F. Regourd, *The colonial machine: French science and overseas expansion in the Old Regime* (Turnhout: Brepols, 2011).

McCollim, G. B. *Louis XIV's assault on privilege. Nicolas Desmaretz and the tax on wealth* (Rochester University Press, 2012).

McCormick, T. *William Petty and the ambitions of political arithmetic* (Oxford University Press, 2009).

McCusker, J. "Worth a war? The importance of the trade between British America and the Mediterranean," in S. Marzagalli et al. (eds.), *Rough waters: American involvement with the Mediterranean in the eighteenth and nineteenth centuries* (St. John's: Maritime Studies Research, 2010), 7–24.

Ménager, D. (ed.), *L'écriture de l'Histoire*, special issue of *Nouvelle Revue du Seizième siècle* 19,1 (2001).

Merimi, M. "Sulaymân al-Hîlâtî: l'information historique et les structures du pouvoir ibadhite à Djerba," in S. Bargaoui and H. Remaoun (eds.), *Savoirs historiques au Maghreb. Construction et usages* (Tunis-Oran: CERES, 2006), 41–52.

Merrett, R. J. *Daniel Defoe: contrarian* (Toronto University Press, 2013).

Meurer, P. H. "Ortelius as the father of historical cartography," in M. van den Broecke et al. (eds.), *Abraham Ortelius and the first Atlas. Essays commemorating the quadricentennial of his death, 1598–1998* (Turdijk: HES, 1998), 133–160.

Meusnier, N. "Vauban: arithmétique politique, Ragot et autre Cochonnerie," in T. Martin (ed.), *Arithmétique politique dans la France du XVIIIe siècle* (Paris: INED, 2003), 91–132.

Meyssonnier, S. "Vincent de Gournay, un intendant du commerce au travail. L'apport du fonds de Saint-Brieuc à l'intelligence de ses textes," L. Charles et al. (eds.), *Le cercle de Vincent de Gournay. Savoirs économiques et pratiques administratives en France au milieu du XVIIIe siécle* (Paris: INED, 2011), 89–110.

Mézin, A. *Les consuls de France au siècle des Lumières (1715–1792)* (Paris: MAE, 1995).

Michaud, F. "Le pauvre transformé: Les hommes, les femmes et la charité à Marseille, du XIIIe siècle jusqu'à la peste noire," in *Revue historique* 650 (2009), 243–290.

Micheau, F. "Le Kâmil d'Ibn al-alAthîr, source principale de l'Histoire des Arabes dans le Mukhtasar de Bar Hebraeus" in A.-M. Eddé and

E. Gannagé (eds.), *Regards croisés sur le moyen âge arabe. Mélanges à la mémoire de Louis Pouzet S.J.* = *Mélanges de l'Université St. Joseph 58* (2005), 425–439.

Michell, J. *Megalithomania: Artists, antiquarians and archaeologists at the old stone monuments* (Ithaca and London: Cornell, 1982).

Middleton, W. E. K. *The experimenters. A study of the Accademia del Cimento* (Baltimore/London: John Hopkins, 1971).

Miele, A. "*Res publica, res communis omnium, res nullius*: Grozio e le fonti romane sul diritto del mare," in *Index. Quaderni camerti di studi romanistici* 26 (1998), 384–387.

Mikhail, A. *Nature and empire in Ottoman Egypt. An environmental history* (Cambridge University Press, 2011).

Milanesi, M. *Tolomeo sostituito. Studi di storia delle conoscenze geografiche nel XVI secolo* (Milan: Unicopli, 1984).

Miller, P. "A philologist, a traveller and an antiquary rediscover the Samaritans in seventeenth-century Paris, Rome and Aix: Jean Morin, Pietro della Valle and N.-C. Fabri de Peiresc," in H. Zedelmaier and M. Mulsow (eds.), *Die Praktiken der Gelehrsamkeit in der Frühen Neuzeit* (Tübingen: Niemeyer, 2001), 123–146.

Miller, P. *Peiresc's Mediterranean world* (Cambridge, MA: Harvard University Press, 2015).

Minard, P. *La fortune du Colbertisme. État et industrie dans la France des Lumières* (Paris: Fayard, 1998).

Mittenhuber, F. *Text- und Kartentradition in der Geographie des Klaudios Ptolemaios. Eine Geschichte der Kartenüberlieferung vom ptolemäischen Original bis in die Renaissance* (Bern: Studies in the History and Philosophy of Science, 2009).

Moennig, U. "Die griechischen Studenten am Hallenser Collegium orientale theologicum," in J. Wallmann and U. Sträter (eds.), *Halle und Osteuropa. Zur europäischen Ausstrahlung des hallischen Pietismus* (Tübingen: Niemeyer, 1998), 299–329.

Mokyr, J. *The enlightened economy. An economic history of Britain 1700–1850* (New Haven and London: Yale University Press, 2009).

Morgan, K. "Mercantilism and the British Empire, 1688–1815," in D. Winch and P. K. O'Brien, *The political economy of British historical experience, 1688–1914* (Oxford University Press, 2002), 165–191.

Mössner, J. M. *Die Völkerrechtspersönlichkeit und die Völkerrechtspraxis der Barbareskenstaaten (Algier, Tripolis, Tunis 1518–1830)* (Berlin: de Gruyter, 1968).

Mousnier, R. "Le Testament politique de Richelieu," *Revue historique* 201 (1940), 55–71.

Muccillo, M. *Platonismo, ermetismo e "prisca theologia". Ricerche di storiografia filosofica rinascimentale* (Florence: Olschki, 1996).

Mueller, R. C. "Merchants and their Merchandise: Identity and Identification in Medieval Italy," in W. Kaiser and C. Moatti (eds.), *Gens de passage en Méditerranée de l'Antiquité à l'époque moderne. Procédures de contrôle et d'identification* (Paris: Maisonneuve & Larose, 2007), 313–344.

Muldoon, J. "Is the sea open or closed? The Grotius-Selden debate renewed," in K. Pennington and M. H. Eichbauer (eds.), *Law as profession and practice in medieval Europe. Essays in honor of James A. Brundage* (Aldershot: Ashgate, 2011), 117–133.

Müller, J.-D. "Literarische und andere Spiele. Zum Fiktionalitätsproblem in vormoderner Literatur," in *Poetica* 36 (2004), 281–312.

Müller, L. *Consuls, corsairs, and commerce. The Swedish consular service and long-distance shipping, 1720–1815* (Stockholm: Uppsala UL, 2004).

Murphy, A. E. *John Law. Economic theorist and policy-maker* (Oxford University Press, 1997).

Neudecker, H. "From Istanbul to London? Albertus Bobovius' appeal to Isaac Basire," in A. Hamilton et al. (eds.), *The republic of letters and the Levant* (Leiden and Boston: Brill, 2005), 173–196.

Neumann, E. "Imagining European community on the title page of Ortelius' Theatrum Orbis Terrarum (1570)," in *Word & Image* 25, 4 (2009), 427–442.

Neveu, B. *Erudition et religion aux XVIIe et XVIIIe siècles* (Paris: A. Michel, 1994).

Nipperdey, J. *Die Erfindung der Bevölkerungspolitik. Staat, politische theorie und population in der Frühen Neuzeit* (Göttingen: V&R, 2012).

North, J. *Stonehenge. Neolithic man and the cosmos* (London: HarperCollins, 1996).

Norwich, O. I. *Maps of Africa. An illustrated and annotated carto-bibliography* (Cape Town: Donker, 1983).

Novak, M. E. "Defoe as a defender of the government, 1727–29: A re-attribution and a new attribution," in *Huntington Library Quarterly* 71, 3 (2008), 503–512.

L'œuvre de Jean Bodin (Paris: Champion, 2004).

O'Reilly, W. "Charles Vallancey and the Military Itinerary of Ireland," in *Proceedings of the Royal Irish Academy* 106C (2006), 125–217.

Overton, J. H. *The nonjurors. Their lives, principles, and writings* (London: Smith, 1902).

Osmond, P. H. *Isaac Barrow. His life and times* (London: SPCK, 1944).

Owen, A. L. *The famous druids; a survey of three centuries of English literature on the Druids* (Oxford: Clarendon, 1962).

Pagano de Divitiis, G. *Mercanti inglesi nell'Italia del Seicento. Navi, traffici, egemonie* (Venice: Marsilio, 1990).

Pagden, A. *The fall of natural man: The American Indian and the origins of comparative ethnology* (Cambridge University Press, 1982).

Pallier, D. *Recherches sur l'imprimerie à Paris pendant la Ligue, 1585–1594* (Geneva: Droz, 1975).

Panzac, D. *'La caravane maritime' – Marins européens et marchands ottomans en Méditerranée (1680–1830)* (Paris: CNRS, 2004).

Panzac, D. *La marine ottomane. De l'apogée à la chute de l'Empire (1572–1923)* (Paris: CNRS, 2009).

Papadopoulos, S. I. *Ἡ κίνησῃ τοῦ δούκα τοῦ Νεβέρ Καρόλου Γονζάγα γὶα τὴν ἀπελευθέρωσιν τὼν Βαλκανικῶν λαῶν (1603–1625)* (Thessaloniki: Hetaireia Makedonikon Spoudon, 1966).

Paris, R. *Histoire du commerce de Marseille, vol. 5: Le Levant (de 1660 à 1789)* (Paris: Plon, 1957).

Parry, G. *The trophies of time. English antiquarians of the seventeenth century* (Oxford University Press, 1995).

Parslow, C. Charles *Rediscovering antiquity: Karl Weber and the excavation of Herculaneum, Pompeii, and Stabiae* (Cambridge University Press, 1995).

Patterson, W. B. "Cyril Lukaris, George Abbot, James VI and I, and the beginning of Orthodox-Anglican relations," in P. Doll (ed.), *Anglicanism and Orthodoxy. 300 years after the "Greek College" in Oxford* (Oxford et al.: Lang, 2006), 39–56.

Pawlisch, H. S. *Sir John Davies and the conquest of Ireland. A study in legal imperialism* (Cambridge University Press, 1985).

Pearson, J. B. *A biographical sketch of the chaplains to the Levant Company, maintained at Constantinople, Aleppo and Smyrna, 1611–1706* (Cambridge University Press, 1883).

Pennell, C. R. *Piracy and diplomacy in seventeenth-century North Africa. The Journal of Thomas Baker, English consul in Tripoli, 1677–1685* (London and Toronto: Associated University Press, 1989).

Perrot, J.-C. *Une histoire intellectuelle de l'économie politique, XVIIe–XVIIIe siècle* (Paris: EHESS, 1992).

Perruso, R. "The development of the doctrine of *res communes* in medieval and early modern Europe," in *Tijdschrift voor Rechtsgeschiedenis* 69 (2002), 69–93.

Peskin, L. A. *Captives and countrymen. Barbary slavery and the American public, 1785–1816* (Baltimore: John Hopkins Press, 2009).

Petersson, R. T. *Sir Kenelm Digby. The ornament of England, 1603–1665* (Cambridge/Mass.: Harvard University Press, 1956).

Petitmengin, P. "Les Patrologies avant Migne," in A. Mandouze and J. Fouilheron (eds.), *Migne et le renouveau des études patristiques. Acte du colloque de Saint-Flour, 7–8 juillet 1975* (Paris: Beauchesne, 1985), 15–38.

Piggott, S. *The druids* (London: Thames and Hudson, 1975).

Piggott, S. *William Stukeley. An eighteenth-century antiquary* (Oxford: Clarendon, 1950).

Piggott, S. *Ancient Britons and the antiquarian imagination: Ideas from the Renaissance to the regency* (London: Thames and Hudson, 1989).

Pippidi, A. *Tradiţia politică bizantină în mările române în secolele XVI– XVIII* (Bucarest: Corint, 2001).

Pippidi, A. *Byzantins, Ottomans, Roumains. Le Sud-Est européen entre l'héritage impérial et les influences occidentales* (Paris: Champion, 2006).

Pitts, J. *A Turn to empire. The rise of imperial liberalism in Britain and France* (Princeton University Press, 2005).

Pizzorusso, G. "La Congrégation de Propaganda fide à Rome: Centre d'accumulation et de production de 'savoirs missionnaires' (XVIIe-début XIXe siècle)," in C. de Castelnau-L'Estoile et al. (eds.), *Missions d'évangélisation et circulation des savoirs, XVIe-XVIIIe siècle* (Madrid: Casa de Velázquez, 2011), 25–40.

Plank, G. "Making Gibraltar British in the eighteenth century," in *History* 98 (2013), 346–369.

Platt, P. G. (ed.), *Wonders, marvels and monsters in early modern culture* (Newark and London: Associated University Presses, 1999).

Podskalsky, G. *Griechische Theologie in der Zeit der Türkenherrschaft 1453–1821* (Munich: Beck, 1988).

Popper, K. *Logik der Forschung*, 9th ed. (Tübingen: Niemeyer, 1989).

Popper, W. *The Cairo Nilometer. Studies in Ibn Taghrī Birdī's chronicles of Egypt I* (Berkeley University Press, 1951).

Porres Alonso, B. *Libertad a los cautivos. Actividad redentora de la orden Trinitaria, vol. 1: Redenciones de cautivos (1198–1785)* (Córdoba-Salamanca: Secretariado Trinitario, 1997).

Porter, A. *Religion versus empire? British Protestant missionaries and overseas expansion, 1700–1914* (Manchester University Press, 2004).

Poumarède, G. "Naissance d'une institution royale: Les consuls de la nation française en Levant et en Barbarie aux XVIe et XVIIe siècles," in *Annuaire-Bulletin de la Société de l'histoire de France* 2001, 65–128.

Poumarède, G. "Le voyage de Tunis et d'Italie de Charles-Quint ou l'exploitation politique du mythe de la croisade (1535–1536)," in *Bibliothèque d'Humanisme et Renaissance* 67 (2005), 247–285.

Powell, J. R. *Robert Blake. General-at-Sea* (London: Collins, 1972).

Preißler, H. "Orientalische Studien in Leipzig vor Reiske," in H.-G. Ebert, T. Hanstein (eds.), *Johann Jacob Reiske – Leben und Wirkung. Ein*

Leipziger Byzantinist und Begründer der Orientalistik im 18. Jahrhundert (Leipzig: Ev. Verlagsanstalt, 2005), 19–43.

Primavesi, O. "Henri II Estienne über philosophische Dichtung: Eine Fragmentsammlung als Beitrag zu einer poetologischen Kontroverse," in O. Primavesi and K. Luchner (eds.), *The Presocratics from the Latin Middle Ages to Hermann Diels* (Stuttgart: Steiner, 2011), 157–196.

Pritchard, J. *In search of empire. The French in the Americas, 1670–1730* (Cambridge University Press, 2004).

Proctor, R. N. and L. Schiebinger (eds.), *Agnotology. The making and unmaking of ignorance* (Stanford University Press, 2008).

Quaglioni, D. "Scienza politica e predizione delle *conversiones rerumpublicarum*: Bodin, Grégoire, Albergati," in *I limiti della sovranità. Il pensiero di Jean Bodin nella cultura politica e giuridia dell'età moderna* (Padua: Cedam, 1992), 169–197.

Quantin, J.-L. "The fathers in seventeenth century Roman Catholic theology," in I. Backus (ed.), *The reception of the church fathers in the West*, 2 vol. (Boston and Leiden: Brill, 2001), 951–986.

Quantin, J.-L. "Anglican scholarship gone mad? Henry Dodwell (1641–1711) and Christian antiquity," in J.-L. Quantin and C. R. Ligota (eds.), *History of scholarship. A selection of papers from the seminar on the history of scholarship held annually at the Warburg Institute* (Oxford University Press, 2006), 305–356.

Quantin, J.-L. *Church of England and Christian antiquity. The construction of a confessional Identity in the 17th century* (Oxford University Press, 2009).

de Quehen, H. "Politics and scholarship in the Ignatian controversy," in *The Seventeenth Century* 13, 1 (1998), 69–84.

Raj, K. *Relocating modern science: Circulation and the construction of knowledge in South Asia and Europe, 1650–1900* (Houndsmills et al.: Palgrave Macmillan, 2007).

Rauchenberger, D. *Johannes Leo der Afrikaner. Seine Beschreibung des Raumes zwischen Nil und Niger nach dem Urtext* (Wiesbaden: Harrassowitz, 1999).

Ravetz, J. R. "Usable knowledge, usable ignorance: incomplete science with policy implications," in W. C. Clark and R. E. Munn (eds.), *Sustainable Development of the Biosphere* (Cambridge University Press, 1986), 415–434.

Ravetz, J. R. *The merger of knowledge with power. Essays in critical science* (London and New York: Mansell, 1990).

Reinert, S. A. *Translating empire. Emulation and the origins of political economy* (Cambridge, MA: Harvard University Press, 2011).

Relaño, F. *The shaping of Africa. Cosmographic discourse and cartographic science in late medieval and early modern Europe* (Aldershot: Ashgate, 2002).

Ressel, M. *Zwischen Sklavenkassen und Türkenpässen. Nordeuropa und die Barbaresken in der Frühen Neuzeit* (Berlin et al.: de Gruyter, 2012).

Ressel, M. and C. Zwierlein: "The ransoming of North European captives from Northern Africa. A comparison of Dutch, Hanseatic and English institutionalization of redemption from 1610–1645," in N. Jaspert, and S. Kolditz (eds.), *Seeraub im Mittelmeerraum. Piraterie, Korsarentum und maritime Gewalt von der Antike bis zur Neuzeit* (Munich: Fink, 2013), 377–406.

Richard, F. "Un érudit à la recherche de textes religieux venus d'Orient, le docteur Louis Picques (1637–1699)," in E. Bury and B. Meunier (eds.), *Les pères de l'église au XVIIe siècle* (Paris: Cerf, 1993), 253–277.

Richthofen, F. v. *China. Ergebnisse eigener Reisen und darauf gegründeter Studien*, vol. 1 (Berlin: Reimer, 1877).

Robinson, C. F. *Islamic historiography* (Cambridge University Press, 2003).

Rodriguez, J. *Captives and their saviors in the medieval crown of Aragon* (Washington: Catholic University of America Press, 2007).

Rodríguez-Picavea, E. "The military orders and hospitaller activity on the Iberian peninsula during the Middle Ages," in *Mediterranean Studies* 18 (2009), 24–43.

Rohr, C. "Man and natural disaster in the Late Middle Ages: The earthquake in Carinthia and Northern Italy on 25 January 1348 and its perception," in *Environment and History* 9 (2003), 127–150.

Rosa, S. "'Il était possible aussi que cette conversion fut sincère': Turenne's Conversion in Context," in *French Historical Studies* 18 (1994), 632–666.

Rose, C. "The origins and ideals of the SPCK 1699–1716," in J. Walsh et al. (eds.), *The Church of England, c. 1689–c. 1833. From Toleration to Tractarianism* (Cambridge University Press, 1993), 172–190.

Rothkrug, L. *Opposition to Louis XIV. The political and social origins of the French Enlightenment* (Princeton University Press, 1965).

Rothman, N. E. *Brokering empire. Trans-imperial subjects between Venice and Istanbul* (Ithaca and London: Cornell, 2012).

Rouighi, R. *The making of a Mediterranean Emirate. Ifriqiya and its Andalusis, 1200–1400* (Philadelphia: Pennstate University Press, 2011).

Rousset, C. *Histoire de Louvois et de son administration politique et militaire jusqu'à la paix de Nimègue*, tome 1 (Paris: Didier, 1862), 35–57.

Roux, F. C. *Les échelles de Syrie et de Palestine au XVIIIe siècle* (Paris: Geuthner, 1928).

Rubiés, J.-P. *Travel and ethnology in the Renaissance. South India through European Eyes, 1250–1625* (Cambridge University Press, 2000).

Rudwick, M. J. S. *The meaning of fossils. Episodes in the history of palaeontology*, 2nd ed. (New York: Science History, 1976).

Runciman, S. *The Great Church in captivity. A study of the patriarchate of Constantinople from the eve of the Turkish Conquest to the Greek War of Independence* (Cambridge University Press, 1968).

Russell, G. A. "Introduction," in G. A. Russell (ed.), *The "Arabick" interest of the natural philosophers in seventeenth-century England* (Leiden and Boston: Brill, 1994), 1–19.

Russell, G. A. (ed.), *The "Arabick" interest of the natural philosophers in seventeenth-century England* (Leiden and Boston: Brill, 1994).

Sadok, B. *La Régence de Tunis au XVIIe siècle: ses relations commerciales avec les ports de l'Europe méditerranéenne Marseille et Livourne* (Zaghouan: Ceroma, 1987).

Safier, N. *Measuring the New World. Enlightenment science and South America* (Chicago University Press, 2008).

Sahlins, P. *Unnaturally French. Foreign citizens in the Old Regime and after* (Ithaca/London: Cornell, 1996).

Said, E. *Orientalism*, with new preface (New York: Vintage, 1979/2003).

dos Santos Lopes, M. *Afrika. Eine neue Welt in deutschen Schriften des 16. und 17. Jahrhunderts* (Stuttgart: Steiner, 1992).

Sasso, G. *Machiavelli e gli antichi*, vol. 1 (Florence: Ricciardi, 1988), 3–65.

Sawyer, J. K. *Printed poison. Pamphlet propaganda, faction politics and the public sphere in early seventeenth-century France* (Berkeley University Press, 1990).

Scattola, M. *Das Naturrecht vor dem Naturrecht. Zur Geschichte des ,ius naturae' im 16. Jahrhundert* (Tübingen: Niemeyer, 1999).

Schaeper, T. J. *The French council of commerce, 1700–1715. A study of mercantilism after Colbert* (Columbus: Ohio State University Press, 1983).

Schlobach, J. *Zyklentheorie und Epochenmetaphorik* (Munich: Fink, 1980).

Schmidt, J. "Between author and library shelf: The intriguing history of some Middle Eastern manuscripts acquired by public collections in the Netherlands prior to 1800," in A. Hamilton et al. (eds.), *The republic of letters and the Levant* (Leiden and Boston: Brill, 2005), 27–51.

Schnakenbourg, É. *Entre la guerre et la paix. Neutralité et relations internationales, XVIIe-XVIIIe siècles* (Rennes University Press, 2013).

Schoonheim, P. L. *Aristotle's meteorology in the Arabico-Latin tradition. A critical edition of the texts, with introduction and indices* (Leiden and Boston: Brill, 2000).

Schulte-Behrbühl, M. *Deutsche Kaufleute in London. Welthandel und Einbürgerung (1600–1800)* (Munich: Oldenbourg, 2007).

Scott, J. "The peace of silence. Thucydides and the English Civil War," in G. A. Rogers and T. Sorell (eds.), *Hobbes and History* (London and New York: Routledge, 2000), 112–136.

Sebag, P. "Sur deux Orientalistes français du XVIIe siècle: F. Petis de la Croix et le sieur de la Croix," in *Revue de l'Occident Musulman et de la Méditerranée* 25 (1978), 89–117.

Sebag, P. *Tunis au XVIIe siècle. Une cité barbaresque au temps de la course* (Paris: L'Harmattan, 1989).

Seifert, A. *Cognitio historica. Die Geschichte als Namensgeberin der frühneuzeitlichen Empirie* (Berlin: Duncker & Humblot, 1976).

Seifert, A. *Der Rückzug der biblischen Prophetie von der neueren Geschiche. Studien zur Geschichte der Reichstheologie des frühneuzeitlichen deutschen Protestantismus* (Cologne: Böhlau, 1990).

Senatore, F. *'Uno mundo de carta'. Forme e strutture della diplomazia sforzesca* (Naples: Liguori, 1998).

Sheehan, J. "Enlightenment, religion, and the enigma of secularization: A review essay," in *AHR* 2003, 1061–1080.

Simmel, G. *Soziologie* (Frankfurt/M: Suhrkamp, 1992).

Simmel, G. "Lebensanschauung. Vier metaphysische Kapitel" (1918), chap. 3: "Tod und Unsterblichkeit," in *Gesamtausgabe*, vol. 16, ed. G. Fitzi and O. Rammstedt (Frankfurt/M: Suhrkamp, 1999).

Sirota, B. S. "The Trinitarian crisis in church and state: Religious controversy and the making of the postrevolutionary Church of England, 1687–1702," in *Journal of British Studies* 52, 1 (2013), 26–54.

Sirota, B. S. "The Church: Anglicanism and the nationalization of maritime space," in P. J. Stern and C. Wennerlind (eds.), *Mercantilism reimagined. Political economy in early modern Britain and its empire* (Oxford University Press, 2014), 196–217.

Skinner, Q. "Liberty and security: The early-modern English debate," in C. Kampmann and U. Niggemann (eds.), *Sicherheit in der Frühen Neuzeit. Norm, Praxis, Repräsentationen* (Cologne et al.: Böhlau, 2013), 30–42.

Smiley, W. "Let *whose* people go? Subjecthood, sovereignty, liberation, and legalism in eighteenth-century Russo-Ottoman relations," in *Turkish Historical Review* 3 (2012), 196–228.

Smithson, M. J. *Ignorance and uncertainty. Emerging paradigms* (New York et al.: Springer, 1989).

Snoeks, R. *L'argument de tradition dans la controverse eucharistique entre catholiques et réformés français au XVIIe siècle* (PU Louvain, 1951).

Sojer, C. "Il manoscritto autografo [della *Graecia Orthodoxia*] di Leone Allacci della Biblioteca Gambalunga di Rimini," in *Schede umanistiche* 20 (2006/07), 119–149.

Solé, J. "Le comparatisme historique et géographique dans la République de Jean Bodin," in *L'œuvre de* Jean Bodin (Paris: Champion, 2004), 411–418.

Soll, J. *The Information Master. Jean-Baptiste Colbert's secret state intelligence system* (Ann Arbor: University Press Michigan, 2009).

Solleveld, F. "Conceptual change in the history of the humanities," in *studium. Tijdschrift voor Wetenschaps- en Universiteitsgeschiedenis* 7 (2014), 223–239.

Somos, M. "Selden's *Mare Clausum*. The secularization of international law and the rise of soft imperialism," in *Journal of the History of International Law* 14 (2012), 287–330.

Sonnino, P. "The dating of Richelieu's *Testament politique*," in *French History* 2005, 262–272.

Spellman, W. M. *The Latitudinarians and the Church of England, 1660–1700* (Athens and London: University of Georgia Press, 1993).

Spengler, J. J. *Économie et population. Les doctrines françaises avant 1800* (Paris: INED, 1954).

Spurr, J. *The Restoration Church of England, 1646–1689* (New Haven and London: Yale University Press, 1991).

Stagl, J. "Die Apodemik oder 'Reisekunst' als Methodik der Sozialforschung vom Humanismus bis zur Aufklärung," in J. Stagl and M. Rassem (eds.), *Statistik und Staatsbeschreibung in der Neuzeit, vornehmlich im 16.-18. Jahrhundert* (Paderborn et al.: Schöningh, 1980), 131–204.

Stanzel, F. K. (ed.), *Europäischer Völkerspiegel. Imagologisch-ethnographische Studien zu den Völkertafeln des frühen 18. Jahrhunderts* (Heidelberg: Winter, 1999).

Starkie, A. *The Church of England and the Bangorian controversy 1716–1721* (Woodbridge: Boydell, 2007).

Stearns, R. P. "Fellows of the Royal Society in North Africa and the Levant, 1662–1800," in *Notes and Records of the Royal Society of London* 11, 1 (1954), 75–90.

Steele, I. K. *Politics of colonial policy. The Board of Trade in colonial administration 1696–1720* (Oxford University Press, 1968).

Stein, T. "Tangier in the Restoration Empire," in *The Historical Journal* 54 (2011), 985–1011.

Steiner, B. *Die Ordnung der Geschichte: Historische Tabellenwerke in der Frühen Neuzeit* (Cologne et al.: Böhlau, 2008).

Steiner, P. "Commerce, commerce politique," in L. Charles et al. (eds.), *Le cercle de Vincent de Gournay. Savoirs économiques et pratiques administratives en France au milieu du XVIIIe siécle* (Paris: INED, 2011), 179–200.

Steitz, G. E. "Die Abendmahlslehre der griechischen Kirche in ihrer geschichtlichen Entwicklung," in *Jahrbücher für deutsche Theologie* 13 (1868), 3–68, 649–700, 672–700.

Stern, P. J. "'Rescuing the age from a charge of ignorance': gentility, knowledge, and the British exploration of Africa in the later eighteenth century," in K. Wilson (ed.), *A New Imperial history. Culture, identity, and modernity in Britain and the Empire, 1660–1840* (Cambridge University Press, 2004), 115–135.

Stern, P. J. and C. Wennerlind (eds.), *Mercantilism reimagined. Political economy in early modern Britain and its empire* (Oxford University Press, 2014).

Stewart-Brown, R. "The avowries of Cheshire," in *The English Historical Review* 29 (1914), 41–55.

Stimson, A. "The longitude problem: The navigator's story," in W.J. H. Andrewes (ed.), *The quest for longitude*, 72–84.

Stolleis, M. *Geschichte des öffentlichen Rechts in Deutschland*. Vol. 1: *Reichspublizistik und Policeywissenschaft 1600–1800* (Munich: Beck, 1988).

Stolleis, M. *Staat und Staatsräson in der frühen Neuzeit. Studien zur Geschichte des öffentlichen Rechts* (Frankfurt/M: Suhrkamp, 1990).

Stolzenberg, D. *Egyptian Oedipus. Athanasius Kircher and the secrets of Antiquity* (Chicago University Press, 2013).

Straumann, B. "Is modern liberty ancient? Roman remedies and natural rights in Hugo Grotius's early works on natural law," in *Law and History Review* 27, 1 (2009), 55–85.

Strong, R. *Anglicanism and the British Empire, c. 1700–1850* (Oxford University Press, 2007).

Stuchtey, B. (ed.), *Science across the European empires, 1800–1950* (Oxford: Oxford University Press, 2005).

Sullivan, G. A. *Memory and forgetting in English Renaissance drama. Shakespeare, Marlowe, Webster* (Cambridge University Press, 2005).

Surun, I. "Le blanc de la carte, matrice de nouvelles représentations des espaces africains," in Laboulais-Lesage (ed.), *Combler les blancs*, 117–144.

Taillemite, É. "Les archives et les archivistes de la Marine des origines à 1870," in *Bibliothèque de l'école des Chartes* 127 (1969), 27–86.

Takeda J. T., *Between crown and commerce. Marseille and the early modern Mediterranean* (Baltimore: J. Hopkins, 2011).

da Terzorio, C. *Le missioni dei minori cappuccini. Sunto storico*, 10 vol. (Rome: Tipografia Pont. Istituto Pio IX, 1913–1938).

Tezcan, B. "The politics of early modern Ottoman historiography," in V. H. Aksan, D. Goffman (eds.), *The early modern Ottomans. Remapping the Empire* (Cambridge University Press, 2007), 167–198.

Théré, C. "Economic publishing and authors, 1566–1789," in G. Faccarello (ed.), *Studies in the history of French political economy* (London and New York: Routledge, 1998), 1–56.

Thomson, A. *Barbary and Enlightenment. European attitudes towards the Maghreb in the 18th century* (Leiden and Boston: Brill, 1987).

Thomson, E. "Commerce, law, and erudite culture: The mechanics of Godefroy's service to Cardinal Richelieu," in *Journal of the History of Ideas* 68, 3 (2007), 407–427.

Thomson, E. "France's Grotian moment? Hugo Grotius and Cardinal Richelieu's commercial statecraft," in *French History* 21, 4 (2007), 377–394.

Thuau, E. *Raison d'État et pensée politique à l'époque de Richelieu*, postface G. Mairet (Paris: A. Michel, 2000).

Tierney, B. *Liberty & Law. The idea of permissive natural law, 1100–1800* (Washington: Catholic University of America, 2014).

Tolmacheva, M. "The medieval Arabic geographers and the beginnings of modern Orientalism," in *International Journal of Middle East Studies* 27, 2 (1995), 141–156.

Tooley, M. J. "Bodin and the Medieval Theory of Climate," in *Speculum* 28, 1 (1953), 64–83.

Toomer, G. J. *Eastern wisedome and learning. The study of Arabic in seventeenth-century England* (Oxford: Clarendon, 1996).

Toomer, G. J. *John Selden. A life in scholarship*, 2 vol. (Oxford University Press, 2009).

Totaro, G. *L'autobiographie d'Athanasius Kircher* (Bern et al.: Lang, 2009).

Tourbet-Delof, G. *L'Afrique barbaresque dans la littérature française aux XVIIe et XVIIIe siècles* (Geneva: Droz, 1973).

Trevor Roper, H. "The Church of England and the Greek church in the time of Charles I," in *From Counter-Reformation to Glorious Revolution* (Chicago University Press, 1992), 83–111.

Trivellato, F. *The familiarity of strangers. The Sephardic Diaspora, Livorno, and cross-cultural trade in the early modern period* (New Haven and London: Yale University Press, 2009).

Turpaud, R. *La juridiction des consuls français dans les échelles du Levant d'après les capitulations* (Paris: Mellottée, 1902).

Tylden-Wright, D. *John Aubrey. A life* (London: HarperCollins, 1991).

Tzirakes N. E., *Ή περὶ μετουσιώσεως (transsubstantiatio) εὐχαριστικὴ ἔρις. Συμβολὴ εἰς τὴν ὀρθόδοξον περὶ μεταβολῆς διδασκαλίαν τοῦ ιζ′ αἰῶνος* (Athens: Diatribe epi diktatoria, 1977).

Ucko, P. J. et al. (eds.), *Avebury reconsidered. From the 1660s to the 1990s* (London and Boston: Unwin Hyman, 1991).

Ulbert, J. and G. Le Bouëdec (eds.), *La fonction consulaire à l'époque moderne. L'Affirmation d'une institution économique et politique (1500–1800)* (PU Rennes, 2006).

Ullmann, M. *Die Medizin im Islam* (Leiden and Boston: Brill, 1970).

Uri, J. (ed.), *Bibliothecae Bodleianae codicum manuscriptorum Orientalium*, pars 1 (Oxford: Clarendon, 1787).

Van den Boogert, M. H. *The capitulations and the Ottoman legal system. Qadis, consuls and beratlis in the 18th Century* (Leiden and Boston: Brill, 2005).

Vandal, A. *L'Odyssée d'un ambassadeur. Les voyages du marquis de Nointel (1670–1680)* (Paris: Plon, 1900).

Veinstein, G. "Le jeune de langues Dantan. Avait-il bien traduit ?" in Hitzel (ed.), *Istanbul*, 319–332.

Verdy du Vernois, F. v. *Die Frage der heiligen Stätten. Ein Beitrag zur Geschichte der völkerrechtlichen Beziehungen der Ottomanischen Pforte* (Berlin: Mittler, 1901).

Villiers, P. *Marine royale, corsaires et trafic dans l'Atlantique de Louis XIV à Louis XVI*, 2 vol. (Dunkerque: Société dunkerquoise d'histoire et d'archéologie, 1991).

Vine, A. *In defiance of time: Antiquarian writing in early modern England* (Oxford University Press, 2010).

Virol, M. *Vauban. De la gloire du roi au service de l'état* (Seyssel: Champ Vallon, 2003).

Völkel, M. *Geschichtsschreibung: eine Einführung in globaler Perspektive* (Cologne et al.: Böhlau, 2006).

Wakefield, C. "Arabic manuscripts in the Bodleian Library: the seventeenth-century collections," in G. A. Russell (ed.), *The "Arabick" interest of the natural philosophers in seventeenth-century England* (Leiden and Boston: Brill, 1994), 128–146.

Wallmann, J. and U. Sträter (eds.), *Halle und Osteuropa. Zur europäischen Ausstrahlung des hallischen Pietismus* (Tübingen: Niemeyer, 1998).

Walton, D. *Arguments from ignorance* (University Park: Penn State University Press, 1994).

Ward, R. *The world of the medieval shipmaster. Law, business and the sea, c. 1350–c. 1450* (Woodbridge: Boydell, 2009).

Ward, R. W. *Early evangelicalism. A global intellectual history, 1670–1789* (Cambridge University Press, 2006).

Ware, T. K. "Orthodox and catholics in the seventeenth century: schism or intercommunion?," in D. Baker (ed.), *Schism, heresy and religious protest* (Cambridge University Press, 1972), 259–276.

Warraq, Ibn *Defending the West. A critique of Edward Said's orientalism* (Amherst: Prometheus, 2007).

Wehling, P. *Im Schatten des Wissens? Perspektiven der Soziologie des Nichtwissens* (Konstanz: UV, 2006).

Weir, D. *The origins of the Federal theology in sixteenth-century reformation thought* (Oxford University Press, 1990).

Wendebourg, D. *Reformation und Orthodoxie. Der ökumenische Briefwechsel zwischen der Leitung der Württembergischen Kirche und Patriarch Jeremias II. von Konstantinopel in den Jahren 1573–1581* (Göttingen: V&R, 1986).

West, M. L. *The Orphic poems* (Oxford University Press, 1983).

White, B. "'Brothers of the string': Henry Purcell and the letter-books of Rowland Sherman," in *Music & Letters* 92, 4 (2011), 519–581.

Wieacker, F. *Römische Rechtsgeschichte*, 2 vol. (Munich: Beck, 1988 and 2006).

Williams, G. *The Orthodox Church of the East in the Eighteenth Century. Being the correspondence between the Eastern patriarchs and the nonjuring bishops* (London: Rivingtons, 1868).

Wilson, K. *The sense of the people. Politics, culture and imperialism in England, 1715–1785* (Cambridge University Press, 1995).

Wilson, R. "Heinrich Wilhelm Ludolf, August Hermann Francke und der Eingang nach Rußland," in J. Wallmann and U. Sträter (eds.), *Halle und Osteuropa. Zur europäischen Ausstrahlung des hallischen Pietismus* (Tübingen: Niemeyer, 1998), 83–108.

Windler, C. *La diplomatie comme expérience de l'autre. Consuls français au Maghreb (1700–1840)* (Geneva: Droz, 2002).

Wollan Teigen, B. *The Lord's supper in the theology of Martin Chemnitz* (Brewster/Mass.: Trinity Lutheran Press, 1986).

Wood, A. C. *A history of the Levant Company* (first ed. 1935) (New York: Barnes & Noble, 1964).

Woolf, D. R. *The social circulation of the past: English historical culture, 1500–1730* (Oxford University Press, 2003).

Woolf, D. R. "From hystories to the historical: Five transitions in thinking about the past, 1500–1700," in *Huntington Library Quarterly* 68 (2005), 33–70.

Woolf, D. R. *A global history of history* (Cambridge University Press, 2011).

Wurm, H. *Der osmanische Historiker Ḥüseyn b. Ga'fer, genannt Hezārfenn, und die Istanbuler Gesellschaft in der zweiten Hälfte des 17. Jahrhunderts* (Freiburg i. Br.: Schwarz, 1971).

Wüstenfeld, F. *Die Geschichtsschreiber der Araber und ihre Werke* (Göttingen: Dieterich, 1882).

Zahedieh, N. "Economy," in D. Armitage and M. J. Braddick (eds.), *The British Atlantic world, 1500–1800*, 2nd ed. (New York: Palgrave Macmillan, 2009), 53–70.

Zwierlein, C. "Reformation als Rechtsreform. Bucers Hermeneutik der lex Dei und sein humanistischer Zugriff auf das römische Recht," in C. Strohm (ed.), *Martin Bucer und das Recht* (Geneva: Droz, 2002), 29–81.

Zwierlein, C. "Consociatio," in F. Ingravalle and C. Malandrino (eds.), *Il lessico della Politica di Johannes Althusius* (Florence: Olschki, 2005), 143–168.

Zwierlein, C. "Das Glück des Bürgers. Eine begriffsgeschichtliche Skizze zur optimistischen Grundmentalität der Aufklärung," in H. E. Friedrich et al. (eds.), *Bürgerlichkeit im 18. Jahrhundert* (Tübingen: Niemeyer, 2006), 71–113.

Zwierlein, C. *Discorso und Lex Dei. Die Entstehung neuer Denkrahmen im 16. Jahrhundert und die Wahrnehmung der französischen Religionskriege in Italien und Deutschland* (Göttingen: V&R, 2006).

Zwierlein, C. "Politik als Experimentalwissenschaft, 1521–1526," in *Philosophisches Jahrbuch* 113 (2006), 31–62.

Zwierlein, C. "'(Ent)konfessionalisierung' (1935) und 'Konfessionalisierung' (1981)," in *Archiv für Reformationsgeschichte* 92 (2007), 199–230.

Zwierlein, C. "Diachrone Diskontinuitäten in der frühneuzeitlichen Informationskommunikation und das Problem von Modellen 'kultureller Evolution'," in A. Brendecke et al. (eds.), *Information in der Frühen Neuzeit. Status, Bestände, Strategien* (Berlin: de Gruyter, 2008), 423–453.

Zwierlein, C. "Machiavellismus / Antimachiavellismus," in H. Jaumann (ed.), *Gelehrtendiskurse der Frühen Neuzeit* (Berlin: de Gruyter, 2010), 903–951.

Zwierlein, C. "Natur/Kultur-Grenzen und die Frühe Neuzeit – Transcodierung von Natur, Klimatheorie und biokulturelle Grenzen," in C. Roll et al. (eds.), *Grenzen und Grenzüberschreitungen. Bilanzen und Perspektiven der Frühneuzeitforschung* (Cologne: Böhlau, 2010), 25–49.

Zwierlein, C. *Der gezähmte Prometheus. Feuer und Sicherheit zwischen Früher Neuzeit und Moderne* (Göttingen: V&R, 2011).

Zwierlein, C. "Forgotten religions, religions that cause forgetting," in I. Karremann et al. (eds.), *Forgetting faith? Negotiating confessional conflict in early modern Europe* (Berlin: de Gruyter, 2012), 117–138.

Zwierlein, C. "Les saints de la communion avec le Christ: Hybridations entre églises et états dans le monde calviniste dans les années 1560," in F. Buttay and A. Guillausseau (eds.), *Des saints d'État ? Politique et sainteté au temps du concile de Trente* (Paris: PU Sorbonne, 2012), 35–50.

Zwierlein, C. "Sicherheitsgeschichte. Ein neues Feld der Geschichtswissenschaften," in *Geschichte & Gesellschaft* 38 (2012), 365–386.

Zwierlein, C. "*Conversiones*, révolutions, guerres civiles: de Bodin au droit international dans la Méditerranée du XVIIIème siècle," in *Il pensiero politico* (forthcoming, fascicolo 3, 2016).

Zwierlein, C. "Early modern history," in A. Lichtenberger et al. (eds.), *Handbuch für Mediterranistik* (Munich: Fink, 2015), 87–105.

Zwierlein, C. "Coexistence and ignorance: What Europeans in the Levant did not read (17th/18th centuries)," in C. Zwierlein (ed.), *The dark side of knowledge. Histories of ignorance, 1400–1800* (Leiden/Boston: Brill, 2016), 225–265.

Zwierlein, C. (ed.), *The dark side of knowledge. Histories of ignorance, 1400–1800* (Leiden/Boston: Brill, 2016).

Zwierlein, C. "Towards a history of ignorance," in C. Zwierlein (ed.), *The dark side of knowledge. Histories of ignorance, 1400–1800* (Leiden/ Boston: Brill, 2016), 1–47.

Zwierlein, C. and M. Ressel, "Zur Ausdifferenzierung zwischen Fiktionalitäts- und Faktualitätsvertrag im Umfeld frühneuzeitlichen pikarischen Erzählens," in M. Waltenberger (ed.), *Das Syntagma des Pikaresken* (Heidelberg: Winter, 2012), 103–129.

Index

Printed in the United States
By Bookmasters